Advances in Experimental Medicine and Biology

Volume 1001

Advances in Experimental Medicine and Biology presents multidisciplinary and dynamic findings in the broad fields of experimental medicine and biology. The wide variety in topics it presents offers readers multiple perspectives on a variety of disciplines including neuroscience, microbiology, immunology, biochemistry, biomedical engineering and cancer research. Advances in Experimental Medicine and Biology has been publishing exceptional works in the field for over 30 years and is indexed in Medline, Scopus, EMBASE, BIOSIS, Biological Abstracts, CSA, Biological Sciences and Living Resources (ASFA-1), and Biological Sciences. The series also provides scientists with up to date information on emerging topics and techniques. 2015 Impact Factor: 1.953

More information about this series at http://www.springer.com/series/5584

Tomohiro Sasanami
Editor

Avian Reproduction

From Behavior to Molecules

Springer

Editor
Tomohiro Sasanami
Department of Applied Biological Chemistry
Shizuoka University
Shizuoka
Japan

ISSN 0065-2598 ISSN 2214-8019 (electronic)
Advances in Experimental Medicine and Biology
ISBN 978-981-10-3974-4 ISBN 978-981-10-3975-1 (eBook)
DOI 10.1007/978-981-10-3975-1

Library of Congress Control Number: 2017955088

Printed on acid-free paper

This Springer imprint is published by Springer Nature
The registered company is Springer Nature Singapore Pte Ltd.
The registered company address is: 152 Beach Road, #21-01/04 Gateway East, Singapore 189721, Singapore

Preface

Our understanding of the reproduction mechanism in animals has advanced significantly in recent years as a result of the remarkable developments in modern technology in the molecular, cellular, biochemical, and genetic sciences. In addition, the refinement of the live-imaging technique, in which the cellular events occurring in this process can be visualized directly in live animals, also has contributed substantially to the answering of long-standing unresolved questions. However, progress remains hampered in non-model animals due to various technical limitations; therefore, currently there remains a less-than-comprehensive understanding of the reproduction mechanism, resulting from the considerable divergence in reproductive tactics.

There are about 10,000 kinds of birds in the world, and they show significant divergence in body size, feather color, dietary habits, behavior, and other characteristics. Recently, the genomes of 48 birds, which cover 32 of the 35 proposed avian orders, were sequenced and analyzed. This will provide valuable information on the genetic and evolutionary mechanisms that control growth, sexual behaviors, and reproductive physiology. In particular, from the point of view of the agricultural sciences, there is a crying need for further increase in meat and egg production in the food industries. Therefore, research on avian species, in any aspect of life science, is now accelerating.

This book, *Avian Reproduction: From Behavior to Molecules*, deals with basic knowledge as well as our recent findings on reproductive physiology leading to more successful fertilization in birds. Fertilization is an indispensable step for zygotic formation in sexual reproduction to ensure the survival of animals. To achieve better fertilization outcomes, diverse reproductive tactics are developed in various vertebrates. Of these, the reproductive strategies in birds are well refined and some phenomena are unique to birds; thus, an understanding of this mechanism will provide new ideas and insights to various scientists in different fields. Application of these reproductive systems will become a powerful tool in a myriad of applications. As very few books cover this subject, our book is also important for scientists in the field of avian biology.

The contents cover many unique aspects of avian reproduction, and I do hope that it will be useful and will spark new interest in avian reproduction for readers including scientists and students. Finally, I would like to express my sincere thanks to all the experts in this field who contributed chapters, and also to take this opportunity to thank the staff of Springer for helping this book come to fruition.

Shizuoka, Japan Tomohiro Sasanami

Contents

Avian Primordial Germ Cells

1

Takahiro Tagami, Daichi Miyahara,
and Yoshiaki Nakamura

Abstract

Germ cells transmit genetic information to the next generation through gametogenesis. Primordial germ cells (PGCs) are the first germ-cell population established during development, and are the common origins of both oocytes and spermatogonia. Unlike in other species, PGCs in birds undergo blood circulation to migrate toward the genital ridge, and are one of the major biological properties of avian PGCs. Germ cells enter meiosis and arrest at prophase I during embryogenesis in females, whereas in males they enter mitotic arrest during embryogenesis and enter meiosis only after birth. In chicken, gonadal sex differentiation occurs as early as embryonic day 6, but meiotic initiation of female germ cells starts from a relatively late stage (embryonic day 15.5). Retinoic acid controls meiotic entry in developing chicken gonads through the expressions of retinaldehyde dehydrogenase 2, a major retinoic acid synthesizing enzyme, and cytochrome P450 family 26, subfamily B member 1, a major retinoic acid-degrading enzyme. The other major biological property of avian PGCs is that they can be propagated in vitro for the long term, and this technique is useful for investigating proliferation mechanisms. The main factor involved in chicken PGC proliferation is fibroblast growth factor 2, which activates the signaling of MEK/ERK and thus promotes the cell cycle and anti-apoptosis. Furthermore, the activation of PI3K/Akt signaling is indispensable for the proliferation and survival of chicken PGCs.

T. Tagami (✉)
Institute of Livestock Grassland Science,
NARO, Ibaraki, Japan
e-mail: tagami@affrc.go.jp

D. Miyahara
Institute of Livestock Grassland Science,
NARO, Ibaraki, Japan

Shinshu University, Ueda, Japan

Y. Nakamura
National Institute for Basic Biology, Okazaki, Japan

T. Sasanami (ed.), *Avian Reproduction*, Advances in Experimental Medicine and Biology 1001,
DOI 10.1007/978-981-10-3975-1_1

Keywords
Primordial germ cells • Germ cell formation • Germ cell markers • Germ cell migration • Germ cell differentiation • Gametogenesis • Germ cell culture • Sexual differentiation • Germline chimeras • Chicken • Quail

1.1 Introduction

Germ cells are the only cell type to convey genetic information to the next generation through gametogenesis. Interestingly, germ cells usually do not arise within the gonads. During early development, germ-cell lineage is set aside from all somatic cells in the embryo. This population of cells is described as primordial germ cells (PGCs), the gamete progenitor cells (Fig. 1.1). Following migration to the genital ridge (also called the gonadal anlage or gonadal ridge), PGCs undergo propagation and meiosis and then differentiate into functional spermatozoa or ova through a complex, multi-step process.

Avian PGCs were first described by Waldeyer in 1870, and the origin, migration, differentiation, and molecular markers of avian PGCs have been investigated, with a particular focus on poultry such as chicken (*Gallus domesticus*) and Japanese quail (*Coturnix japonica*). The major biological properties of avian PGCs are as follows; firstly, their unique migration pathway to the genital ridge through blood circulation, and secondly, their ability to propagate for the long term in vitro, which remains restricted to the germ line. These have yielded the establishment of unique developmental technologies on the basis of PGC manipulation in birds to generate genetically modified poultry and conserve poultry genetic resources. Therefore, this chapter will deal with our current understanding of the origin, migration, and differentiation of chicken PGCs including the historical background of chicken PGC identification. A comparison with mammals, particularly the mouse, will be described. This chapter will also introduce the in-vitro culture of chicken PGCs.

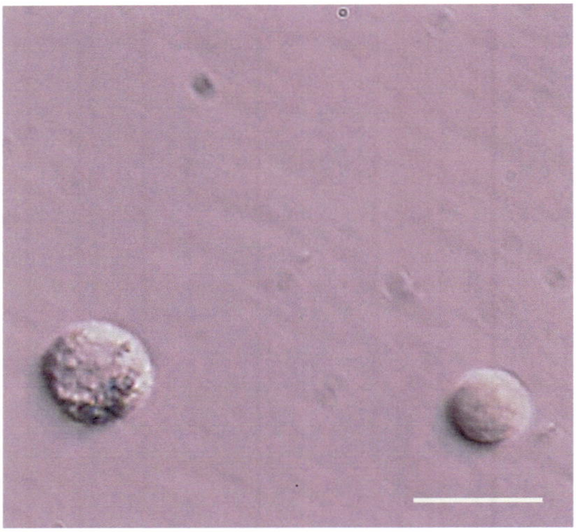

Fig. 1.1 The morphological characteristic of a chicken primordial germ cell (PGC). A chicken PGC collected from embryonic blood. Chicken PGCs have an eccentrically placed nucleus and a considerable amount of lipids in the cytoplasm. A PGC (*left*) and an erythrocyte (*right*). *Bar* = 25 μm

1.2 Determination of PGCs in Birds

In many vertebrate and invertebrate animals, germ cells are separated clearly from somatic cell types in the early stage of development. In these organisms, germ cells often do not arise within the gonads, rather arising from elsewhere. They then migrate to the developing gonads. Germ cells can acquire specification either autonomously, the so-called "preformation" model, or by induction the so-called "epigenesis" model.

In the preformation model, a certain region of the egg cytoplasm contains maternally inherited determinants that specify the cells as PGCs before, or immediately after, fertilization (Eddy 1975; Illmensee and Mahowald 1974; Olsen et al. 1997; Venkatarama et al. 2010). The cytoplasmic region containing such determinants is often referred to as the germ plasm. One of the most well-studied model organisms in the preformation model is the arthropod *Drosophila melanogaster* (common fruit fly). Before blastoderm formation in *Drosophila melanogaster*, precocious cellularisation at the posterior pole of the embryo produces four to five pole cells (Huettner 1923), which are the exclusive progenitors of the PGCs (Technau and Campos-Ortega 1986; Williamson and Lehmann 1996). Transplantation of pole cells (Illmensee and Mahowald 1974, 1976; Illmensee et al. 1976) or forced assembly of pole plasm in ectopic sites, such as at the anterior of the oocyte (Ephrussi and Lehmann 1992), results in PGC formation at these sites. The experimental evidence indicates that the pole plasm is a true germ-cell determinant, and not simply a germ-cell marker. Nematodes (*C. elegans*), fish (*Danio rerio*), and anuran amphibian (*Xenopus laevis*) also employ a similar preformation mechanism with respect to germ cells.

In contrast, in the urodele amphibian axolotl (*Ambystoma mexicanum*) and the mouse (*Mus musculus*), there is no obvious appearance showing the presence of germ plasm in the embryo. Rather, PGCs arise as a result of inductive signals from neighboring cells in the developing embryo. In mice, *bone morphogenetic protein 4* (*Bmp4*) is expressed in the extraembryonic ectoderm on embryonic day 5.5 and 6.25. Epiblast cells that receive this signal express *B-lymphocyte-induced maturation protein-1* (*Blimp-1*), and the signal induces the differentiation of the progenitor cells to PGCs (Ohinata et al. 2005). A single-cell level analysis of gene expression has indicated that the genes *fragilis* and *stella* are upregulated in a subset of the proximal epiblast cells. The expression of these two genes appears to increase the competency of the cells to respond to BMP signals, which directs them to differentiate into PGCs (Saitou et al. 2002). However, even cells of the distal epiblast, which do not normally express *fragilis* or *stella*, can be induced to differentiate into PGCs if placed next to the source of the BMP signals (Tam and Zhou 1996). These results have demonstrated that specification of PGCs in mice is clearly epigenetic and does not depend on maternally inherited determinants. The investigations have been extended to induce mouse PGC-like cells from pluripotent stem cells, including embryonic stem (ES) cells, as well as induced pluripotent stem (iPS) cells, through an epiblast-like state (Hayashi et al. 2011, 2012).

Chicken germ cells were initially thought to originate from the hypoblast (Swift 1914). In 1981, Eyal-Giladi et al. (1981) experimentally showed that avian germ cells are derived from epiblast, not from hypoblast. Since then, avian PGCs have been thought to specify through a gradual epigenetic process around the blastoderm stages (Karagenç et al. 1996). However, studies on the chicken homolog expression pattern of a *Drosophila vasa*, a germline specific RNA-binding protein, have shown that chicken VASA protein forms part of the mitochondrial cloud in oocytes, and localizes in the cleavage furrows during cleavage stages (Tsunekawa et al. 2000). These findings suggest that preformation may be the mechanism for germ-cell determination in chickens, although functional studies have yet to be performed. In the future, ectopic transplantation and functional inhibition of germ plasm is expected to determine the avian germ cell specification mechanism.

1.3 Identification of Avian PGCs

The morphological characteristics of chicken PGCs are as follows: an eccentrically positioned, large, and spherical nucleus with a distinct nucleolus, and the presence of multiple lipid granules in the cytoplasm. PGCs, which circulate temporally through bloodstream to migrate to the genital ridge, are easily distinguished from erythrocytes due to characteristic morphology and larger size (diameter = 10–20 μm: Fig. 1.1). These morphological characteristics of PGCs are widely conserved among avian species (e.g., quail, ducks, pigeons, pheasants, etc.).

Chicken PGCs contain a large amount of glycogen in the cytoplasm; hence, they have been classically detected by staining with periodic acid-Schiff (PAS) reaction (Meyer 1961, 1964). PAS staining efficiently detects chicken PGC only after primitive streak formation from embryonic stage 4 (18–19 h of incubation; Arabic numerals refer to the staging system of Hamburger and Hamilton 1951). In contrast, Japanese quail PGCs have almost no glycogen in the cytoplasm and thus cannot be detected by PAS staining (Swartz and Domm 1972). Although the detailed role of glycogen in avian PGCs remains unclear, this difference in PAS reactivity has been applied to identify the origin of PGCs in the interspecific germline chimeras between the chicken and the Japanese quail. Chicken PGCs have also been immunohistochemically identified with monoclonal antibodies that raised against cell surface antigens such as embryonic mouse antigen-1 (EMA-1), stage-specific embryonic antigen-1 (SSEA-1), SSEA-3, and SSEA-4 (Urven et al. 1988; Karagenç et al. 1996; Jung et al. 2005). The monoclonal antibody 2C9, produced by immunizing the gonad extracts of 6-day-old chick embryos, can recognize male and female PGCs up to day 8 and day 6 respectively (Maeda et al. 1994; Ono et al. 1998). However, these markers are not specific enough to investigate the origin of chicken PGCs. Moreover, these markers do not always recognize entire PGCs during development. For example, EMA-1 recognized only 24% of VASA-positive PGCs in the germinal crescent region at stage 8 (26–29 h of incubation), whereas the percentage of both EMA-1 and VASA-positive PGCs in the genital ridge at stages 20–22 (3–3.5 days of incubation) is drastically increased to 89% (Hen et al. 2014). Although SSEA-1 has been used widely as an appropriate marker to identify and purify the PGCs of chicken embryos, it is an inadequate marker of PGCs, at least at stage 5–7 (19–26 h of incubation) because SSEA-1 expresses in many other cells and tissues in the chicken embryos. Furthermore, SSEA-1 is only expressed by a fraction, albeit increasing, of VASA-positive PGCs between stages 8–19, and not in the entire PGC population (De Melo Bernardo et al. 2012).

Japanese quail–chicken chimeras have been used extensively in the field of developmental biology. To detect the cellular processes of quail cells in Japanese quail–chicken chimeras, several monoclonal antibodies specific to some quail tissues have been produced. Since the monoclonal antibodies QH1, QCR1 and QB2 can recognize PGCs in the Japanese quail, but not in the chicken (Pardanaud et al. 1987; Aoyama et al. 1992; Ono et al. 1996), these have been used to trace xenotransplanted quail PGCs in chicken recipients (Atsumi et al. 2008, 2009; Ishiguro et al. 2009). Lectins such as lycopersicon esculentum agglutinin (LEA) and solanum tuberosum agglutinin (STA) recognize carbohydrate chains expressed on the surface of the PGCs of Japanese quail and chicken respectively. Recent investigations into cell surface markers of avian PGCs have shown they can be purified by antigen–antibody reaction-based systems such as magnetic-activated cell sorting (MACS) or florescence-activated cell sorting (FACS) (Ono and Machida 1999; Mozdziak et al. 2005; Kang et al. 2008).

Although the mechanisms for the segregation of germ cells vary widely across species, the proteins expressed in the germ cells are remarkably conserved. Up to now, several germline specific molecular markers for avian germ cells have been investigated as described below. The *vasa* gene, a DEAD-box RNA helicase, originally discovered in *Drosoplila*, is widely conserved among various organisms, including *C. elegans*, *Xenopus laevis*, *Danio rerio*, *Oncorhynchus mykiss*, *Mus musculus*, and *Homo sapiens* (Lasko and Ashburner 1988; Roussell and Bennett 1993; Fujiwara et al. 1994; Olsen et al. 1997; Yoon et al. 1997; Castrillon et al. 2000; Yoshizaki et al. 2000; Sengoku et al. 2006). In mouse, it has been shown that Vasa participated in the biosynthesis of piRNA (Kuramochi-Miyagawa et al. 2010). In 2000, Tsunekawa et al. isolated the *vasa* homologue gene in chickens, and found its products in the chicken germ cells. With the exception of the male elongated spermatid stage and spermiogenesis, chicken VASA is expressed in the cytoplasm of germ cells in both sexes. Hence, chicken VASA has been widely used as a specific marker for chicken germ cells. The vasa is crucial for the initiation of germ-cell differentiation and meiosis in *Drosophila melanogaster* (Ghabrial and Schupbach 1999), but its role in chicken remains unclear. Other chicken germ-cell-specific molecular markers include the RNA-binding protein, chicken dead end homolog (also known as DND1) (Aramaki et al. 2007, 2009), and deleted in azoospermia-like (DAZL) (Kito et al. 2010). Chicken DND1 is localized in the nuclei of PGCs, while chicken DAZL is present in the nuclei and cytoplasm of PGCs. Chicken VASA and DND1 are expressed in both male and female germ cells, whereas DAZL expression is not detected in premature chicken ovaries that are composed of oocytes and gonadal somatic cells. DND1 is critical for germ-cell development in zebrafish and mice (Weidinger et al. 2003; Youngren et al. 2005). On the other hand, DAZL functions in germ-cell differentiation are variable across species (Houston and King 2000; Eberhart et al. 1996; Karashima et al. 2000; Ruggiu et al. 1997). Recent development of PGC-mediated gene-manipulation techniques will allow us to investigate the functions of these germline-specific genes in chicken germ cells.

1.4 Development and Migration of Chicken PGCs

In birds, fertilization takes place in the infundibulum part of the oviduct, after which a yolky zygote undergoes blastoderm formation along with egg formation (deposition of egg albumen, shell membrane, and egg shell) through the passage of the oviduct. Chicken development in the first 24 h after fertilization takes place in the oviduct, and in the remaining 21 days in the egg. The distribution of chicken VASA protein in cleavage-stage chicken embryos was described by Tsunekawa et al. (2000). Avian embryos exhibit discoidal meroblastic cleavage; the first cleavage furrow appears centrally in the blastodisc, and subsequent cleavages follow to form a single-layered tissue. At first cleavage (5–6 h after ovulation), a germplasm-like structure containing the aforementioned chicken VASA protein in the fertilized chicken ovum is shown

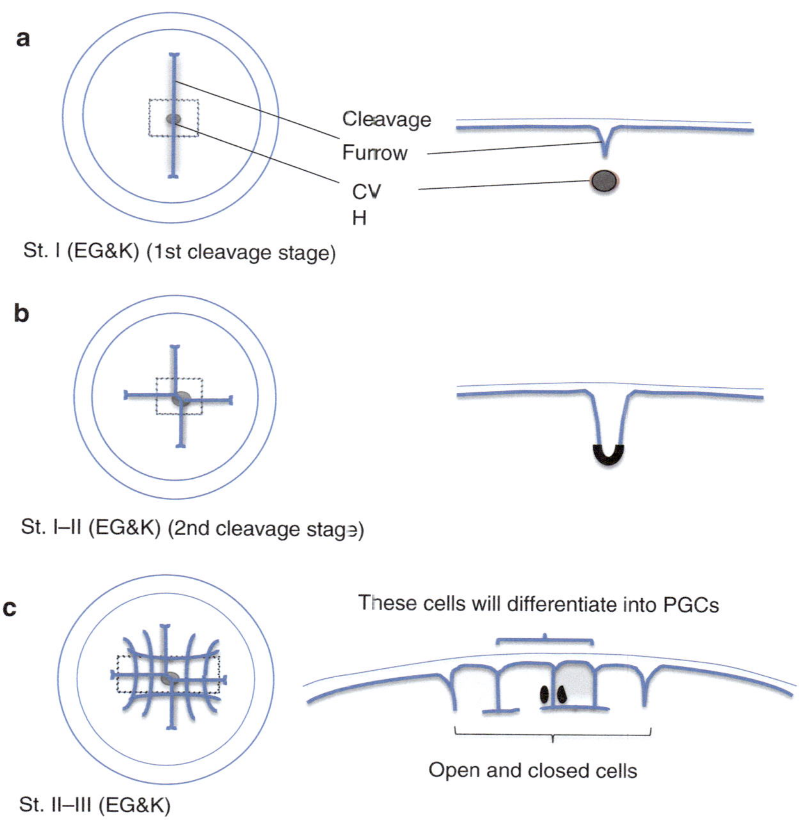

Fig. 1.2 Localization of chicken VASA protein in early cleavage blastoderms. It is thought that the chicken VASA protein is maternally inherited to a zygote, then cells that incorporate a germplasm including VASA protein give rise to PGCs. **a** A chicken embryo at stage I, viewed from the dorsal side (*left*), and viewed from the lateral side (*right*). **b** and **c** A chicken embryo at stage I–II (**b**) and at stage II–III (**c**) respectively. St. I (EG&K): developmental stage I classified by Eyal-Giladi and Kochav (1976). (These figures were reconstructed based on the articles of Tsunekawa et al. (2000) Development 127:2741–2750. and Eyal-Giladi and Kochav (1976) Dev Biol 49:321–337)

(Fig. 1.2a). This VASA-positive structure is spherical, with a diameter of around 20 μm, and is located in the basal part of the cleavage furrow. At the second cleavage, 20 min after the first cleavage, the VASA-positive structure changes to a "V" shape at the basal part of the cleavage furrow (Fig. 1.2b). Chicken VASA protein is localized to patch-like structures near the ventral cleavage furrow at the 15-blastomere stage. At stage III [(8–9 h after ovulation), Roman numerals refer to the staging system of Eyal-Giladi and Kochav 1976], the VASA-positive structure is locally observed in the center of a specific blastomere or in specific cells formed at the center of the blastodisc by horizontal cleavage (Fig. 1.2c). Thereafter, equatorial and vertical cleavages divide the blastodisc into a 5–6 cell layered-tissue, the so-called blastoderm. Between the central portion of the blastoderm and the yolk, a thin fluid-filled cavity, called the subgerminal cavity, appears. At stage VI (approximately 14 h after ovulation), granular spherical structures containing chicken VASA protein are localized in the ventral cytoplasm of 6–8 cells among a total of approximately 300 blastoderm cells, which reside in the center of the blastoderm, as observed in stage X embryos (freshly oviposited). These VASA-positive cells are hypothesized to be the precursor cells for PGCs. Thereafter, the deep cells in the central portion of the blastoderm collapse and die, leaving behind the 1-cell-thick area pellucida, which forms most of the actual embryo. In contrast, the peripheral ring of blastoderm cells persist, and their deep cells give rise to the area opaca, which comprises 2–3 layers of cells.

In stage X chicken embryos, which consist of 20,000–60,000 cells, approximately 30–130 VASA-positive PGCs are scattered centrally in the area pellucida of the blastoderm (Fig. 1.3) (Ginsburg and Eyal-Giladi 1987; Tsunekawa et al. 2000; Nakamura et al. 2007). At this developmental stage, most of the area pellucida cells remain at the surface, forming an upper layer, the so-called epiblast, while other cells migrate into the subgerminal cavity to form the primary hypoblast. Immediately after primary hypoblast formation, a sheet of cells, derived from the deep cells at the posterior margin of the blastoderm, migrates anteriorly beneath the surface to form a lower layer, called the secondary hypoblast. The gastrulation process starts with the formation of the primitive streak initially formed by the accumulation of the cells between the epiblast and hypoblast, followed by a thickening of the epiblast at the posterior marginal zone at stage 2 (6–7 h of incubation). The epiblast cells migrating through the anterior end of the primitive streak pass down into the blastocoel and migrate anteriorly, forming the endoderm, head meso-

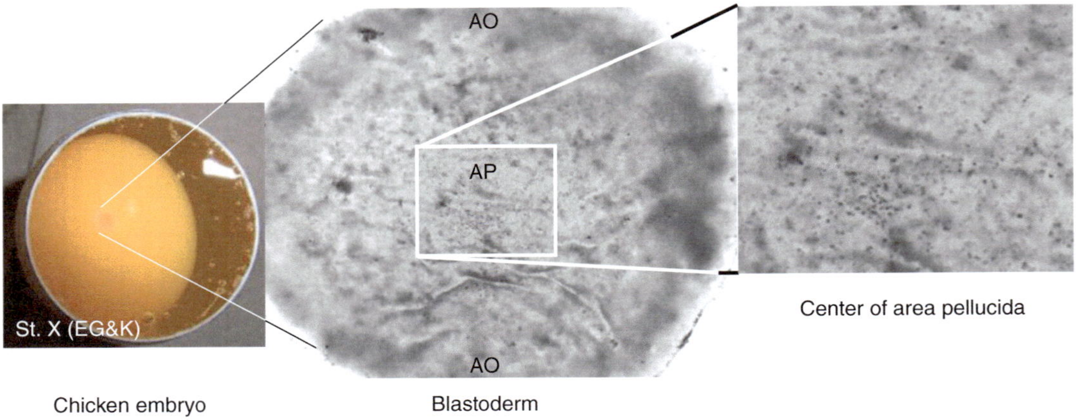

Fig. 1.3 Chicken PGCs are scattered at the center of the blastoderm at stage X. A chicken embryo at stage X just after oviposition (*left*). The whole of the chicken blastoderm was immunostained with anti-chicken VASA antibodies, viewed from the dorsal side (*middle*), and a high magnification view of the boxed zone (*right*). *AO* area opaca, *AP* area pellucida. (Revised Nakamura et al. (2007) Poult Sci 86:2182–2193)

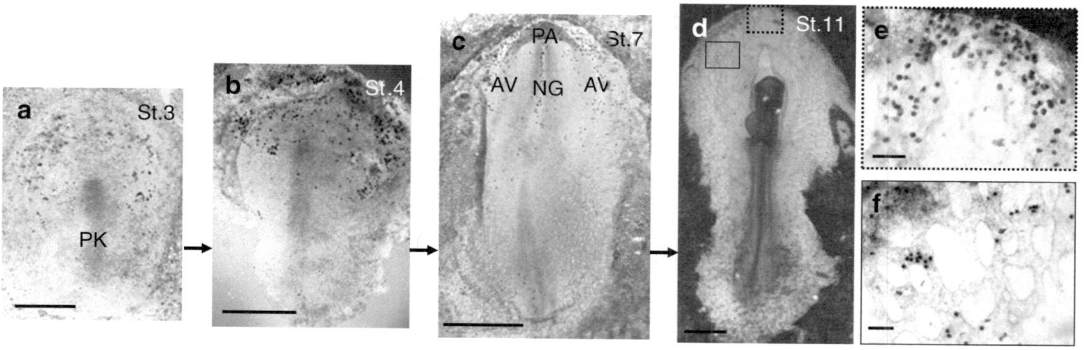

Fig. 1.4 Distribution of chicken PGCs during early development [stage 3–11 (HH)]. Whole embryos at stage (St.) 3(**a**), St. 4(**b**), St. 7(**c**) and St. 11(**d**) were immunostained with anti-chicken VASA antibodies. The *black and dotted boxed zone* in **d** corresponds to **e** and **f** respectively. A portion of the region anterior to the head of St.11 (**e**). A portion of the blood vessels of St.11 (**f**). *PK* primitive streak, *PA* proamnion, *AV* amniocardiac vesicle, *NG* neural groove. *Arrows* indicate the vicinity of the anterior vitelline veins where PGCs transit from the extraembryonic region to the intraembryonic vascular system. *Bars* = 1 mm (**a**, **b**, **c**, **d**), 100 μm (**e**, **f**). (Revised Nakamura et al. (2007) Poult Sci 86:2182–2193)

derm, and notochord. Endodermal cells migrate anteriorly and eventually displace the hypoblast cells. These hypoblast cells are confined to a region in the anterior extraembryonic region known as the germinal crescent. Translocation of PGCs from the central region of the area pellucida to the germinal crescent is thought to be due to passive movement caused by formation of the primitive streak. At stage 3 (12 h of incubation), VASA-positive PGCs distribute with a horseshoe-shaped pattern in the anterior region of the primitive streak (Fig. 1.4a). Since VASA-positive PGCs are predominantly located on the dorsal surface of the hypoblast at this developmental stage, they migrate from the epiblast toward the hypoblast by stage 3. At stage 4 (18–19 h of incubation), the dense accumulation of PGCs in the presumptive amniocardiac vesicle of the germinal crescent region occurs (Fig. 1.4b). During stages 7–9 (23–33 h of incubation), a major part of the PGCs distribute in the proamnion and the amniocardiac vesicle (Fig. 1.4c). Chicken PGCs during the primitive streak formation and somitogenesis are highly proliferative because the total PGC number increases dramatically (approximately threefold increase in the cell numbers during stage X to stage 9 progression).

At stage 10 (33–38 h of incubation), PGCs begin to concentrate in the splanchnopleura, anterior to the head of the embryo. At stage 11 (40–45 h of incubation), PGCs begin to transit from the extraembryonic region to the intraembryonic vascular system, and this coincides with the formation of the anterior vitelline veins (Fig. 1.4d). The majority of PGCs accumulate in the vicinity of the anterior vitelline veins at early stage 11 (Fig. 1.4e). At late stage 11, PGCs begin to be found in the sinusoids (Fig. 1.4f) with a decrease in the number of PGCs in the vicinity of the anterior vitelline veins. These observations indicate that chicken PGCs mainly infiltrate the vascular system from this region (Nakamura et al. 2007). The majority of PGCs concentrate axially in the sinus terminalis, and favor transport axially via the anterior vitelline veins into the embryonic circulation during stages 12–14 (De Melo Bernardo et al. 2012).

Avian PGCs utilize the vascular system as a vehicle to transport themselves to the genital ridge, and these PGCs are often called circulating PGCs (referred to as "cPGCs"). This unique biological property is only conserved in birds and some reptiles, including snakes, and some lizards. The concentration of cPGCs reaches a peak in the bloodstream at stage 14 (50–53 h of incubation; Tajima et al. 1999). At stage 15 (54 h of incubation), chicken cPGCs begin to concentrate to the vitelline artery in the lateral plate mesoderm

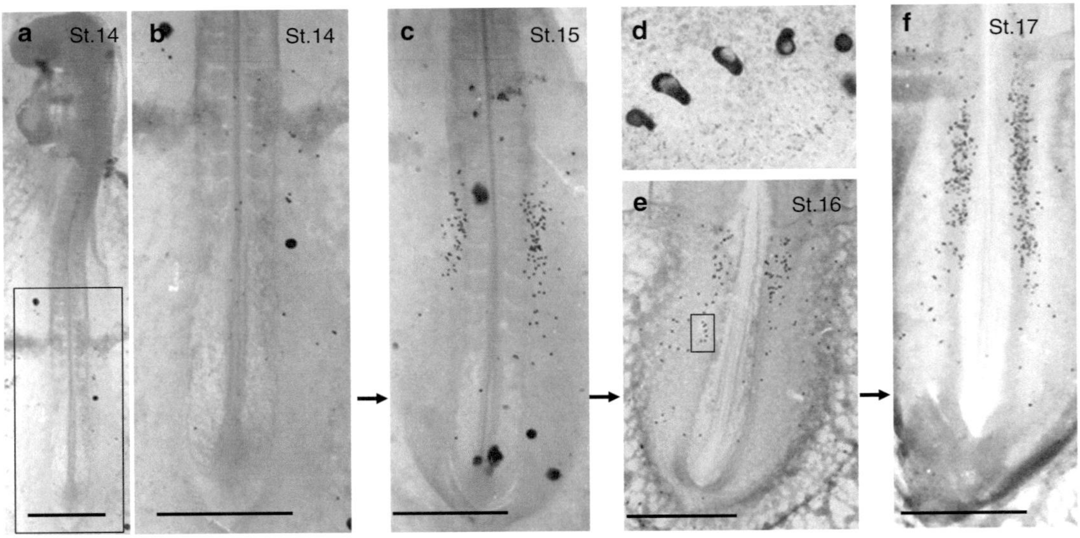

Fig. 1.5 Migration of chicken primordial germ cells (PGC) from the blood vessel to the intermediate mesoderm (stage 14–17 (HH)). Whole embryos at stage (St.) 14(**a**), St. 15(**c**), St. 16(**e**) and St. 17(**f**) were immunostained with anti-chicken VASA antibodies. *Black dot* PGCs. The *black boxed zone* in **a** corresponds to **b**. The *black boxed zone* in **e** corresponds to **d**. *Bars* = 1 mm. (Revised Nakamura et al. (2007) Poult Sci 86:2182–2193)

(Fig. 1.5a–c). Thereafter, PGCs migrate to the intermediate mesoderm-derived coelomic epithelium by extending lamellipodia (Fig. 1.5d, e). The coelomic epithelium then thickens and becomes the genital ridge (Swartz 1982; Didier et al. 1990; Yoshinaga et al. 1992). Interestingly, it is known that chicken PGCs preferentially colonize the left genital ridge both in males and females as shown in Fig. 1.5 (Van Limborgh 1968; Nakamura et al. 2007; Intarapat and Stern 2013; De Melo Bernardo et al. 2012). It has been calculated that the colonization efficiency of chicken cPGCs into the genital ridge is over 60% (Yasuda et al. 1992; Ishiguro et al. 2009). Until stage 20 (72 h of incubation), a majority of PGCs accumulate at the genital ridge (Swift 1914; Nieuwkoop and Sutasurya 1979; Nakamura et al. 2007) while around 10% of PGCs ectopically localize to the extra-gonadal region, including the embryonic head, mesonephros, and intestinal membrane. It is believed that these ectopically localized PGCs are eventually eliminated, mostly by apoptosis. The detailed mechanisms of the former (from the extraembryonic region to the intravascular system) and latter (from the intravascular system to the genital ridge) stages of PGC migration are largely unknown. In mice and zebrafish, involvement of the interaction of peptide chemokine, stromal cell-derived factor 1 (SDF-1), also known as C-X-C motif chemokine ligand 12 (CXCL12) with its receptor, C-X-C chemokine receptor type 4 (CXCR4), also known as fusin or cluster of differentiation 184 (CD184), has been suggested in the directional migration of PGCs toward the future gonad (Doitsidou et al. 2002; Knaut et al. 2003; Ara et al. 2003; Molyneaux et al. 2003). Chicken PGCs would also use the same SDF-1/CXCR4 chemokine signaling to migrate toward the genital ridge, since the chicken genital ridge and PGCs express *SDF-1* and *CXCR4* respectively (Stebler et al. 2004). Indeed, ectopically expressed chicken SDF-1α has shown an ability to accumulate PGCs at those positions (Stebler et al. 2004). In addition, it has been suggested that stem-cell factor (SCF), also referred to as kit-ligand, and its receptor c-KIT, also known as CD117, may be involved in the normal migration and proliferation of PGCs in the mouse (Mintz and Russell 1957; Donovan 1994). On the other hand, in the chicken, the putative secreted form of SCF (SCF1) functions as a chemoattractant in the migration of PGCs, whereas the membrane-bound form of SCF (SCF2) enhances proliferation of PGCs cooperatively with fibroblast growth factor 2 (FGF2) (Srihawong et al. 2015; Miyahara et al. 2016). As in the mouse, chicken PGCs are used for chemotaxis during the latter stage of PGC migration. It therefore appears that SDF-1 and SCF1 gradients guide chicken PGCs toward the genital ridge after they leave the vascular network.

1.5 Differentiation of Chicken Germ Cells

Once chicken PGCs settle in the genital ridge, they expand in number as the gonad enlarges (Fig. 1.6). The PGCs colonizing in the gonads are often called gonadal PGCs (gPGCs) or gonadal germ cells (GGCs). The undifferentiated gonads at stages 27–28 (5.0–5.5 days of incubation) form the primary sex cords, which develop after the rete cords (Romanoff 1960). The primary sex cord, which contains the gPGCs, protrudes into the mesenchymal region of the gonad. Gonadal somatic sex differentiation starts at stage 29 (6 days of incubation) and this gonadal sex differentiation is characteristic in birds; the gonads develop symmetrically in males, but asymmetrically in females, with only the left gonad developing. In males, the primary sex cord becomes the seminiferous tubule, and the rete cord becomes the tubules that connect the Wolffian duct (also known as the mesonephric duct) with the seminiferous tubule. The primary sex cord of the left ovary regresses in females at stages 31–38 (7–12 days of incubation) and the germinal epithelium proliferates, leading to the formation of the cortex via the formation of the secondary sex cord containing germ cells. In females, only the left ovary and left Müllerian duct (also known as the paramesonephric ducts) develop, the latter eventually differentiating into the oviduct. Meanwhile, the right ovary and Müllerian duct mostly disappear before hatching.

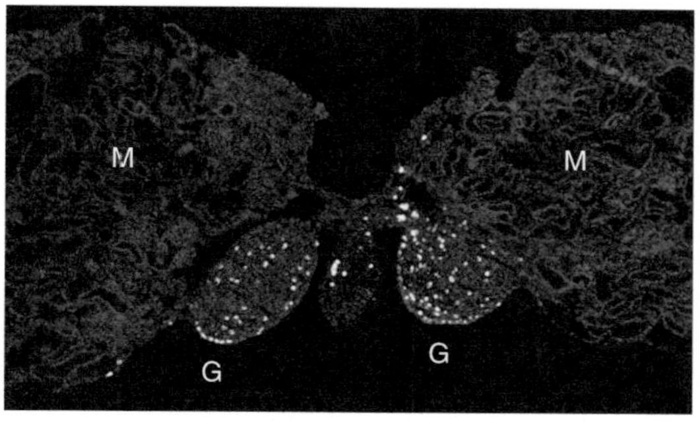

Fig. 1.6 Chicken germ cells colonized in the gonads of a male 7-day-old embryo. A section of the gonads from a male chicken embryo incubated for 7 days was immunostained with anti-chicken VASA antibodies. *White dots* PGCs, *G* a gonad, *M* mesonephros (photograph by Takahiro Tagami)

Male germ cells do not proliferate actively from the time of gonadal sex differentiation (Mendez et al. 2005), while female germ cells proliferate considerably from at least as early as stage 35 (9 days of incubation) (Hughes 1963). It has been estimated that the total population of female germ cells in the left gonadal cortex increases about twenty-five fold between stages 35 and 43 (9–17 days of incubation) (Hughes 1963). In contrast, female germ cells in the right gonad, where the cortex is ill developed, proliferate in the underlying medulla, but they undergo apoptosis until hatching. Meiosis is initiated only in females at a late stage of embryonic development (at stage 41: 15.5 days of incubation).

Retinoic acid (RA) is the key factor controlling meiotic initiation in many animal species. RA synthesis in vertebrate embryos is tightly regulated by the tissue-specific expression of three enzymes, retinaldehyde dehydrogenase (RALDH) including RALDH1, RALDH2, and RALDH3. In chicken embryos, RALDH2 is the major enzyme responsible for RA synthesis, and sites of *RALDH2* gene expression correlate with chicken embryo RA production/release, as also observed in the mouse (Swindell et al. 1999). In the chicken embryo, *RALDH2* is robustly expressed in the gonads of both sexes from stage 30 (6.5 days of incubation), and expression persists throughout development. *RALDH2* expression localizes to the seminiferous cords in males, while in females it localizes in the proliferating cortex of the left gonad and in the medulla of the right gonad (Smith et al. 2008). In the chicken, the expression of RALDH2 is coordinated with the specific expression of a RA-degrading enzyme, cytochrome P450 family 26, subfamily B member 1 (CYP26B1). In males, *CYP26B1* is strongly expressed in the seminiferous cords of developing-testes from stage 30, and expression continues throughout development. In females, *CYP26B1* is initially expressed in the medulla of both gonads from stage 30, as in males, but expression becomes concentrated in the juxtacortical medulla, at the border between the cortex and medulla, from stage 38 (12.5 days of incubation), and expression declines thereafter. The developing ovarian cortex, at the site of germ-cell meiosis, did not show CYP26B1 expression throughout development (Smith et al. 2008). Two classes of receptors, RA receptors (RARs) and retinoid X receptors (RXRs), which belong to a large family of nuclear hormone receptors, mediate RA signaling. Three subtypes of the gene for RAR α, β, and γ, have been identified in chicken and are expressed in distinct spatial and temporal patterns during embryogenesis. In the chicken embryo, *RARβ* has been shown to prominently express in the germ cells in the gonads of both sexes from stage 30, and expression persists throughout development, except in females, where expression declines in the right gonad (Smith et al. 2008). RA signaling induced the expression of the Stimulated by retinoic acid 8 (STRA8) gene, a premeiotic germ cell marker, only in the left female gonads from stage 38. No STRA8 expression was detectable in the right female gonad, consistent with the lack of meiotic cells. Subsequently, synaptonemal complex 3 (SYCP3), a meiosis structural protein, expresses in the germ cells in the developing cortex in the female left ovary from stage 41 (15.5 days of incubation) but not the male developing testes (Smith et al. 2008). Based on the expression patterns of the two RA metabolism enzymes together with the timing of gene expression involving meiosis in chicken germ cells, it could be concluded that RALDH2 and CYP26B1 strongly co-express in the seminiferous cords in the developing male gonads, with no accumulation of RA to influence germ cells (Fig. 1.7). In contrast, high expression of RALDH2 persisted in the cortex with a low level of CYP26B1 in the left female gonad, leading to RA accumulation to induce germ cell meiosis (Fig. 1.7). Unlike in the chicken, in the mouse, *RALDH2* is expressed in the mesonephric kidney, not in the gonads. RA enters the gonad from the mesonephric kidney, and is degraded in males by CYP26B1 that is up-regulated in the gonads just prior to germ cell mitotic arrest (Bowles et al. 2006; Menke et al. 2003), whereas the absence of CYP26B1 allows RA to induce germ cell meiosis. It therefore appears that a role for RA is conserved between birds and mammals.

1.6 Sexual Differentiation of Avian Germ Cells

In avian species, sex is determined at the time of fertilization with the combination of the sex chromosomes in the fertilized egg (ZZ are males and ZW are females, respectively). Accordingly, after colonization in the genital ridge, male PGCs (ZZ) in the testes differentiate into spermatozoa, while female PGCs (ZW) in the left ovary differentiate into ova. Generally, the sexual differentiation of germ cells is induced by various factors from the surrounding gonadal somatic cells. To what extent does the sex of PGCs—male (ZZ) or female (ZW)—contribute to the production of sperm and ova? To answer this question requires histological and test-cross analysis of germ cells in the gonads of chimeric embryo, where both ZZ and ZW or opposite sex of somatic cells present. To demonstrate whether PGCs could differentiate into functional gametes in the mixed-sex gonads, Kagami et al. (1995, 1997) transplanted chicken blastodermal cells from stage X into the opposite sex of stage X chicken embryos. As mentioned above, chicken blastodermal cells at this stage include 30–150 PGCs per embryo. A progeny test revealed that PGCs (ZZ) from genetically male blastoderms could differentiate into functional ova with a Z chromosome (Z-bearing ovum) in a mixed-sex ovary of female chickens. It was also demonstrated that Z-derived, but not W-derived, offspring were obtained from male mixed-sex chimeras by test-cross analysis. In the

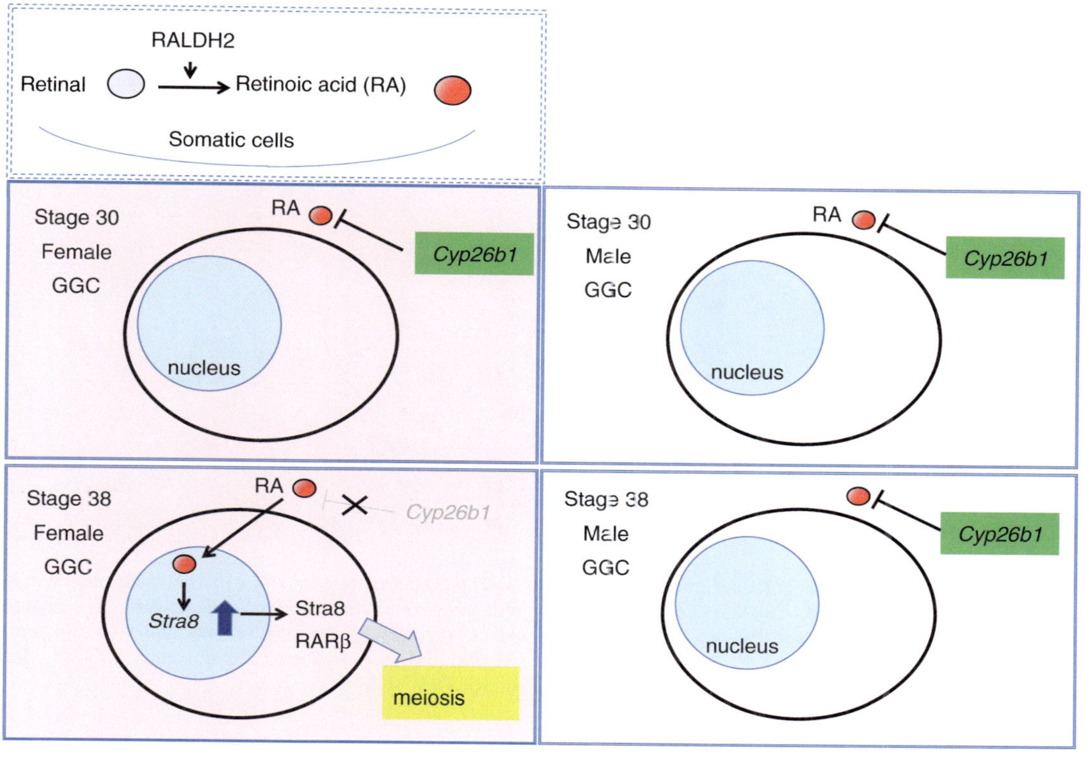

Fig. 1.7 A comparison of retinoic acid (RA) metabolism in the female and male chicken embryos. RA regulates meiotic entry in developing chicken gonads through the expressions of RA-synthesizing enzyme RALDH2 and RA-degrading enzyme CYP26B1. Both RALDH2 and CYP26B1 express in both sexes of the gonads of the chicken embryos at stage 30. In females, persistent expression of RALDH2 in the cortex of the left female gonad, where germ cells are present, but exclusion and down-regulation of CYP26B1 expression, allows the exposure of RA to germ cells that are expressing the RA receptor. RA signaling induces stra8 expression at stage 38, then meiotic initiation occurs. In males, co-expression of RALDH2 and CYP26B1 in the seminiferous cords in the gonads potentially allow no accumulation of RA that influences germ cells

semen of these mixed-sex male chimeras, the W-specific sequence was hardly detected. Therefore, it can be concluded that chicken PGCs obtained from the blastodermal stage have the ability to differentiate into functional gametes in the mixed-sex gonads. Since the fertility of chicken W-bearing spermatozoa was confirmed by intra-cytoplasmic sperm injection (Takagi et al. 2007), the infertility of W-bearing spermatozoa in mixed-sex chimeras may be caused by no or low motility, low survivability, and/or impaired differentiation during spermatogenesis. Meanwhile, germline chimeric chickens have been produced by the intravascular transplantation of purified female cPGCs (ZW) into male recipients. Histological analysis showed that female cPGCs (ZW) could differentiate into spermatogonia in the testes and could pass through meiosis I and II in adapting themselves to a male environment; however, the spermiogenesis (morphological formation of spermatozoa from spermatid) was markedly impaired as shown in Fig. 1.8 (Tagami et al. 1997, 2007; Tagami and Kagami 1998). A similar phenotype was reported in the testes of masculinized chickens, in which all cells are initially ZW in chromosome constitution, by inhibition of aromatase (an enzyme involved in estrogen synthesis) before gonadal sex differentiation (Abinawanto et al. 1998). It, therefore, seems that the chromosomal sex of testicular somatic cells does not affect the spermatogenesis of ZW germ cells in the testes. Test-cross

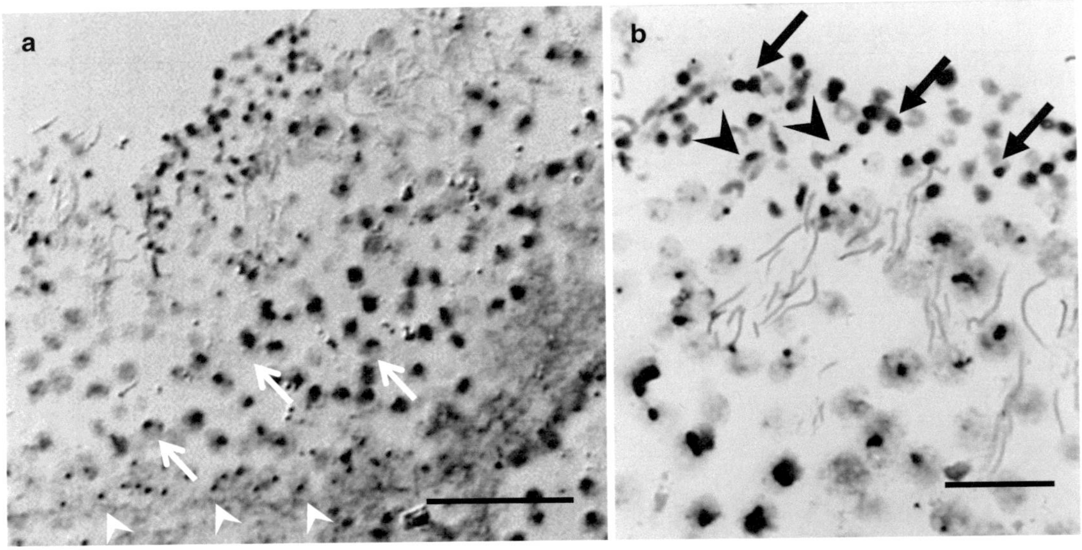

Fig. 1.8 Histological analysis of female germ cells (ZW) in the testes of a male recipient chicken (ZZ). This germline chimeric chicken was made by injection of female PGCs (ZW) into a male 2.5 days embryo (ZZ). W chromosome-bearing germ cells are detected in these testis sections. **a** W-bearing spermatogonia (*white arrowhead*) and spermatocytes (*white arrow*) were observed. **b** W-bearing round (*black arrow*) and elongating spermatids (*black arrowhead*) were observed. *Bars* = 50 mm. (Reproduced from Takahiro Tagami et al. (2007): Mol Reprod Dev 74:68–75)

analysis has demonstrated that both male and female PGCs obtained from the blood circulatory stage barely differentiate into functional gametes in the opposite sex gonads (Tagami et al. 2007; Naito et al. 1999). Considered together, chicken PGCs may have sexual bipotency at the blastoderm stage, but this is lost by the blood circulatory stage.

1.7 Culture of Chicken PGCs

In 1993, Tajima et al. (1993) reported the successful generation of a germline chimeric chicken by transplanting 100 cPGCs into a recipient embryo. However, only a small number of cPGCs (10–100) can be obtained in a limited time period from incubated, 2.5-day-old embryos, which has imposed limitations on the use of this technique. Consequently, there is demand for a technique to allow the mass propagation of PGCs in vitro. In 1995, Chang et al. (1995) cultivated chicken PGCs in vitro for 4 days using stroma cells derived from the genital ridge of 5-day-old embryos as feeder cells in a 199 culture medium containing insulin-like growth factor 1 (IGF-1), FGF2, leukemia inhibitory factor (LIF), and serum. Although chicken PGCs have been able to maintain themselves in an undifferentiated state without losing germline competency (Chang et al. 1997), this culture method has rarely been applied to genetic resource conservation and transgenic chicken production because the in-vitro growth of PGCs is very slow and transient.

In 2006, van de Lavoir et al. (2006) succeeded in the long-term culture of chicken PGCs in vitro using a culture condition similar to that used for ES cells. In their study, cPGCs isolated from embryonic blood were expanded on the feeder of either Sandoz inbred mouse-derived thioguanine-resistant and ouabain-resistant (STO) fibroblast or buffalo rat liver (BRL) cells in knockout-Dulbecco's modified Eagle's medium conditioned with BRL cells, which are known to secrete LIF, SCF, and IGF-1. 7.5% fetal calf serum, 2.5% chicken serum, SCF, and human recombinant FGF2 (Fig. 1.9) was also included. In this culture condition, male PGCs could be

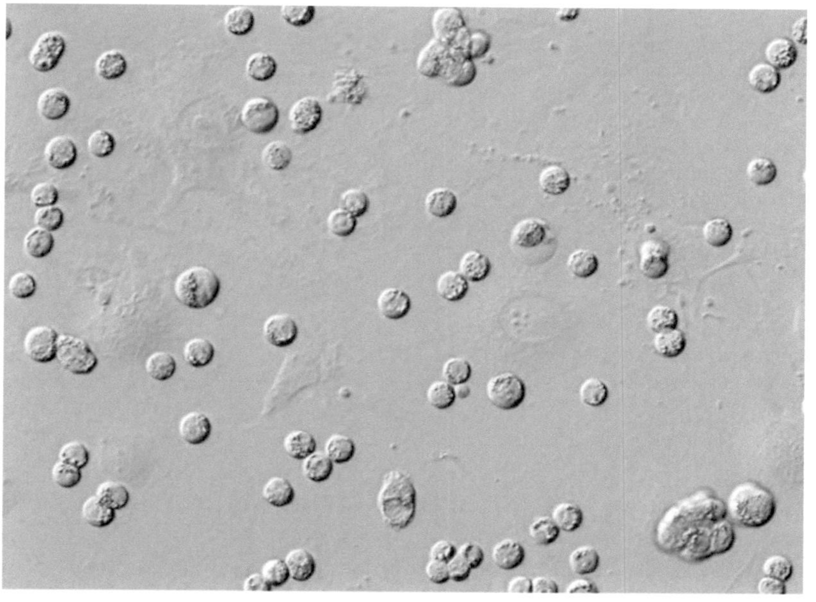

Fig. 1.9 Chicken PGCs cultured on BRL feeder cells. These PGCs were propagated in vitro in the culture condition described in Miyahara et al. (2014) J Poult Sci 51:87–95. *Bar* = 100 μm (photograph by Daichi Miyahara)

maintained with their restriction to the germline for over 100 days, whereas female PGCs were not (maximum 77 days culture). Establishment of a long-term culture of chicken PGCs further enabled the generation of transgenic chickens through the transgenesis of PGCs in vitro. In contrast to the chicken, mouse PGCs can only be expanded as lineage-restricted germ cells for short periods in vitro. Mouse PGCs isolated from the genital ridge transformed into multipotent/unipotent stem cells, called embryonic germ cells (EG cells), in a culture containing SCF, LIF, and FGF2 (Matsui et al. 1992; Resnick et al. 1992). The biological characteristics of PGCs that enable maintenance of linage specificity and germline transmission competency in vitro for the long term are unique to avian species among vertebrates. Favorable culture conditions for the survival of chicken PGCs in vitro are low osmotic pressure (up to 250 mosm/kg) and low calcium concentrations (up to 0.15 mM) (Whyte et al. 2015). As this culture system can induce clonal propagation from the primary cultures of PGCs in both sexes with an exceptionally high efficiency, it is likely to be one of the most appropriate culture conditions for the proliferation and survival of chicken PGCs. In addition to the above-mentioned culture system, PGCs have also been cultured using a KAv-1 medium developed by Kuwana et al. (1996; Naito et al. 2015).

1.8 Factors and Mechanisms Involved in Chicken PGC Proliferation

The establishment of long-term chicken PGC cultures has also initiated the investigation of the mechanisms for chicken PGC proliferation in vitro (Fig. 1.10). The main factor involved in chicken PGC proliferation is FGF2 (Choi et al. 2010; Macdonald et al. 2010; Miyahara et al. 2014). In chicken PGCs, FGF2 activates the signaling of mitogen-activated protein kinase (MAPK)/extracellular signal-regulated kinase (ERK) kinase (MEK)/ERK, a member of the mitogen-activated protein kinase family, which promotes the cell cycle and anti-apoptosis. ERK activation does not occur when the FGF2 receptor and MEK are pharmacologically inhibited by PD173074 and PD0325901 respectively. If MEK/ERK signaling is not activated, PGCs are rapidly eliminated in vitro (Choi et al. 2010; Macdonald et al. 2010).

During early development, the SMAD family is strongly phosphorylated in chicken PGCs.

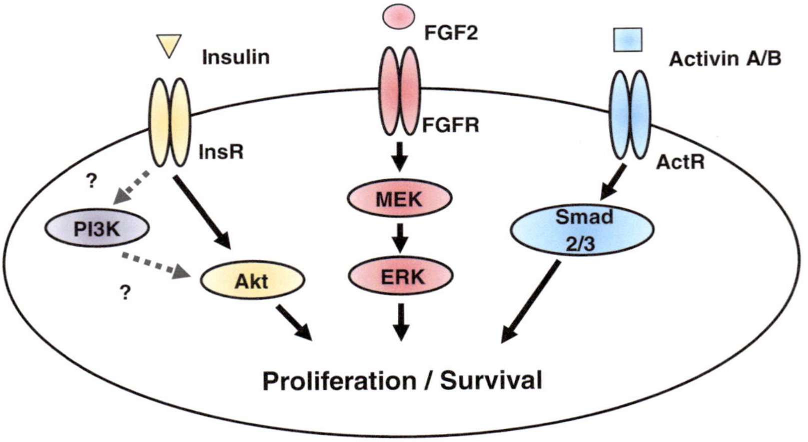

Fig. 1.10 Model for in-vitro proliferation of chicken primordial germ cells (PGCs). A schema of growth factors, receptors, and signaling molecules involved in the proliferation of chicken PGCs. *InsR* insulin receptor, *FGFR* FGF receptor, *ActR* activin receptor. *Dashed arrows* with PI3K indicate that PI3K might activate the Akt directly. (See text)

However, the role of SMAD in chicken PGCs remains unclear, although the loss of SMAD signaling markedly inhibits in-vitro proliferation. Consequently, the elevated phosphorylation of SMAD family members is indispensable for chicken PGC proliferation. The main cytokines that participate in the activation of SMAD signaling include bone morphogenetic protein 4 (BMP4) and activin, which belong to the transforming growth factor beta (TGF-β) superfamily. Of these, activin most strongly activates SMAD signaling in chicken PGCs in vitro (Whyte et al. 2015). Activin regulates the gene that encodes the pluripotency factor NANOG in human ES cells and mouse epiblast stem cells (Vallier et al. 2009; Xu et al. 2008). NANOG is required for the specification and survival of mouse germ cells (Chambers et al. 2007; Yamaguchi et al. 2009). NANOG is also strongly expressed in chicken PGCs; therefore, activin–SMAD interactions are believed to participate in the survival and maintenance of PGCs in an undifferentiated state through regulation of NANOG.

The activation of the serine/threonine kinase Akt (also known as protein kinase B) by phosphorylation is necessary for the proliferation and the survival of mammalian endogenous PGCs and the in-vitro establishment of EG cells (Kimura et al. 2008; Matsui et al. 2014). In chicken PGCs, the inhibition of phosphoinositide 3 kinase (PI3K)-dependent phosphorylation of Akt arrests in-vitro proliferation and promotes apoptosis (Macdonald et al. 2010; Whyte et al. 2015). This suggests that the activation of PI3K/Akt signaling is indispensable for the proliferation and survival of chicken PGCs. Insulin and SCF is the principal activator of PI3k/Akt signaling. The addition of insulin to chicken PGC culture media enhances Akt phosphorylation, supporting the in-vitro propagation and survival of chicken PGCs. Chicken SCF2, but not mouse SCF2, was able to greatly improve the proliferation of chicken PGCs in vitro (Miyahara et al. 2016). However, the proliferative effect of chicken SCF2 on chicken PGCs is cooperative with that of FGF2, and may contribute to PGC division by enhancing the proliferative effect of FGF2.

Conclusion

On Earth, there are currently 10,424 avian species, of which 1375 (13%) are thought to be endangered species (The IUCN Red List of Threatened Species: 2015. Ver 4. http://www.iucnredlist.org/about/summary-statistics#Tables_1_2). Additionally, each country in the world has numerous breeds, as well as species of poultry, that were bred with original directionality, and these are precious genetic resources. In recent years, massive fatalities, caused by highly pathogenic avian influenza, for example, have become problematic, and there is a demand to develop efficient methods to cryopreserve and regenerate poultry genetic resources. Furthermore, techniques practically applying genetically modified chickens to produce useful proteins in egg whites for pharmaceuticals are actively being developed to generate a whole new poultry industry. Such techniques for preserving the genetic resources and modifying the genes of birds have been slow to develop compared to industrial livestock and experimental animals, as micromanipulation of fertilized eggs and ova is extremely difficult because avian ova are contained within masses of yolk.

Avian PGCs will become a potent tool for the preservation of avian genetic resources, in contrast to the preservation of mammalian sperm, ova, and fertilized eggs. Additionally, the development of chicken PGC culturing techniques mentioned here has rapidly improved techniques for producing genetically modified chickens.

References

Abinawanto, Zhang C, Saito N, et al. Identification of sperm-bearing female-specific chromosome in the sex-reversed chicken. J Exp Zool. 1998;280:65–72.

Aoyama H, Asamoto K, Nojyo Y, et al. Monoclonal antibodies specific to quail embryo tissues: their epitopes in the developing quail embryo and their application to identification of quail cells in quail–chick chimeras. J Histochem Cytochem. 1992;40:1769–77.

Ara T, Nakamura Y, Egawa T, et al. Impaired colonization of the gonads by primordial germ cells in mice lacking a chemokine, stromal cell-derived factor 1 (SDF-1). Proc Natl Acad Sci U S A. 2003;100:5319–23.

Aramaki S, Sato F, Kato T, et al. Molecular cloning and expression of dead end homologue in chicken primordial germ cells. Cell Tissue Res. 2007;330:45–52.

Aramaki S, Kubota K, Soh T, et al. Chicken dead end homologue protein is a nucleoprotein of germ cells including primordial germ cells. J Reprod Dev. 2009;55:214–8.

Atsumi Y, Tagami T, Kagami H, et al. Restriction of germline proliferation by soft X-ray irradiation of chicken embryos and its application to chimera production. J Poult Sci. 2008;45:292–7.

Atsumi Y, Yazawa S, Usui F, et al. Depletion of primordial germ cells (PGCs) by X-irradiation to extraembryonic region of chicken embryos and expression of xenotransplanted quail PGCs. J Poult Sci. 2009;46:136–43.

Bowles J, Knight D, Smith C, et al. Retinoid signaling determines germ cell fate in mice. Science. 2006;312:596–600.

Castrillon DH, Quade BJ, Wang TY, et al. The human VASA gene is specifically expressed in the germ cell lineage. Proc Natl Acad Sci U S A. 2000;97:9585–90.

Chambers I, Silva J, Colby D, et al. Nanog safeguards pluripotency and mediates germline development. Nature. 2007;450:1230–4.

Chang IK, Tajima A, Chikamune T, et al. Proliferation of chick primordial germ cells cultured on stroma cells from the germinal ridge. Cell Biol Int. 1995;19:143–9.

Chang IK, Jeong DK, Hong YH, et al. Production of germline chimeric chickens by transfer of cultured primordial germ cells. Cell Biol Int. 1997;21:485–9.

Choi JW, Kim S, Kim TM, et al. Basic fibroblast growth factor activates MEK/ERK cell signaling pathway and stimulates the proliferation of chicken primordial germ cells. PLoS One. 2010;5(9):e12968.

De Melo Bernardo A, Sprenkels K, Rodrigues G, et al. Chicken primordial germ cells use the anterior vitelline veins to enter the embryonic circulation. Biol Open. 2012;1:1146–52.

Didier E, Didier P, Fargeix N, et al. Expression and distribution of carbohydrate sequences in chick germ cells: a comparative study with lectins and the NC-1/HNK-1 monoclonal antibody. Int J Dev Biol. 1990;34:421–31.

Doitsidou M, Reichman-Fried M, Stebler J, et al. Guidance of primordial germ cell migration by the chemokine SDF-1. Cell. 2002;111:647–59.

Donovan PJ. Growth factor regulation of mouse primordial germ cell development. Curr Top Dev Biol. 1994;29:189–225.

Eberhart CG, Maines JZ, Wasserman SA. Meiotic cell cycle requirement for a fly homologue of human Deleted in Azoospermia. Nature. 1996;381:783–5.

Eddy EM. Germ plasm and the differentiation of the germ cell line. Int Rev Cytol. 1975;43:229–80.

Ephrussi A, Lehmann R. Induction of germ cell formation by *oskar*. Nature. 1992;358:387–92.

Eyal-Giladi H, Ginsburg M, Farbarov A. Avian primordial germ cells are of epiblastic origin. J Embryol Exp Morphol. 1981;65:139–47.

Eyal-Giladi H, Kochav S. From cleavage to primitive streak formation: a complementary normal table and a new look at the first stage of the development of the chicken I. General morphology. Dev Biol. 1976;49:321–37.

Fujiwara Y, Komiya T, Kawabata H, et al. Isolation of a DEAD-Family protein gene that encodes a murine homolog of Drosophila vasa and its specific expression in germ cell lineage. Proc Natl Acad Sci U S A. 1994;91:12258–62.

Ghabrial A, Schupbach T. Activation of a meiotic checkpoint regulates translation of Gurken during Drosophila oogenesis. Nat Cell Biol. 1999;1:354–7.

Ginsburg M, Eyal-Giladi H. Primordial germ cells of the young chick blastoderm originate from the central zone of the area pellucida irrespective of the embryo-forming process. Development. 1987;101:209–19.

Hamburger V, Hamilton HL. A series of normal stages in the development of the chick embryo. J Morphol. 1951;88:49–92.

Hayashi K, Ogushi S, Kurimoto K, et al. Offspring from oocytes derived from in vitro primordial germ cell-like cells in mice. Science. 2012;338:971–5.

Hayashi K, Ohta H, Kurimoto K, et al. Reconstitution of the mouse germ cell specification pathway in culture by pluripotent stem cells. Cell. 2011;146:519–32.

Hen G, Friedman-Einat M, Sela-Donenfeld D. Primordial germ cells in the dorsal mesentery of the chicken embryo demonstrate left–right asymmetry and polarized distribution of the EMA1 epitope. J Anat. 2014;224:556–63.

Houston DW, King ML. A critical role for Xdazl, a germ plasm-localized RNA, in the differentiation of primordial germ cells in Xenopus. Development. 2000;127:447–56.

Huettner AF. The origin of the germ cells in *Drosophila melanogaster*. J Morphol. 1923;2:385–422.

Hughes GC. The population of germ cells in the developing female chick. J Embryol Exp Morphol. 1963;11:513–36.

Illmensee K, Mahowald AP. Transplantation of posterior polar plasm in *Drosophila*. Induction of germ cells at the anterior pole of the egg. Proc Natl Acad Sci U S A. 1974;71:1016–20.

Illmensee K, Mahowald AP. The autonomous function of germ plasm in a somatic region of the *Drosophila* egg. Exp Cell Res. 1976;97:127–40.

Illmensee K, Mahowald AP, Loomis MR. The ontogeny of germ plasm during oogenesis in *Drosophila*. Dev Biol. 1976;49:40–65.

Intarapat S, Stern CD. Chick stem cells: current progress and future prospects. Stem Cell Res. 2013;11:1378–92.

Ishiguro S, Minematsu T, Naito M, et al. Migratory ability of chicken primordial germ cells transferred into quail embryos. J Reprod Dev. 2009;55:183–6.

Jung JG, Kim DK, Park TS, et al. Development of novel markers for the characterization of chicken primordial germ cells. Stem Cells. 2005;23:689–98.

Kagami H, Clark ME, Verrinder Gibbins AM, et al. Sexual differentiation of chimeric chickens containing ZZ and ZW cells in the germline. Mol Reprod Dev. 1995;42:379–87.

Kagami H, Tagami T, Matsubara Y, et al. The developmental origin of primordial germ cells and the transmission of the donor-derived gametes in mixed-sex germline chimeras to the offspring in the chicken. Mol Reprod Dev. 1997;48:501–10.

Kang SJ, Choi JW, Kim SY, et al. Reproduction of wild birds via interspecies germ cell transplantation. Biol Reprod. 2008;79:931–7.

Karagenç L, Ginsburg M, Eyal-Giladi H, et al. Origin of primordial germ cells in the prestreak chick embryo. Dev Genet. 1996;19:290–301.

Karashima T, Sugimoto A, Yamamoto M. Caenorhabditis elegans homologue of the human azoospermia factor DAZ is required for oogenesis but not for spermatogenesis. Development. 2000;127:1069–79.

Kimura T, Tomooka M, Yamano N, et al. AKT signaling promotes derivation of embryonic germ cells from primordial germ cells. Development. 2008;135:869–79.

Kito G, Aramaki S, Tanaka K, et al. Temporal and spatial differential expression of chicken germline-specific proteins cDAZL, CDH and CVH during gametogenesis. J Reprod Dev. 2010;56:341–6.

Knaut H, Werz C, Geisler R, et al. A zebrafish homologue of the chemokine receptor Cxcr4 is a germ-cell guidance receptor. Nature. 2003;421:279–82.

Kuramochi-Miyagawa S, Watanabe T, Gotoh K, et al. MVH in piRNA processing and gene silencing of retrotransposons. Genes Dev. 2010;24:887–92.

Kuwana T, Hashimoto K, Nakanishi A, et al. Long-term culture of avian embryonic cells *in vitro*. Int J Dev Biol. 1996;4:1061–4.

Lasko PF, Ashburner M. The product of the Drosophila gene *vasa* is very similar to eukaryotic initiation factor-4A. Nature. 1988;335:611–7.

Macdonald J, Glover JD, Taylor L, et al. Characterisation and germline transmission of cultured avian primordial germ cells. PLoS One. 2010;5(11):e15518.

Maeda S, Ohsako S, Kurohmaru M, et al. Analysis for the stage specific antigen of the primordial germ cells in the chick embryo. J Vet Med Sci. 1994;56:315–20.

Matsui Y, Takehara A, Tokitake Y, et al. The majority of early primordial germ cells acquire pluripotency by AKT activation. Development. 2014;141:4457–67.

Matsui Y, Zsebo K, Hogan BL. Derivation of pluripotential embryonic stem cells from murine primordial germ cells in culture. Cell. 1992;70:841–7.

Mendez C, Carrasco E, Pedernera E. Adenohypophysis regulates cell proliferation in the gonads of the developing chick embryo. J Exp Zool A Comp Exp Biol. 2005;303:179–85.

Menke DB, Koubova J, Page DC. Sexual differentiation of germ cells in XX mouse gonads occurs in an anterior-to-posterior wave. Dev Biol. 2003;262:303–12.

Meyer DB. The intra-embryonic migration of primordial germ cells in staged chick embryos. Anat Rec. 1961;139:314–5.

Meyer DB. The migration of primordial germ cells in the chick embryo. Dev Biol. 1964;10:154–90.

Mintz B, Russell ES. Gene-induced embryological modifications of primordial germ cells in the mouse. J Exp Zool. 1957;134:207–37.

Miyahara D, Mori T, Makino R, et al. Culture conditions for maintain propagation, long-term survival and germline transmission of chicken primordial germ cell-like cells. J Poult Sci. 2014;51:87–95.

Miyahara D, Oishi I, Makino R, et al. Chicken stem cell factor enhances primordial germ cell proliferation cooperatively with fibroblast growth factor 2. J Reprod Dev. 2016;61:143–9.

Molyneaux KA, Zinszner H, Kunwar PS, et al. The chemokine SDF1/CXCL12 and its receptor CXCR4 regulate mouse germ cell migration and survival. Development. 2003;130:4279–86.

Mozdziak PE, Angerman-Stewart J, Rushton B, et al. Isolation of chicken primordial germ cells using fluorescence-activated cell sorting. Poult Sci. 2005;84:594–600.

Nakamura Y, Yamamoto Y, Usui F, et al. Migration and proliferation of primordial germ cells in the early chicken embryo. Poult Sci. 2007;86:2182–93.

Naito M, Matsubara Y, Harumi T, et al. Differentiation of donor primordial germ cells into functional gametes in the gonads of mixed-sex germline chimaeric chickens produced by transfer of primordial germ cells isolated from embryonic blood. J Reprod Fertil. 1999;117:291–8.

Naito M, Harumi T, Kuwana T. Long-term culture of chicken primordial germ cells isolated from embryonic blood and production of germline chimaeric chickens. Anim Reprod Sci. 2015;153:50–61.

Nieuwkoop PD, Sutasurya LA. The migration of the primordial germ cells. In: Nieuwkoop PD, Sutasurya LA, editors. Primordial germ cells in the chordates. London: Cambridge University Press; 1979. p. 113–27.

Ohinata Y, Payer B, O'Carroll D, et al. Blimp1 is a critical determinant of the germ cell lineage in mice. Nature. 2005;436:207–13.

Olsen LC, Aasland R, Fjose A. A vasa-like gene in zebrafish identifies putative primordial germ cells. Mech Dev. 1997;66:95–105.

Ono T, Machida Y. Immunomagnetic purification of viable primordial germ cells of Japanese quail (*Coturnix japonica*). Comp Biochem Physiol A Mol Integr Physiol. 1999;122:255–9.

Ono T, Yokoi R, Aoyama H. Transfer of male or female primordial germ cells of quail into chick embryonic gonads. Exp Anim. 1996;45:347–52.

Ono T, Yokoi R, Maeda S, et al. Settlement of quail primordial germ cells in chicken gonads. Anim Sci Technol. 1998;69:546–55.

Pardanaud L, Buck C, Dieterlen-Liever F. Early germ cell segregation and distribution in the quail blastodisc. Cell Differ. 1987;22:47–60.

Resnick JL, Bixler LS, Cheng L, et al. Long-term proliferation of mouse primordial germ cells in culture. Nature. 1992;359:550–1.

Romanoff AL. The urogenital system. In: Romanoff AL, editor. The avian embryo. New York: Macmillan; 1960. p. 783–862.

Roussell DL, Bennett KL. glh-1, a germ-line putative RNA helicase from Caenorhabditis, has four zinc fingers. Proc Natl Acad Sci U S A. 1993;90:9300–4.

Ruggiu M, Speed R, Taggart M, et al. The mouse Dazla gene encodes a cytoplasmic protein essential for gametogenesis. Nature. 1997;389:73–7.

Saitou M, Barton SC, Surani MA. A molecular programme for the specification of germ cell fate in mice. Nature. 2002;41:293–300.

Sengoku T, Nureki O, Nakamura A, et al. Structural basis for RNA unwinding by the DEAD-box protein *Drosophila vasa*. Cell. 2006;125:287–300.

Smith CA, Roeszler KN, Bowles J, et al. Onset of meiosis in the chicken embryo; evidence of a role for retinoic acid. BMC Dev Biol. 2008;8:85.

Srihawong T, Kuwana T, Siripattarapravat K, et al. Chicken primordial germ cell motility in response to stem cell factor sensing. Int J Dev Biol. 2015;59:453–60.

Stebler J, Spieler D, Slanchev K, et al. Primordial germ cell migration in the chicken and mouse embryo: the role of the chemokine SDF-1/CXCL12. Dev Biol. 2004;272:351–61.

Swartz WJ, Domm LV. A study on division of primordial germ cells in the early chick embryo. Am J Anat. 1972;135:51–70.

Swartz WJ. Acid and alkaline phosphatase activity in migrating primordial germ cells of the early chick embryo. Anat Rec. 1982;202:379–85.

Swindell EC, Thaller C, Sockanathan S, et al. Complementary domains of retinoic acid production and degradation in the early chick embryo. Dev Biol. 1999;216:282–96.

Swift CH. Origin and early history of the primordial germ-cells in the chick. Am J Anat. 1914;15:483–516.

Tagami T, Matsubara Y, Hanada H, et al. Differentiation of female chicken primordial germ cells into spermatozoa in male gonads. Dev Growth Differ. 1997;39:267–71.

Tagami T, Kagami H. Developmental origin of avian primordial germ cells and its unique differentiation in the gonads of mixed-sex chimeras. Mol Reprod Dev. 1998;50:370–6.

Tagami T, Kagami H, Matsubara Y, et al. Differentiation of female primordial germ cells in the male testes of chicken (*Gallus Gallus Domesticus*). Mol Reprod Dev. 2007;74:68–75.

Tajima A, Hayashi H, Kamizumi A, et al. Study on the concentration of circulating primordial germ cells (cPGCs) in early chick embryos. J Exp Zool. 1999;284:759–64.

Tajima A, Naito M, Yasuda Y, Kuwana T. Production of germ line chimera by transfer of primordial germ cells in the domestic chicken (Gallus domesticus). Theriogenology. 1993;40:509–19.

Takagi S, Ono T, Tsukada A, et al. Fertilization and blastoderm development of quail oocytes after intracytoplasmic injection of chicken sperm bearing the W chromosome. Poult Sci. 2007;86:937–43.

Tam PP, Zhou SX. The allocation of epiblast cells to ectodermal and germ-line lineages is influenced by the position of the cells in the gastrulating mouse embryo. Dev Biol. 1996;178:124–32.

Technau GM, Campos-Ortega JA. Lineage analysis of transplanted individual cells in embryos of *Drosophila melanogaster* Part III. Commitment and proliferative capabilities of pole cells and midgut progenitors. Rouxs Arch Dev Biol. 1986;195:489–98.

Tsunekawa N, Naito M, Sakai Y, et al. Isolation of chicken vasa homolog gene and tracing the origin of primordial germ cells. Development. 2000;127:2741–50.

Urven LE, Erickson CA, Abbott UK, et al. Analysis of germline development in the chicken embryo using an anti mouse EC cell antibody. Development. 1988;103:299–304.

Vallier L, Mendjan S, Brown S, et al. Activin/Nodal signalling maintains pluripotency by controlling Nanog expression. Development. 2009;136:1339–49.

van de Lavoir MC, Diamond JH, Leighton PA, et al. Germline transmission of genetically modified primordial germ cells. Nature. 2006;441:766–9.

Van Limborgh J. Le premier indice de la différenciation sexuelle des gonades chez l'embryon de poulet. Arch Anat Microsc Morphol Exp. 1968;57:79–90.

Venkatarama T, Lai F, Luo X, et al. Repression of zygotic gene expression in the Xenopus germline. Development. 2010;137:651–60.

Waldeyer W. Eierstock und Ei. Ein Beitrag zur Anatomie und Entwicklungsgeschichte der Sexualorgane. Leipzig: W. Engelmann; 1870.

Weidinger G, Stebler J, Slanchev K, et al. Dead end, a novel vertebrate germ plasm component, is required for zebrafish primordial germ cell migration and survival. Curr Biol. 2003;13:1429–34.

Whyte J, Glover JD, Woodcock M, et al. FGF, Insulin, and SMAD signaling cooperate for avian primordial germ cell self-renewal. Stem Cell Reports. 2015;5:1171–82.

Williamson A, Lehmann R. Germ cell development in *Drosophila*. Annu Rev Cell Dev Biol. 1996;12:365–91.

Xu RH, Sampsell-Barron TL, Gu F, et al. NANOG is a direct target of TGFbeta/activin-mediated SMAD signaling in human ESCs. Cell Stem Cell. 2008;7:196–206.

Yasuda Y, Tajima A, Fujimoto T, et al. A method to obtain avian germ-line chimaeras using isolated primordial germ cells. J Reprod Fertil. 1992;96:521–8.

Yamaguchi S, Kurimoto K, Yabuta Y, et al. Conditional knockdown of Nanog induces apoptotic cell death in mouse migrating primordial germ cells. Development. 2009;136:4011–20.

Yoon C, Kawakami K, Hopkins N. Zebrafish vasa homologue RNA is localized to the cleavage planes of 2- and 4-cell-stage embryos and is expressed in the primordial germ cells. Development. 1997;124:3157–65.

Yoshinaga K, Fujimoto T, Nakamura M, et al. Selective binding sites of primordial germ cells in chick and quail embryos. Anat Rec. 1992;233:625–32.

Yoshizaki G, Sakatani S, Tominaga H, et al. Cloning and characterization of vasa-like gene in rainbow trout and its expression in the germ cell lineage. Mol Reprod Dev. 2000;55:364–71.

Youngren KK, Coveney D, Peng X, et al. The Ter mutation in the dead end gene causes germ cell loss and testicular germ cell tumours. Nature. 2005;435:360–4.

Sex-Determining Mechanism in Avians

2

Asato Kuroiwa

Abstract

The sex of birds is determined by inheritance of sex chromosomes at fertilization. The embryo with two Z chromosomes (ZZ) develops into a male; by contrast, the embryo with Z and W chromosomes (ZW) becomes female. Two theories are hypothesized for the mechanisms of avian sex determination that explain how genes carried on sex chromosomes control gonadal differentiation and development during embryogenesis. One proposes that the dosage of genes on the Z chromosome determines the sexual differentiation of undifferentiated gonads, and the other proposes that W-linked genes dominantly determine ovary differentiation or inhibit testis differentiation. Z-linked *DMRT1*, which is a strong candidate avian sex-determining gene, supports the former hypothesis. Although no candidate W-linked gene has been identified, extensive evidence for spontaneous sex reversal in birds and aneuploid chimeric chickens with an abnormal sex chromosome constitution strongly supports the latter hypothesis. After the sex of gonad is determined by a gene(s) located on the sex chromosomes, gonadal differentiation is subsequently progressed by several genes. Developed gonads secrete sex hormones to masculinize or feminize the whole body of the embryo. In this section, the sex-determining mechanism as well as the genes and sex hormones mainly involved in gonadal differentiation and development of chicken are introduced.

Keywords

Sex determination • Z chromosome • W chromosome • Sex hormone • Sex reversal

A. Kuroiwa
Hokkaido University, Sapporo, Hokkaido, Japan
e-mail: asatok@sci.hokudai.ac.jp

T. Sasanami (ed.), *Avian Reproduction*, Advances in Experimental Medicine and Biology 1001,
DOI 10.1007/978-981-10-3975-1_2

2.1 The Z and W Chromosomes in Birds

The chicken is the most useful experimental model for birds. Draft genome sequences of chicken have been available since 2004 (International Chicken Genome Sequencing Consortium 2004), and the estimated genome size is approximately 1 billion bp. The chromosomes of a female chicken (*Gallus gallus domesticus*) are shown in Fig. 2.1. The chromosome number of chickens is $2n = 78$, and ten pairs of macrochromosomes (Chromosomes 1–9, Z, and W) and 29 pairs of microchromosomes can be observed. As with the chicken, the typical karyotype of birds consists of several pairs of macrochromosomes and many microchromosomes. The macrochromosomes are distinguishable by their size, morphology, and banding pattern, which can be obtained by treatment with enzymes or salt solutions. By contrast, microchromosomes are too small to distinguish individually.

The sex of birds is genetically determined. Females have the heterogametic sex chromosomes ZW, whereas males have the homogametic ZZ. This sex-determining mechanism is highly conserved in avians, which include nearly ten thousand species.

The Z chromosome is relatively large. The estimated size of the Z chromosome is 82 million bp, and the number of genes reported on the Z chromosome is 1137 (NCBI, Gnome: http://www.ncbi.nlm.nih.gov/genome/111?genome_assembly_id=22848). The W chromosome has degenerated during evolution, and its estimated size is 5.16 million bp. At least 50 genes are estimated on the W chromosome.

2.2 Gonadal Development in Chickens

Development of the chicken embryo takes 21 days. The developmental stage is described by 45 distinct stages under the Hamburger and Hamilton (HH) system (Hamburger and Hamilton 1951). The development of embryo can also be staged according to days of embryonic development post-lay (embryonic day [E]).

The developmental process of the urogenital system in chickens is similar to that in other amniotes. Urogenital tissues arise from the intermediate mesoderm, and the first evidence of gonadal development is observed at E3.5, characterized by a thickening of the coelomic epithelium ventral to the mesonephros. Until this embryonic stage, primordial germ cells (PGCs) of extragonadal origin migrate into the gonads through the blood stream.

The sex of birds is determined by genes located on the sex chromosome (see Sect. 2.3). In chickens, it is thought that sex determination occurs at E4.5. After sex determination, the gonads differentiate to testes or ovaries according to the sex chromosome constitution of cells, ZZ or ZW. However, until around E6.5, the gonads are considered "bipotential," which means they are able to differentiate into either testes or

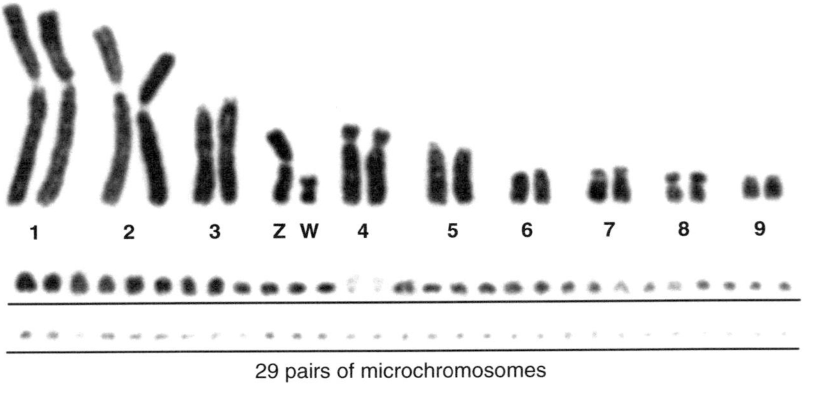

Fig. 2.1 Giemsa-stained chromosomes of female chicken (*Gallus gallus domesticus*), $2n = 78$. The karyotype of chickens includes ten pairs of macrochromosomes and 29 pairs of microchromosomes

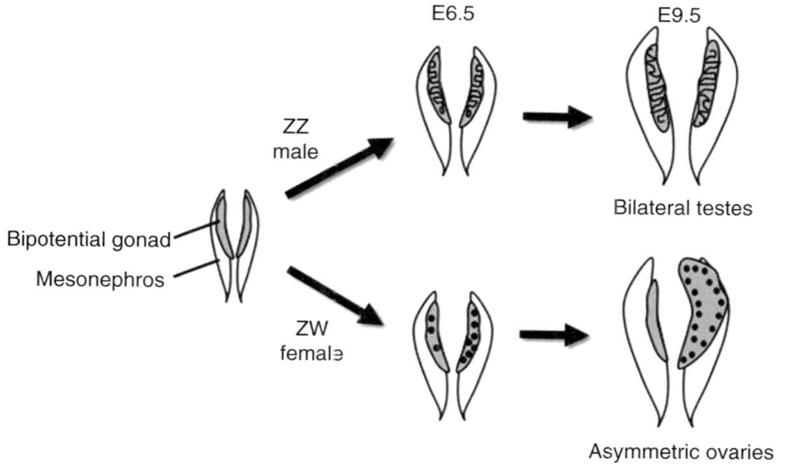

Fig. 2.2 A schematic image of gonadal development of chicken. After sex-determining, bipotential gonads of ZZ embryos differentiate to bilateral testes, whereas ZW embryos appear asymmetric ovaries

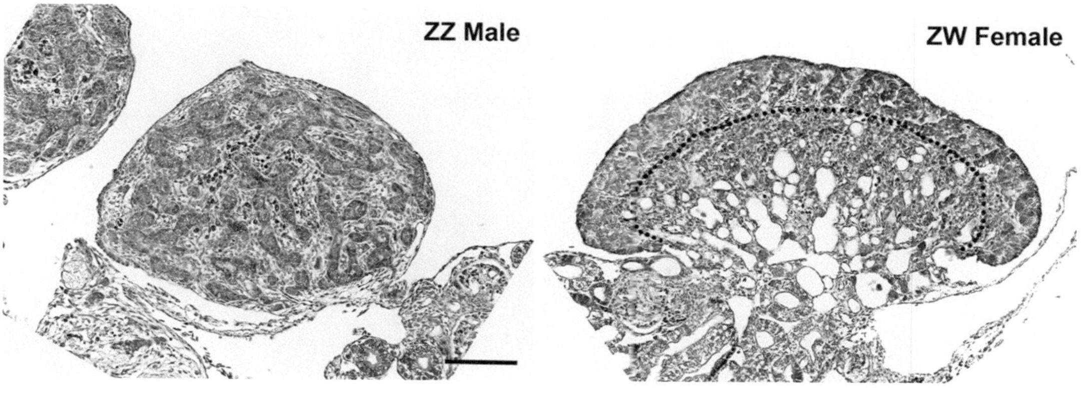

Fig. 2.3 Histological section of embryonic gonads with hematoxylin and eosin (HE)-staining. *Left*: a ZZ male gonad exhibits a developed medulla characterized by seminiferous tubules with Sertoli cells and prospermatogonia. *Right*: a left gonad of ZW female exhibits a diagnostic thickened cortex and lacunae in the cortex by becoming vacuolated. *Scale bar* means 100 μm

ovaries. After E6.5, histological differentiation of the gonads can be observed between sexes. In ZZ embryos, the gonads are differentiated to bilateral testes (Fig. 2.2). The medulla is developed and is characterized by seminiferous tubules with Sertoli cells and prospermatogonia (Fig. 2.3). The pre-Sertoli cells produce anti-Müllerian hormone (AMH), which regresses embryonic oviducts (Müllerian ducts). Leydig cells adjacent to the seminiferous tubules in the testicle secrete testosterone to differentiate around cords (Wolffian ducts) to internal genitalia. The PGCs become enclosed in developing seminiferous cords, and undergo mitotic arrest in males. Meiosis only occurs after hatching.

The right gonads are gradually depressed and fail to develop in ZW female embryos. The left gonads rapidly develop into ovaries with diagnostic thickened cortexes (Fig. 2.2). This asymmetric morphology observed between right and left gonads in females becomes apparent at E6.5 (Fig. 2.2). The proliferating germ cells exhibit a cortical distribution and begin to enter meiosis at E15.5 in the left gonad. PGCs in the right gonad undergo some proliferation, but do not enter meiosis (Ukeshima 1996). The Müllerian duct on the right side also regresses to form a dysfunctional vestige (Carlon and Stahl 1985). The medullary cords in the cortex of the female left gonad form lacunae by becoming vacuolated during development (Fig. 2.3). The left ovary finally develops into a functional organ, in which follicles are formed.

2.3 The Two Hypotheses for Sex Determination

Two hypotheses have been proposed for the mechanism of avian sex determination (Fig. 2.4). One of them is called the "Z dosage" model. This model explains that the dosage of a Z-linked gene mediate sex determination, whereby two copies are required for male development (ZZ). This model is supported by the observation that birds have no dosage compensation system for the Z chromosome, such as X inactivation observed in mammals (McQueen et al. 2001; Kuroda et al. 2001; Kuroiwa et al. 2002; Itoh et al. 2010). Therefore, it is thought that a high expression level of a Z-linked gene in the gonads of ZZ embryo triggers testis development. A strong candidate gene for sex determination under this hypothesis is the Z-linked *DMRT1* (doublesex and mab-3-related transcription factor 1) gene (see Fig. 2.5).

The other hypothesis is the "W dominant" model. According to this model, the W chromosome carries a dominant-acting ovary determinant or an

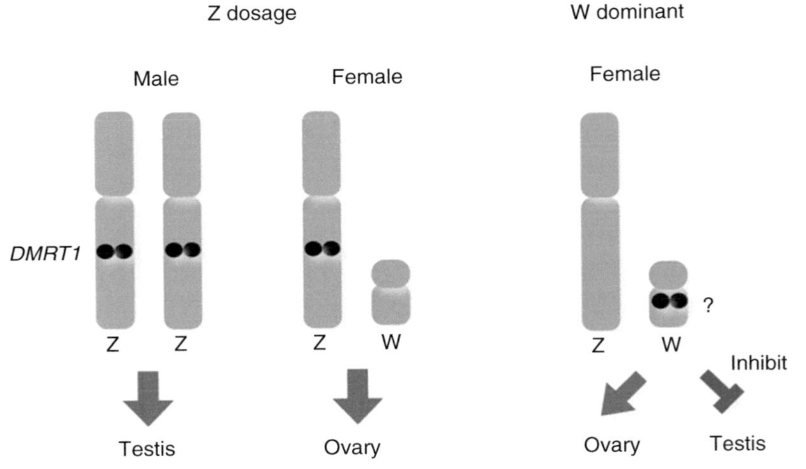

Fig. 2.4 Two hypotheses of sex-determining mechanisms in birds

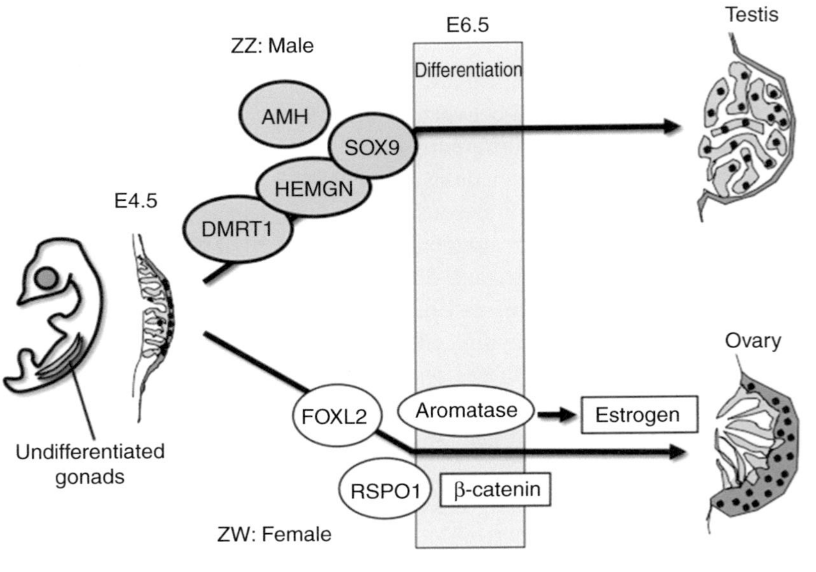

Fig. 2.5 Molecular mechanisms model underlying gonadal differentiation in the chicken embryo

inhibitor of testis differentiation. Two W-linked genes have been reported as candidate sex-determining genes in chickens.

HINTW (histidine triad nucleotide-binding protein W) was reported as the best candidate W-linked ovary-determining gene (also known as *WPKCI* and *ASW*) at approximately the same time by two research groups (Hori et al. 2000; O'Neill et al. 2000). This gene is an ortholog of *HINT1* (histidine triad nucleotide binding protein 1) on autosome in mammals. The *HINT* gene encodes an aberrant form of a nucleotide hydrolase enzyme (HINT). HINT proteins generally have endogenous adenosine 5′ monophosphoramidate enzyme activity. *HINTZ* which is a Z homologue of chicken *HINTW*, has a functional catalytic domain, the HIT motif, like other HINT proteins. By contrast, this motif is absent in *HINTW*. Several in-vitro biochemical experiments have shown that HINTZ function can be inhibited via the formation of HINTZ/HINTW heterodimers (Pace and Brenner 2003). However, ZZ embryos that overexpress *HINTW* develop to normal males with bilateral testes (Smith et al. 2009a). This provides genetic evidence against a role for *HINTW* in avian sex determination.

FET1 (female-expressed transcript 1) is another candidate W-linked ovary determinant (Reed and Sinclair 2002). This gene is found only in chicken genome, thus there are no orthologs in another bird species. The expression was almost exclusively observed in the female urogenital system. In particular, it is strongly expressed in female left gonads leading up to sexual differentiation, at E4.5–E6.5. However, genome sequencing analyses revealed that the gene is located on chromosome 4. Therefore, there are no candidate W-linked ovary-determining genes at present.

2.4 ZO and ZZW Chickens

Abnormal sex chromosome constitutions are useful to understand the sex-determining mechanism of a species. Very nice examples are mammals and fruit flies (*Drosophila melanogaster*). They have XX/XY sex chromosome constitutions; however, their mechanism for sex determination differs. To distinguish between these, sex chromosome aneuploids could be particularly helpful. Mammals have a male-dominant Y chromosome, whereas the sex of fruit flies is determined by X chromosome dosage. Therefore, XO individuals are female in mammals, but male in fruit flies. By contrast, XXY animals are male in mammals, but female in fruit flies. From these findings, distinguishing between the mechanisms of sex determination in birds would be straightforward if sex chromosome aneuploids were available (Graves 2003).

Unfortunately, despite intensive studies of chickens with aberrant sex chromosome constitutions, there are no reports of ZO chickens (Graves 2003). This means that ZO chromosomes may indeed be embryonic lethal. By contrast, several reports have described ZZW triploid chickens. Thorne and Sheldon (1991) reported these chickens were sterile intersex with both ovarian and testicular tissues. Furthermore, Lin et al. (1995) gave a detailed description of the gonads of 63 ZZW triploid chickens ranging in age from 1 day to 4.5 years. ZZW triploid fowls developed similarly to normal ZW hens until about 20 weeks of age, when their plumage, comb, and wattles developed like a male (Lin et al. 1995). Both left and right gonads were found in 59 chickens, and only a left gonad was found in the remaining four chickens. The development of right and left gonads was separately described by Lin et al. (1995), because they developed differently.

In the right gonads of 1-day-old chickens (chicks), seminiferous tubules were well developed, as observed in normal diploid ZZ males. After 3 months, the development of seminiferous tubules was retarded in the right gonads. The slow growth in the diameter of seminiferous tubules up to 7 weeks of age was associated with a twofold increase in the number of Sertoli cells from hatching (Lin et al. 1995). In 6-month-old chickens, few spermatozoa among a large number of round, condensed spermatid nuclei in the seminiferous tubules were observed. In 9-month-old chickens, the seminiferous tubules were degenerated. Primary spermatocytes were found in some tubules by 9 weeks of age, their formation being retarded by about 2 weeks by comparison with ZZ males.

By 15 weeks of age, the most advanced germ cells were primary spermatocytes, whereas spermatozoa were present in ZZ males (Lin et al. 1995).

ZZW triploid chickens differed with respect to the development of the left gonads. In 1-day-old chickens, the left gonads showed ovotestis features: oocytes in the cortex and seminiferous tubules in the medulla. The cortex was less developed in ZZW chickens than in ZW chickens. After 1 week, more than 50% of oocytes in cortical cords degenerated and contained no nucleus or a poorly defined nucleus. Seminiferous tubules were present in the medullae of all left gonads. After 3 weeks old, the cortex was structurally distinct from the medulla. Seminiferous tubules were developing in the medulla, whereas the development remained less advanced than that of those in the right gonads. The cortex continued to degenerate, and was infiltrated by leukocytes that were mainly small lymphocytes in 5-week-old fowls. By 6 months of age, leukocytic infiltration of the cortical region stopped, and there were no ovarian components in the gonads. By contrast, the structure of seminiferous tubules in the medulla was similar to that of those in the right gonads. A second phase of gonadal degeneration began at 9 months old, and all left gonads degenerated, leaving a large portion of parenchyma composed of loose connective tissue. In chickens older than 1 year, seminiferous tubules could not be observed in the left gonads.

These observations in ZZW triploid chickens, which are complete triploids with all autosomes present in triplicate, indicated that the W chromosome is associated with ovarian development to some degree; however, ovarian features are ultimately overridden by two Z chromosomes (reviewed in Lambeth and Smith 2012). Interestingly, chimeric chickens with a mixture of diploid and triploid cells are also informative. Although the estimated ratio of 2AZZ/3AZZW chimeras (2A: two sets of autosome, 3A: three sets of autosome) is only 5%, the left gonads consistently develop to an ovary (Thorne and Sheldon 1993). This means that the ovary-determining gene is located on the W chromosome, because a small number of cells including the W chromosome is sufficient to induce ovarian development (reviewed in Lambeth and Smith 2012). This observation supports the "W dominant" model.

2.5 *DMRT1*: A Z-Linked Candidate for Sex Determination in Avians

In birds and lower vertebrates, DM domain genes that encodes transcription factors with a zinc-finger-like DM domain are highly conserved. One of these genes, *DMRT1* located on the Z chromosome in birds, is a strong candidate avian sex determinant under the "Z dosage" model. This gene is highly conserved in vertebrate and non-vertebrate species, and is involved in the development of male reproductive organs (Raymond et al. 1999). In vertebrates, *DMRT1* expression is essential for testis differentiation. The overexpression of *DMRT1* in XX mouse fetal gonads drives the development of testes and represses the expression of key markers of ovarian development (Zhao et al. 2015). *DMRT1* paralogs were identified as sex-determining genes in medaka (*Oryzias latipes*, *DMY*/*Dmrt1bY* in the Y chromosome) (Matsuda et al. 2002, 2003; Nanda et al. 2002) and African clawed frogs (*Xenopus laevis*, *DMW* in the W chromosome) (Yoshimoto et al. 2008). These DM domain genes have acquired new roles in gonadal sex differentiation via gene duplication and translocation (medaka), duplication and truncation (African clawed frogs), or loss of function of one allele (birds) (Cutting et al. 2013).

The chicken homolog of *DMRT1* is located on the Z chromosome. It is expressed more highly in male undifferentiated gonads than in females (Smith et al. 1999). *DMRT1* knockdown via RNA interference results in the feminization of embryonic gonads in genetically male (ZZ) embryos (Smith et al. 2009b). The feminized left gonad shows female-like histology, disorganized testis cords, and a decline in the testicular marker *SOX9* (SRY-box 9). The ovarian marker aromatase is ectopically activated. The feminized right gonad shows a more variable loss of *DMRT1* and ectopic aromatase activation, suggesting differential sensitivity to *DMRT1* between the left and right gonads. Germ cells also show a female pattern of distribution in feminized male gonads. Furthermore, overexpressing of *DMRT1* by infection of retroviruses in the left and right gonads of ZW embryo induces masculinization, characterized by increased expression of male marker genes and reduced expression of female marker genes (Lambeth et al. 2014). These reports indicate that *DMRT1* is a master regulator for testis determination in the chicken and also support the "Z dosage" model for avian sex determination.

2.6 Genes Involved in Male Sexual Differentiation

DMRT1 expression begins at ~E3.5. The expression is observed in the medulla of gonads and is higher in ZZ males than in ZW females. After high *DMRT1* expression, *SOX9* functions in testis development in ZZ chicken embryos (Fig. 2.5). In placental mammals, the sex determination gene *SRY* (sex-determining region Y) directly activates *SOX9* expression by binding to the *SOX9* enhancer together with the NR5A1 (nuclear receptor subfamily 5, group A, member 1) protein in the undifferentiated gonads of XY embryos (Sekido and Lovell-Badge 2008). However, in chickens, there is a time-lag between the initial expression of *DMRT1* and *SOX9*, which occur at days 3.5 and 6.5 respectively. Therefore, other factors, which are probably chicken-specific, must be components of the molecular cascade between *DMRT1* and *SOX9*.

HEMGN (hemogen) was firstly reported as a chicken-specific factor for testis differentiation that mediates SOX9 regulation under DMRT1 (Nakata et al. 2013, Fig. 2.5). In mice, *Hemgn* (also known as *EDAG* in humans) is a recently characterized hematopoietic tissue-specific gene encoding a nuclear protein (Yang et al. 2001). *Hemgn* expression is restricted to the blood islands of the yolk sac and the fetal liver during embryogenesis, the adult spleen, and bone marrow. *EDAG* shows a similar expression pattern in humans. High *EDAG* expression is observed in bone-marrow cells in acute myeloid leukemia, suggesting that *EDAG* plays a regulatory role in acute myeloid leukemia (An et al. 2005). However, the gene is not expressed in the gonads during embryogenesis in mammals. In chickens, *HEMGN* is located on the Z chromosome and expressed not only in hematopoietic tissues, but also in the early embryonic gonads of male chickens (Nakata et al. 2013). Male-specific expression has been observed in the nuclei of (pre-)Sertoli cells after the sex determination period and prior to the expression of *SOX9*. In ZW embryonic gonads masculinized by aromatase inhibitor treatment (see Sect. 2.8), the expression of *HEMGN* was induced. ZW embryos overexpressing *HEMGN*, generated by infection with a retrovirus carrying *HEMGN*, had masculinized gonads: the expressions of male marker genes, *DMRT1* and *SOX9*, are increased, whereas female marker genes, aromatase and *FOXL2* (forkhead box L2) (see Sect. 2.7) are decreased. Furthermore, distribution of germ cells showed a testis-like pattern. These findings suggest that *HEMGN* is a transcription factor that is specifically involved in chicken sex determination.

AMH is a glycoprotein belonging to the transforming growth factor-β (TGF-β) superfamily. This hormone is secreted by the gonads, and plays a role in sexual differentiation of reproductive organs. AMH is synthesized and secreted by Sertoli cells of the embryonic testis, and directly acts to regress the paired Müllerian ducts of males (Josso and Picard 1986; Josso et al. 1993; Vigier et al. 1983). In mammals, SOX9 directly regulates *Amh* transcription together with NR5A1, GATA4 (GATA binding protein 4), and WT1 (Wilms tumor protein homolog). However, in chickens, *AMH* mRNA expression is expressed prior to *SOX9* mRNA (Fig. 2.5). *AMH* mRNA is also present in the female gonads of embryonic chickens, and acts to regress the right female Müllerian ducts (Hutson et al. 1981). It is thought that estrogens protect the left duct from regression by AMH (Josso et al. 2001; Hutson et al. 1982; Tran and Josso 1977). However, the exact mechanism is not known.

AMH is widely conserved in vertebrates. Its function is primarily related to Müllerian duct regression, whereas Y-linked duplicated *AMH* functions in male sex differentiation in the Patagonia pejerrey (*Odontesthes hatcheri*). In chickens, *AMH* expression precedes that of *SOX9*, indicating that *AMH* plays a more central role in avian testis development, similar to fish species. Lambeth et al. (2015) suppressed *AMH* expression in embryonic gonads of chickens using RNA interference, and did not observe an effect on the expression of the key testis pathway genes *DMRT1* and *SOX9*; and male embryos exhibited normal testicular structure. However, the sizes of the mesonephros and gonads were reduced, with normal phenotypes in male and female embryos. These findings indicate that *AMH* is required for proper cell proliferation and urogenital system growth, irrespective of sex, whereas AMH does not directly contribute to testicular or ovarian differentiation (Lambeth et al. 2015). The same research group continuously characterized embryos overexpressing *AMH* generated by infection with a retrovirus carrying *AMH* (Lambeth et al. 2016). The embryos overexpressing AMH showed small and undeveloped structures in gonads of both sexes at embryonic and adult stages. ZW female gonads developed to testis-like cords lacking Sertoli cells, and were incapable of steroidogenesis. In ZZ males, a similar phenotype is observed: complete loss of both Sertoli cells and gonadal steroidogenesis. These observations suggest that AMH does not operate as an early testis activator in the chicken but can affect sex hormone production (Lambeth et al. 2016).

2.7 Genes Involved in Female Differentiation

In most vertebrates including birds, gonadal sex differentiation in females is sensitive to the sex steroid hormone estradiol. This hormone is only detected in female embryonic gonads, and is required and sufficient for ovarian development (Elbrecht and Smith 1992). The enzyme aromatase is responsible for converting androgens to estradiol. The aromatase protein is expressed in the medullae of female gonads from E6.5 onwards, and its expression increases during ovarian development (Smith et al. 2005).

Unfortunately, there are no candidates for W-linked ovary-determining genes. However, several genes of autosomes have been known to be involved in female sex determination. *FOXL2* is an essential factor that is widely conserved in vertebrates, including chickens (Loffler et al. 2003; Wang et al. 2007; Pisarska et al. 2011, Fig. 2.5). The expression patterns of *FOXL2* and aromatase transcripts are highly correlated in the developing ovary at E4.7–12.7 (Govoroun et al. 2004). The proteins encoded by both genes are colocalized in the nuclei of medullar cord cells, and FOXL2 is mainly expressed in the granulosa cells of developing follicles. The expression timing of *FOXL2* is just prior to that of aromatase, suggesting that FOXL2 directly or indirectly regulates aromatase transcription. The *FOXL2* expression is reduced by aromatase inhibitor treatment in vivo, suggesting that there is a feedback regulator loop between *FOXL2* and aromatase (Hudson et al. 2005).

RSPO1 (R-spondin 1) and *WNT4* (wingless-type MMTV integration site family, member 4) activation of β-catenin signaling plays an important role in the developing ovary in several vertebrate species (Biason-Lauber and Konrad 2008; Liu et al. 2010; Chue and Smith 2011), including chickens (Fig. 2.5). *RSPO1* mRNA is expressed in the left and right gonads of ZW chicken embryos from E4.5, and increases strongly from E8.5 onward (Smith et al. 2008). By contrast, its expression remains low in the gonads of ZZ embryos. *WNT4* expression is observed in the bipotential gonads of ZZ and ZW embryos at E4.5. However, in the left gonads of ZW embryos, it is upregulated during sexual differentiation, at E6.5–8.5. *RSPO1* and *WNT4* are strongly expressed in the cortex of the developing ovary (Ayers et al. 2013). They may act synergistically to activate β-catenin.

2.8 Sex Reversal in Birds

Sex-reversed animals have been particularly found in many species of fishes. Both cases, that of male-to-female and female-to-male, have been reported in wild fish species. The phenomenon of sex-reversal had also often observed in birds. Aristotle (384–322 BC) seems to have been the first to record the phenomenon of abnormal sex development in poultry. He observed hens that changed into cockerels (reviewed in Taber 1964). Examples of spontaneous sex reversal in chickens and many other bird species have been reported. Interestingly, only the masculinization of female birds (i.e., female-to-male) has been documented, and there are no reports of male-to-female sex reversal. Aristotle recorded the reciprocal event, cockerel to feminine behavior, but this has not been confirmed (Taber 1964).

Sex-reversed chickens can be generated experimentally. ZZ embryos treated with exogenous estrogen prior to gonadal differentiation show feminization of the left gonads, resulting in ovaries or ovotestes. Additionally, the administration of the anti-estrogen tamoxifen disturbs normal female development (Scheib 1983). Aromatase inhibitors, such as fadrozole, can effectually lead to feminization in ZW embryos (Elbrech and Smith 1992). These inhibitors induce female-to-male sex reversal in ZW females when applied between E0 and E7.5. Bilateral gonads develop, and a testis-like structure with a thick cortex and dense medulla is observed, although the embryo is genetically female (ZW) (Fig. 2.6). Furthermore, genes involved in testis differentiation are unregulated; by contrast, female marker genes are downregulated.

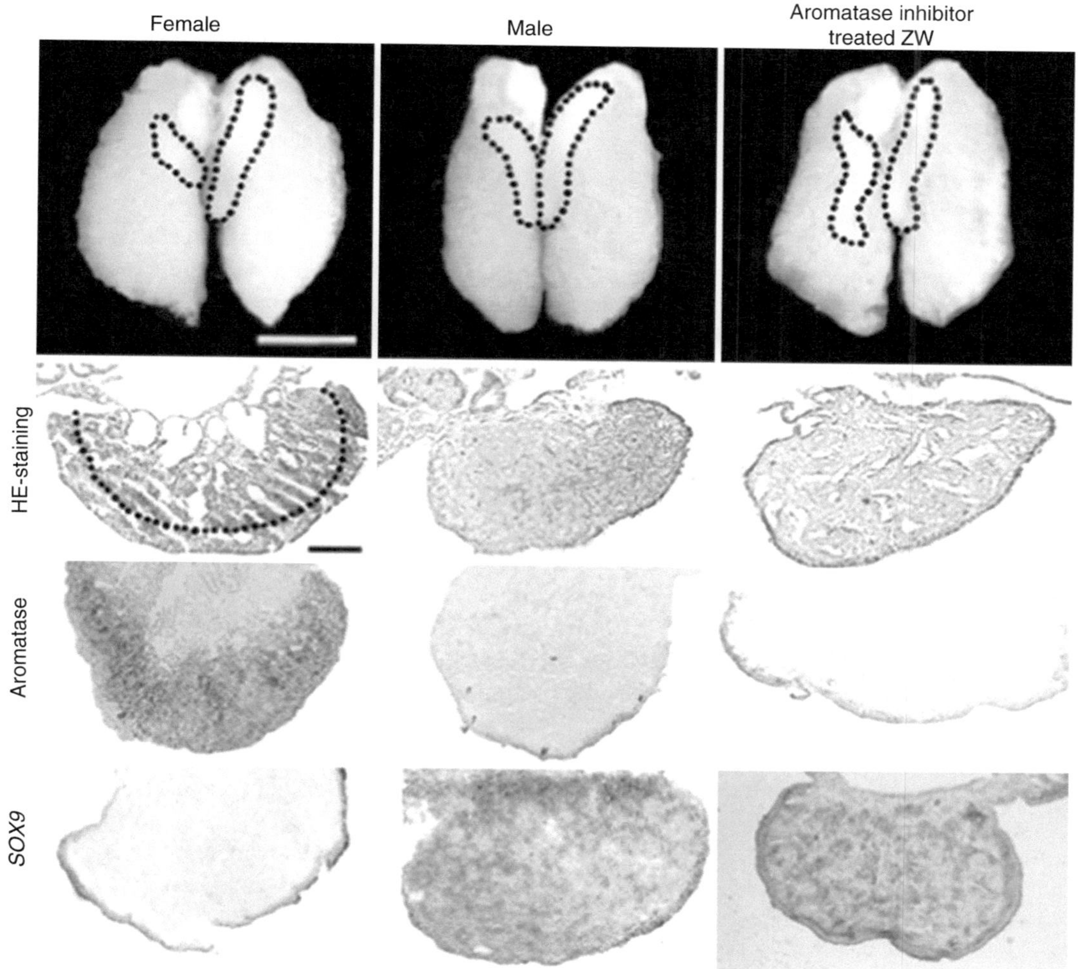

Fig. 2.6 The gonads of ZW female embryo, ZZ male embryo, and masculinized ZW embryo induced by aromatase inhibitor treatment. *Upper*: gonads on top of the mesonephros of female, male, and masculinized ZW embryos at E10.5. The gonads of masculinized ZW embryos showed bilateral development, similar to male gonads. *Dashed lines* indicate gonads. *Scale bar* means 1 mm. *Middle*: HE-staining of gonad sections from female, male, and masculinized ZW embryos. The left masculinized ZW gonad has a testis-like phenotype with a dense medulla and thin cortex, although a slightly fragmented medulla was observed. The *dashed line* indicates the border between the cortex and medulla in the female gonad. *Scale bar* means 100 μm. *Lower*: aromatase and *SOX9* in situ hybridization in male, female, and masculinized ZW gonad frozen sections at day 10.5. Aromatase is detected in female gonads, but no expression is observed in male or masculinized ZW gonads. By contrast, *SOX9* expression is not detected in female

Evidence obtained from experimental sex-reversed chickens suggests that sex hormones and their enzymatic pathway are very important for ovary development in chickens, and potentially in all bird species. Furthermore, there are no examples of male-to-female sex reversal in birds under natural conditions, indicating that birds cannot be female without the W chromosome. This inference strongly supports the "W dominant" hypothesis, which maintains that W-linked genes dominantly determine ovary differentiation (or inhibit testis differentiation).

2.9 Sex Chimera Birds: Gynandromorphs—Genes or Hormones?

Gynandromorphs have been generally observed in insects and crustaceans. The individuals have the physical characteristics of both genders, usually displaying a bilateral difference. Interestingly, gynandromorphs are found only in birds in vertebrates; one side of the animal appears male and the other female. These birds are rare, but researchers have focused on this interesting phenomenon as a "genetic mosaicism" (Hollander 1944). Gynandromorph birds have been observed in pigeons, zebra finches, and especially domestic fowls (Hollander 1975; Agate et al. 2003; Cock 1955), among other species. A long-standing theory on the etiology of gynandromorphs proposes that a single sex chromosome is lost at the two-cell stage on one side of the animal (Cock 1955). However, it is now understood that gynandromorph birds arise as a result of a failure in the extrusion of a polar body during meiosis and subsequent fertilization of both a Z- and W-bearing female pronucleus (Hollander 1975; Zhao et al. 2010).

Gynandromorphs are spontaneously generated, namely the experimental samples are rare. Zhao et al. (2010) reported valuable data examined three lateral gynandromorph chickens. All chickens were ISA brown commercial hybrids with sex-linked coloration in which males show white plumage and females show brown plumage. Gynandromorph chickens show a marked bilateral asymmetry: half of the body is phenotypically female and the other side is phenotypically male. The female side with brown plumage appears a small wattle and small leg spur. By contrast, the male side, which is white, shows like a typical cockerel with a large wattle, a large leg spur, a heavier leg structure, and an obviously greater mass of breast muscle (Zhao et al. 2010).

In their report, to identify the sex chromosome constitutions of somatic cells, fluorescence in-situ hybridization using Z and W chromosome probes was performed using chromosome preparations obtained from cells in blood and skin samples from both sides of three gynandromorph birds. All three birds were composed of a mixture of normal diploid male and female cells. Tissues of the male side were mainly composed of ZZ (male) cells, whereas tissues of the female side mainly contained ZW (female) cells.

Understanding gonadal differentiation in gynandromorph chickens is highly complicated, because gonadal structure does not correspond to external appearance (Zhao et al. 2010). Two gynandromorph chickens (G1 and G2) appeared female on the left side and male on the right, whereas the remained one, G3, showed the reverse external appearance, left is male and right is female. The left gonad differed in appearance between these three gynandromorph chickens. G1 contained a testis-like gonad composed primarily of sperm containing seminiferous tubules. G2 had an ovary-like gonad composed predominantly of large and small follicles. G3 showed an ovotestis comprised of a mixture of empty tubules and small follicular-like structures. The morphological appearance of the gonads reflected the cellular composition (ZZ or ZW) of the individual organs. Testis-like and ovary-like gonads were composed principally of ZZ- and ZW-containing cells respectively, whereas the ovotestis comprised a mixture of ZZ- and ZW-containing cells.

In the traditional view of sexual development in vertebrates including birds, the gonads develop into either ovaries or testis during the embryonic stage, and then release sex hormones to masculinize or feminize the rest of the body. However, the phenomenon of gynandromorphy cannot be explained by this process, because hormones are expected to flow equally to both sides of the body. Nevertheless, the organs exhibit male or female phenotypes depending on the cellular composition, ZZ or ZW.

Gonadal chimeras generated by transplantation of presumptive mesoderm exhibit a similar pattern. Zhao et al. (2010) transplanted sections of presumptive mesoderm from green fluorescent protein (GFP)-expressing embryos at developmental stage 12 to the equivalent tissue on non-GFP embryos at the same stage of development between sexes. The transplanted embryos were allowed to develop until stage 35, and the expression patterns of the male marker AMH and female marker aromatase were examined. Interestingly, donor male cells expressed AMH and donor female cells expressed aromatase in mixed-sex chimeras. Donor cells appear to be incapable of contributing to specialized compartments of the host gonad. Female donor cells in the testis of a male host cannot be recruited into the functional male Sertoli cell compartment, and male donor cells in the ovary of a female host are excluded from the functionally female compartment.

These observations seem paradoxical. Hormones must play a role in early sexual development because genetically female chicken embryos develop as males with testes when treated with an aromatase inhibitor (see Sect. 2.8). However, studies on gynandromorphy and gonadal chimeras provide evidence that all somatic cells recognize their sex, ZZ male or ZW female. This observation led to the idea that male and female chicken somatic cells possess a cell-autonomous sex identity (CASI) (Zhao et al. 2010; Clinton et al. 2012). Based on gynandromorph chicken studies, Zhao et al. (2010) proposed that Z-linked genes underlie sex determination throughout the avian body. This idea is supported by the fact that birds have no chromosome-wide dosage compensation mechanism (see Fig. 2.3). The dosage of most Z-linked genes is twofold, higher in male than in female cells, and this might determine the sex identities of each cell.

Conclusion

The sex-determining process is essential for reproduction. In recent years, understanding the molecular mechanism of sex-determination in birds has been progressed, however, it is lagging behind that of mammals and fishes. Although genes and regulatory networks that govern the fate of gonads were recently identified, many gaps in knowledge remain. There is conflicting evidence concerning the importance of sex hormones in sex differentiation. Additional investigations of the role of genes involved in sex determination and differentiation, and the relative contribution of the genetic sex of each somatic cell and hormones to sexual differentiation, are expected in the future.

References

Agate RJ, Grisham W, Wade J, Mann S, Wingfield J, Schanen C, Palotie A, Arnold AP. Neural, not gonadal, origin of brain sex differences in a gynandromorphic finch. Proc Natl Acad Sci U S A. 2003;100:4873–8.

An LL, Li G, KF W, Ma XT, Zheng GG, Qiu LG, Song YH. High expression of EDAG and its significance in AML. Leukemia. 2005;19:1499–502.

Ayers KL, Sinclair AH, Smith CA. The molecular genetics of ovarian differentiation in the avian model. Sex Dev. 2013;7:80–94.

Biason-Lauber A, Konrad D. WNT4 and sex development. Sex Dev. 2008;2:210–8.

Carlon N, Stahl A. Origin of the somatic components in chick embryonic gonads. Arch Anat Microsc Morphol Exp. 1985;74:52–9.

Chue J, Smith CA. Sex determination and sexual differentiation in the avian model. FEBS J. 2011;278:1027–34.

Clinton M, Zhao D, McBride D. Evidence for avian cell autonomous sex identity (CASI) and implications for the sex-determination process? Chromosome Res. 2012;20:177–90.

Cock AG. Half-and-half mosaics in the fowl. J Genet. 1955;53:49–80.

Cutting A, Chue J, Smith CA. Just how conserved is vertebrate sex determination? Dev Dyn. 2013;242:380–7.

Elbrecht A, Smith RG. Aromatase enzyme activity and sex determination in chickens. Science. 1992;255:467–70.

Govoroun MS, Pannetier M, Pailhoux E, Cocquet J, Brillard JP, Couty I, Batellier F, Cotinot C. Isolation of chicken homolog of the FOXL2 gene and comparison of its expression patterns with those of aromatase during ovarian development. Dev Dyn. 2004;231:859–70.

Graves JAM. Sex and death in birds: a model of dosage compensation that predicts lethality of sex chromosome aneuploids. Cytogenet Genome Res. 2003;101:278–82.

Hamburger V, Hamilton HL. A series of normal stages in the development of the chick embryo. J Morphol. 1951;88:49–92.

Hollander WF. Mosaic effects in domestic birds. Q Rev Biol. 1944;19:285–307.

Hollander WF. Sectorial mosaics in the domestic pigeon: 25 more years. J Hered. 1975;66:197–202.

Hori T, Asakawa S, Itoh Y, Shimizu N, Mizuno S. Wpkci, encoding an altered form of PKCI, is conserved widely on the avian W chromosome and expressed in early female embryos: implication of its role in female sex determination. Mol Biol Cell. 2000;11:3645–60.

Hudson QJ, Smith CA, Sinclair AH. Aromatase inhibition method reduces expression of FOXL2 in the embryonic chicken ovary. Dev Dyn. 2005;233:1052–5.

Hutson J, Ikawa H, Donahoe PK. The ontogeny of Müllerian inhibiting substance in the gonads of the chicken. J Pediatr Surg. 1981;16:822–7.

Hutson JM, Ikawa H, Donahoe PK. Estrogen inhibition of Müllerian inhibiting substance in the chick embryo. J Pediatr Surg. 1982;17:953–9.

International Chicken Genome Sequencing Consortium. Sequence and comparative analysis of the chicken genome provide unique perspectives on vertebrate evolution. Nature. 2004;432(7018):695–716.

Itoh Y, Replogle K, Kim YH, Wade J, Clayton DF, Arnold AP. Sex bias and dosage compensation in the zebra finch versus chicken genomes: general and specialized patterns among birds. Genome Res. 2010;20:512–8.

Josso N, Cate RL, Picard JY, Vigier B, di Clemente N, Wilson C, Imbeaud S, Pepinsky RB, Guerrier D, Boussin L, et al. Anti-Müllerian hormone: the Jost factor. Recent Prog Horm Res. 1993;48:1–59.

Josso N, di Clemente N, Gouédard L. Anti-Müllerian hormone and its receptors. Mol Cell Endocrinol. 2001;179:25–32.

Josso N, Picard JY. Anti-Müllerian hormone. Physiol Rev. 1986;66:1038–90.

Kuroda Y, Arai N, Arita M, Teranishi M, Hori T, Harata M, Mizuno S. Absence of Z-chromosome inactivation for five genes in male chickens. Chromosome Res. 2001;9:457–68.

Kuroiwa A, Yokomine T, Sasaki H, Tsudzuki M, Tanaka K, Namikawa T, Matsuda Y. Biallelic expression of Z-linked genes in male chickens. Cytogenet Genome Res. 2002;99:310–4.

Lambeth LS, Ayers KL, Cutting AD, Doran TJ, Sinclair AH, Smith CA. Anti-Müllerian hormone is required for chicken embryonic urogenital system growth but not sexual differentiation. Biol Reprod. 2015; doi:10.1095/biolreprod.115.131664. [Epub ahead of print].

Lambeth LS, Morris K, Ayers KL, Wise TG, O'Neil T, Wilson S, Cao Y, Sinclair AH, Cutting AD, Doran TJ, Smith CA. Overexpression of anti-Müllerian hormone disrupts gonadal sex differentiation, blocks sex hormone synthesis, and supports cell autonomous sex development in the chicken. Endocrinology. 2016;157:1258–75.

Lambeth LS, Raymond CS, Roeszler KN, Kuroiwa A, Nakata T, Zarkower D, Smith CA. Over-expression of DMRT1 induces the male pathway in embryonic chicken gonads. Dev Biol. 2014;389:160–72.

Lambeth LS, Smith CA. Disorder of sexual development in poultry. Sex Dev. 2012;6:96–103.

Lin M, Thorne MH, Martin IC, Sheldon BL, Jones RC. Development of the gonads in the triploid (ZZW and ZZZ) fowl, Gallus domesticus, and comparison with normal diploid males (ZZ) and females (ZW). Reprod Fertil Dev. 1995;7:1185–97.

Liu CF, Liu C, Yao HH. Building pathways for ovary organogenesis in the mouse embryo. Curr Top Dev Biol. 2010;90:263–90.

Loffler KA, Zarkower D, Koopman P. Etiology of ovarian failure in blepharophimosis ptosis epicanthus inversus syndrome: FOXL2 is a conserved, early-acting gene in vertebrate ovarian development. Endocrinology. 2003;144:3237–43.

Matsuda M, Nagahama Y, Shinomiya A, Sato T, Matsuda C, Kobayashi T, Morrey CE, Shibata N, Asakawa S, Shimizu N, Hori H, Hamaguchi S, Sakaizumi M. DMY is a Y-specific DM-domain gene required for male development in the medaka fish. Nature. 2002;417:559–63.

Matsuda M, Sato T, Toyazaki Y, Nagahama Y, Hamaguchi S, Sakaizumi M. Oryzias curvinotus has DMY, a gene that is required for male development in the medaka, O. latipes. Zool Sci. 2003;20:159–61.

McQueen HA, McBride D, Miele G, Bird AP, Clinton M. Dosage compensation in birds. Curr Biol. 2001;11:253–7.

Nakata T, Ishiguro M, Aduma N, Izumi H, Kuroiwa A. Chicken hemogen homolog is involved in the chicken-specific sex-determining mechanism. Proc Natl Acad Sci U S A. 2013;110:3417–22.

Nanda I, Kondo M, Hornung U, Asakawa S, Winkler C, Shimizu A, Shan Z, Haaf T, Shimizu N, Shima A, Schmid M, Schartl M. A duplicated copy of DMRT1 in the sex-determining region of the Y chromosome of the medaka, Oryzias latipes. Proc Natl Acad Sci U S A. 2002;99:11778–83.

O'Neill M, Binder M, Smith C, Andrews J, Reed K, Smith M, Millar C, Lambert D, Sinclair A. ASW: a gene with conserved avian W-linkage and female specific expression in chick embryonic gonad. Dev Genes Evol. 2000;210:243–9.

Pace HC, Brenner C. Feminizing chicks: a model for avian sex determination based on titration of Hint enzyme activity and the predicted structure of an Asw–Hint heterodimer. Genome Biol. 2003;4:R18.

Pisarska MD, Barlow G, Kuo FT. Mini review: roles of the forkhead transcription factor FOXL2 in granulosa cell biology and pathology. Endocrinology. 2011;152:1199–208.

Raymond CS, Kettlewell JR, Hirsch B, Bardwell VJ, Zarkower D. Expression of Dmrt1 in the genital ridge of mouse and chicken embryos suggests a role in vertebrate sexual development. Dev Biol. 1999;215:208–20.

Reed KJ, Sinclair AH. FET-1: a novel W-linked, female specific gene up-regulated in the embryonic chicken ovary. Mech Dev. 2002;119:S87–90.

Scheib D. Effects and role of estrogens in avian gonadal differentiation. Differentiation. 1983;23(Suppl):S87–92.

Sekido R, Lovell-Badge R. Sex determination involves synergistic action of SRY and SF1 on a specific Sox9 enhancer. Nature. 2008;453:930–4.

Smith CA, McClive PJ, Hudson Q, Sinclair AH. Male-specific cell migration into the developing gonad is a conserved process involving PDGF signalling. Dev Biol. 2005;284:337–50.

Smith CA, McClive PJ, Western PS, Reed KJ, Sinclair AH. Conservation of a sex-determining gene. Nature. 1999;402:601–2.

Smith CA, Roeszler KN, Sinclair AH. Genetic evidence against a role for W-linked histidine triad nucleotide binding protein (HINTW) in avian sex determination. Int J Dev Biol. 2009a;53:59–67.

Smith CA, Roeszler KN, Ohnesorg T, Cummins DM, Farlie PG, Doran TJ, Sinclair AH. The avian Z-linked gene DMRT1 is required for male sex determination in the chicken. Nature. 2009b;461:267–71.

Smith CA, Shoemaker CM, Roeszler KN, Queen J, Crews D, Sinclair AH. Cloning and expression of R-Spondin1 in different vertebrates suggests a conserved role in ovarian development. BMC Dev Biol. 2008;8:72.

Taber E. Intersexuality in birds. In: Armstrong CN, Marshall AJ, editors. Intersexuality of vertebrates including man. New York: Academic Press; 1964. p. 285–310.

Thorne MH, Sheldon BL. Cytological evidence of maternal meiotic errors in a line of chickens with a high incidence of triploidy. Cytogenet Cell Genet. 1991;57:206–10.

Thorne MH, Sheldon BL. Triploid intersex and chimeric chickens: useful models for studies of avian sex determination. In: Reed KC, Graves JAM, editors. Sex chromosomes and sex-determining genes. Chur: Harwood Academic Publishers; 1993. p. 199–205.

Tran D, Josso N. Relationship between avian and mammalian anti-Müllerian hormones. Biol Reprod. 1977;16:267–73.

Ukeshima A. Germ cell death in the degenerating right ovary of the chick embryo. Zool Sci. 1996;13:559–63.

Yang LV, Nicholson RH, Kaplan J, Galy A, Li L. Hemogen is a novel nuclear factor specifically expressed in mouse hematopoietic development and its human homologue EDAG maps to chromosome 9q22, a region containing breakpoints of hematological neoplasms. Mech Dev. 2001;104:105–11.

Vigier B, Tran D, du Mesnil du Buisson F, Heyman Y, Josso N. Use of monoclonal antibody techniques to study the ontogeny of bovine anti-Müllerian hormone. J Reprod Fertil. 1983;69:207–14.

Wang DS, Kobayashi T, Zhou LY, Paul-Prasanth B, Ijiri S, Sakai F, Okubo K, Morohashi K, Nagahama Y. Foxl2 up-regulates aromatase gene transcription in a female-specific manner by binding to the promoter as well as interacting with ad4 binding protein/steroidogenic factor 1. Mol Endocrinol. 2007;21:712–25.

Yoshimoto S, Okada E, Umemoto H, Tamura K, Uno Y, Nishida-Umehara C, Matsuda Y, Takamatsu N, Shiba T, Ito M. A W-linked DM-domain gene, DM-W, participates in primary ovary development in Xenopus laevis. Proc Natl Acad Sci U S A. 2008;105:2469–74.

Zhao D, McBride D, Nandi S, McQueen HA, McGrew MJ, Hocking PM, Lewis PD, Sang HM, Clinton M. Somatic sex identity is cell autonomous in the chicken. Nature. 2010;464:237–42.

Zhao L, Svingen T, Ng ET, Koopman P. Female-to-male sex reversal in mice caused by transgenic overexpression of Dmrt1. Development. 2015;142:1083–8.

Female Reproductive System and Immunology

3

Yukinori Yoshimura and Animesh Barua

Abstract

Health of the reproductive organs is essential for formation and production of high quality and hygienic eggs. It is of importance to review the structures and functions of female reproductive system for better understanding of the mechanism by which the eggs are formed. The unique functions of ovarian cells for follicular growth and differentiation as well as steroidogenesis and oocyte maturation are regulated by gonadotropins and gonadal steroids. The oviduct is responsible for egg formation, while the unique function to store sperms for a prolonged period takes place in the specific tissue of this organ. The unique innate and adaptive immuno-defense systems that play essential role to prevent infection are developed in the ovary and oviduct. Toll-like receptors (TLRs) that recognize the molecular pattern of microbes and initiate the immunoresponse are expressed in those organs. Avian β-defensins (AvBDs), a member of antimicrobial peptides, are synthesized by the ovarian and oviductal cells. Challenge of those cells by TLR ligands upregulates the expression of proinflammatory cytokines, which in turn stimulate the expression of AvBDs. The adaptive immune system in the ovary and oviduct is also unique, since the migration of lymphocytes is enhanced by estrogens. In contrast to the development of immuno-defense system, spontaneous ovarian cancer and uterine fibroids appear more frequently in chickens than in mammals, and thus chickens could be used as a model for studying these diseases. Thus the avian reproductive organs have unique functions not only for egg formation but also for the immuno-defense system, which is essential for prevention of infection and production of hygienic eggs.

Y. Yoshimura (✉)
Graduate School of Biosphere Science, Hiroshima University, Higashi-Hiroshima 739-8528, Japan
e-mail: yyosimu@hiroshima-u.ac.jp

A. Barua
Departments of Pharmacology, Obstetrics & Gynecology and Pathology, Rush University Medical Center, Chicago, IL, USA
e-mail: Animesh_Barua@rush.edu

T. Sasanami (ed.), *Avian Reproduction*, Advances in Experimental Medicine and Biology 1001,
DOI 10.1007/978-981-10-3975-1_3

Keywords
Ovarian follicles • Oviduct • Oogenesis • Oocyte maturation • Sperm storage • Immuno-defense • Toll-like receptors • Antimicrobial peptides • Ovarian cancer

3.1 Introduction

The avian egg consists of the yolk, albumen, eggshell membrane and eggshell, and contains all nutrients required for embryo development. The yolk mass is formed as a sequential layers of yellow and white yolk, and the latebra extends from the nucleus of Pander at the bottom of the germinal disc to the middle of the yolk. The yolk is attached with a chalazal structure and surrounded by the albumen. The eggshell membrane consists of the inner and outer layers surrounds the albumen, and the air sac is formed between the inner and outer layers at the blunt end of eggs. The eggshell formed on the eggshell membrane consisting of three major layers of mammillae, palisades, and crystal layers, and is highly rich in calcite (calcium carbonate). The pigments on the surface of the eggshell appear only in eggs laid by birds able to synthesize them. The surface of the eggshell is coated by the cuticle, which may play a role in preventing the invasion of microbes.

The ovary and oviduct are developed only in the left side of the abdominal cavity in hens. Those in the right side are regressed during the embryonic phase by the effects of anti-Müllerian factors. In female embryos, the gonadal cortex undergoes differentiation into the major ovarian tissue where the follicles and steroidogenic cells develop, whereas the medulla differentiates into connective tissue including nerve and blood-vessel systems. In laying hens, primordial follicles are embedded in the ovarian cortex. The growing follicles are protruded on the ovarian surface, and differentiate into numerous white follicles and several larger yellow follicles (Fig. 3.1). The yellow follicles undergo rapid growth and show a hierarchal order in size among each follicle. Postovulatory follicles do not form corpus luteum, unlike mammals, and regress gradually after ovulation. Small follicles that are not recruited to enter the rapid growth phase undergo atresia, even during the laying phase (Johnson 2000).

The oviducts consist of five segments including the infundibulum, magnum, isthmus, uterus (shell gland), and vagina (Fig. 3.1). The ovum remains in these segments for approximately 15–30 min, 2–3 h, 1.5–2 h, 20 h, and a few minutes respectively. The infundibulum forms the funnel and tubular regions. The funnel region engulfs the

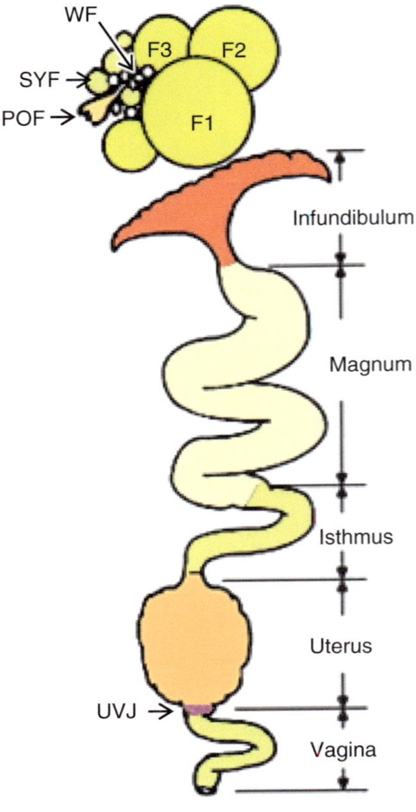

Fig. 3.1 Chicken ovary and oviduct. *F1–F3* the largest to third largest preovulatory follicles, *SYF* small yellow follicle, *WF* white follicle, *POF* postovulatory follicle. The oviduct consists of the infundibulum, magnum, isthmus, uterus, and vagina. The tissue between the uterus and vagina is the utero-vaginal junction (UVJ). (Reproduced with agreement of Japanese Association of Veterinary Anatomists)

ovulated ovum, whereas the tubular region secretes the substances that form the outer vitelline layer and chalaza on the surface of the ovum, which is surrounded by the perivitelline layer. Fertilization occurs in the infundibulum before the outer perivitelline layer is formed. Then, the fertilized ovum descends the magnum and isthmus, where the albumen and eggshell membrane are formed. In the uterus, water and minerals are incorporated in the albumen, followed by calcification forming the eggshell. The vagina is at the caudal end of the oviduct which opens to the cloaca. The part between the uterus and vagina is called the utero-vaginal junction (UVJ), where sperm are kept alive for a few weeks.

Health of the reproductive organs is essential for formation and production of high quality and hygienic eggs. In this chapter, the general structures and functions of female reproductive system will be described first, for better understanding of the mechanism by which the eggs are formed. Then, the unique immuno-defense system in the ovary and oviduct is described, which is essential to prevent infection and maintain the health of these organs. Finally, the recent research topic on the spontaneous ovarian cancer and uterine fibroids is described, since these diseases appear more frequently in the chickens than in mammals, and thus chickens could be used as a model for the studies of them.

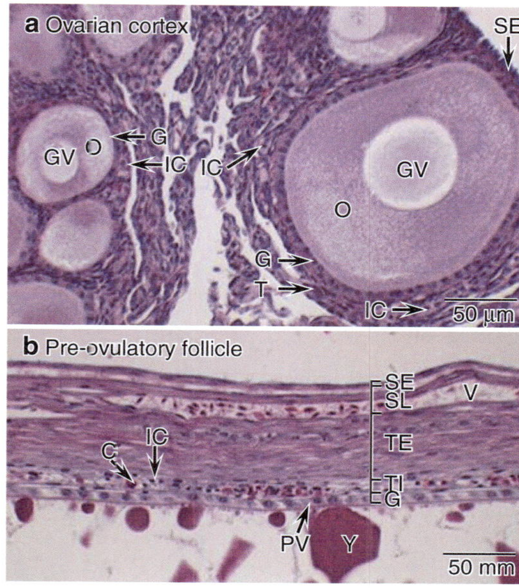

Fig. 3.2 Chicken ovarian follicles. **a** Primordial follicles embedded in the ovarian cortex. The oocytes (O) with germinal vesicle (GV) are surrounded by the granulosa (G) and thin theca layers (T), and the interstitial cells (IC) are localized in the outer part of the theca layer. **b** The follicular wall of preovulatory follicle consists of the granulosa later (G), theca interna and externa (TI and TE), superficial layer and epithelium (SL and SE). Interstitial cells (IC) are localized and capillary beds (C) are developed in the theca interna. *Y* yolk, *PV* perivitelline layer, *V* vein. (Reproduced with agreement of Japanese Association of Veterinary Anatomists)

3.2 The Ovary

3.2.1 Structure of Follicles

The primary follicles are embedded in the ovarian cortex (Fig. 3.2a). The oocyte contains a large and prominent germinal vesicle (nucleus), and is surrounded by a single layer of follicular epithelium (pre-granulosa cells). As the follicles grow, the granulosa layer become multi-layered, and the thecal layer develops by incorporating the interstitial cells located near the follicular surface.

White follicles are formed by further development and differentiation of follicular cells, and protrude from the ovarian surface. The germinal vesicle (nucleus) is located at the animal pole of the oocyte. During the growth of white follicles, the granulosa layer structure changes from a multi-layer to a single-cell layer. The thecal layer differentiates into the theca interna and externa layers, and the interstitial cells are located in the theca interna at the start of this developmental stage.

In the yellow follicles, the oocyte containing a large amount of yolk is surrounded by the perivitelline layer and the follicular wall, which consists of the granulosa layer, basal lamina, theca interna and externa, superficial tissue and epithelium (Fig. 3.2b). These follicles grow rapidly with the accumulation of yolk precursors and other materials circulating in the blood. The germinal disc containing germinal vesicle and cytoplasmic organelles is located at the animal pole. The perivitelline layer surrounding the outer surface of the oocyte is

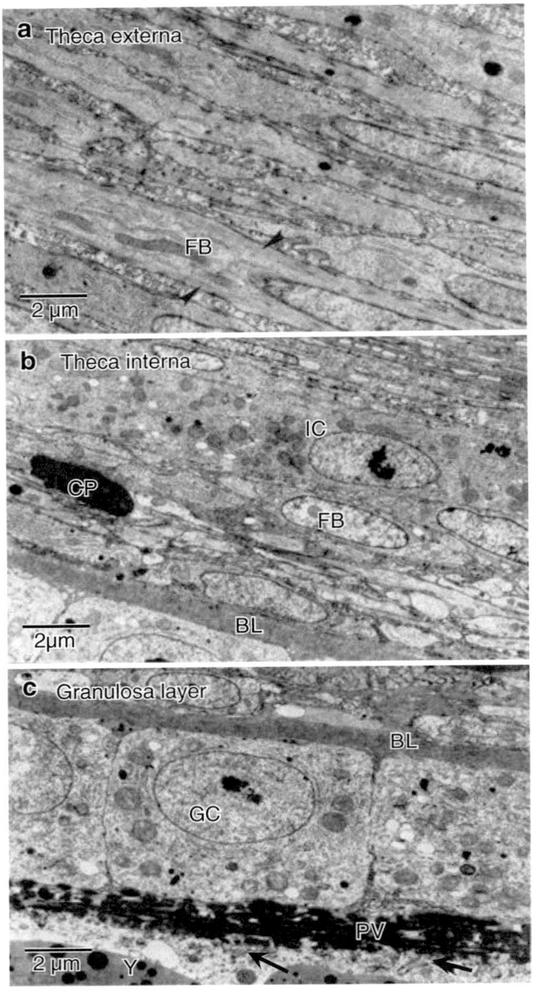

Fig. 3.3 Electron micrographs of the follicular wall. **a** Theca externa. Fibroblasts contain microfilaments (*arrow heads*). **b** Theca interna. The thecal interstitial cells (IC) and fibroblasts (FB) are localized. *BL* basal lamina, *CP* blood capillary. **c** Granulosa layer. Granulosa cells (GC) are cuboidal in shape, and packed between perivitelline layer (PV) and basal lamina (BL). *Arrows* show the cytoplasmic process of the granulosa cells. *Y* yolk. (Reproduced with agreement of Japanese Association of Veterinary Anatomists)

formed from the meshwork of fibers (Fig. 3.3c), which contains molecules responsible for binding the sperm to the ovum during the process of fertilization, namely ZP1, ZP2, and ZP4. Among them, ZP2 is localized in the germinal disc region of mature eggs, where sperm bind to the egg coat at high density, suggesting some role for ZP2 in the preferential binding and penetration of sperm in the germinal disc region (Nishio et al. 2014).

The granulosa cells show a cuboidal shape, and are rich in cytoplasmic organelles including mitochondria with lamellar cristae, rough endoplasmic reticulum and dense granules (Fig. 3.3c). Microvilli are well developed on their lateral and apical surfaces, and they form interdigitations with adjacent cells. The large cytoplasmic processes extend from their apical surface to penetrate the perivitelline layer and interact with the oolemma. At the germinal disc region, the cell membranes of granulosa cells and oolemma form gap junctions (Yoshimura et al. 1993b). The basal lamina underlining the granulosa layer is thick, approximately 1 µm in the largest of follicles.

In the theca interna, the interstitial cells, spindle-shaped fibroblasts and relatively small amount of collagenous fibers exist (Fig. 3.3b). The blood capillary network is well developed in this tissue, suggesting that yolk materials in the blood circulation are released from the capillary and provided to the oocyte. The interstitial cells have a typical structure of steroidogenic cells, namely mitochondria with tubular cristae, smooth endoplasmic reticulum, and lipid droplets, and synthesize progesterone and androgens.

The theca externa is a dense connective tissue where fibroblast-like cells, aromatase cells and rich collagenous fibers have been identified (Fig. 3.3a). The fibroblast-like cells may be categorized as myo-fibroblasts, because they contain microfilaments in the peripheral region of the cytoplasm, and the theca layer shows contraction activity that may facilitate follicular rupture at ovulation. Aromatase cells are identified by the presence of immunoreactive aromatase molecules, and have been suggested to contribute to aromatizing testosterone for estrogen synthesis (Nitta et al. 1991).

The superficial tissue surrounding the theca layer is a loose connective tissue extending from the ovarian cortex and follicular stalk, and the vascular and nervous systems enter the follicle through this tissue. At the stigma region, this tissue becomes thin, while theca externa increases in thickness, and thus blood vessels except for capillaries in the theca interna are not developed here. The superficial epithelium consists of flattened epithelial cells. The microvilli on their surface are not well developed, unlike in mammalian follicles.

3.2.2 Follicular Growth, Differentiation, and Steroidogenesis

The primary follicles in the ovarian cortex undergo growth into white follicles, followed by pre-hierarchal follicles (6–8 mm in diameter). Then, one pre-hierarchal follicle is recruited in 1 day to enter the hierarchal stage, namely the pre-ovulatory follicles that undergo rapid growth. During the process of follicular growth, yolk accumulates with increasing the size of follicles, and there is noticeable differentiation of the granulosa and theca cells. The white follicles incorporate the protein-rich yolk, whereas pre-hierarchal yellow and pre-ovulatory follicles accumulate lipid-rich yolk (Johnson 2000). The vasculature is well developed for transportation of yolk materials in the superficial layer of pre-ovulatory follicles, compared with white follicles.

In the pre-ovulatory follicles of hens, three types of cells including the granulosa cells, thecal interstitial cells, and aromatase cells in the theca externa participate in steroidogenesis (Nitta et al. 1991). Granulosa cells express Cytochrome P450 side chain cleavage (P450scc), an enzyme responsible for synthesizing pregnenolone from cholesterols, and 3β-hydroxysteroid dehydrogenase (3β-HSD) for synthesizing progesterone from pregnenolone. The thecal interstitial cells also express the 3β-HSD and P45017α that is responsible for testosterone synthesis from progesterone. The aromatase cells in the theca externa can be identified by immunostaining for aromatase which synthesizes estradiol from testosterone (Nitta et al. 1991). Thus, progesterone synthesized in the granulosa and thecal interstitial cells is used for synthesis of testosterone in the later cells, which is converted to estradiol in the aromatase cells.

It is reported that the selection of pre-hierarchal follicles entering the rapid growth phase represents a process in which a single undifferentiated follicle is predicted to escape an inhibitory mechanism each day to begin rapid growth and final maturation before ovulation (Johnson 2015). The report described that several processes initiated within the granulosa cells at selection are dependent on G protein-coupled receptors signaling via cyclic adenosine monophosphate (cAMP). Another research group suggested that SLIT protein ligand and their receptors (ROBOs) are implicated in the pre-hierarchal follicular development of the hen ovary, and are influenced by activin A and inhibin A (Qin et al. 2015). However, the precise mechanism by which a pre-hierarchal follicle is selected and initiate for rapid growth remains to be investigated further.

3.2.3 Regulation of Follicular Functions by Gonadotropins and Gonadal Steroids

The functions of follicular cells are changed when follicular growth is regulated by gonadotropins and gonadal steroids. It is reported that the response to follicle-stimulating hormone (FSH) for the synthesis of cAMP declines, whereas that of luteinizing hormone (LH) increases with the growth of follicles (Calvo and Bahr 1983). Such changes in the response to gonadotropins are likely to be associated with the changes in the steroidogenic functions. Estrogen synthesis in the theca stimulated by FSH is greater in smaller follicles than in larger follicles; whereas progesterone synthesis in the granulosa cells stimulated by LH is greater in the larger follicle (Bahr et al. 1983).

We have localized the receptors for progesterone (PR), androgen (AR) and estrogen (ER) in the follicular cells, suggesting that these steroids regulate the functions of follicular cells in an autocrine and/or paracrine manner (Yoshimura and Bahr 1991; Yoshimura et al. 1993a, 1995). PR has been localized in the nuclei of theca externa fibroblasts and superficial epithelial cells within the growing pre-ovulatory follicle. In addition, PR has also appeared in the granulosa cells of the largest pre-ovulatory follicle (Fig. 3.4a; Yoshimura and Bahr 1991). In the majority of granulosa cells of primordial follicles AR was negligible, whereas all of the granulosa cells of pre-ovulatory follicles exhibited a strong AR immunoreaction. Thecal interstitial cells and fibroblasts also stained positive for AR in pre-ovulatory follicles (Yoshimura et al. 1993a). The primordial follicles do not express immunoreactive ER, whereas the interstitial cells in the outer theca externa were ER positive

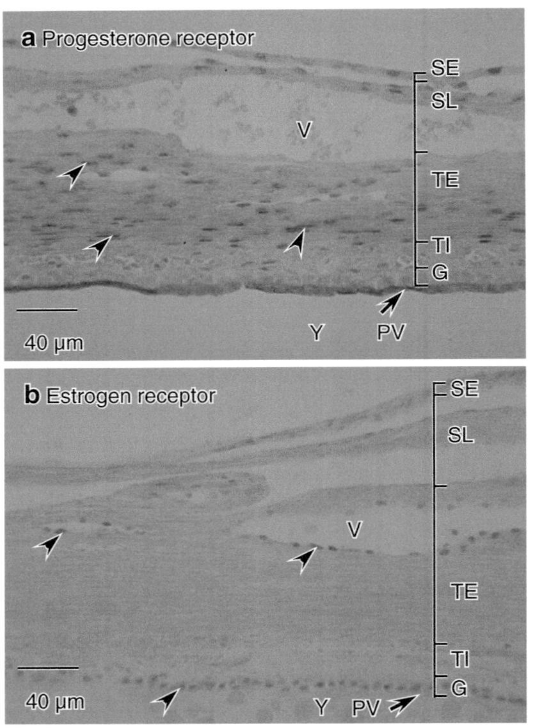

Fig. 3.4 Progesterone and estrogen receptors in the wall of growing follicle. **a** Immunoreactive progesterone receptors are localized in the thecal fibroblasts (Yoshimura and Bahr 1991). **b** Immunoreactive estrogen receptors are localized in the granulosa cells (G) and cells of veins (V) (Yoshimura et al. 1995). *Arrow heads* indicate the cells immunoreactive for progesterone and estrogen. *TI and TE* theca interna and exprena, *SL and SE* superficial layer and epithelium, *Y* yolk, *PV* perivitelline layer

in the pre-hierarchal follicles. The islets of epithelial cells (identical to estrogen-producing cells) in the outer theca externa and granulosa cells were ER-positive in the growing pre-ovulatory follicles, but they were remarkably reduced in the largest pre-ovulatory follicles (Fig. 3.4b; Yoshimura et al. 1995). The expression of these receptors suggests that ovarian tissues are the target for progesterone, androgen, and estrogen, and it is assumed that these steroids play a role in regulating the proliferation and differentiation of follicular cells, follicular maturation, and ovulation through receptor-mediated pathways. Furthermore, the profiles of the receptor expression suggest that the effects of progesterone may be greater in the larger follicles than in smaller ones, those of estrogen may be reduced in the largest follicles, and the androgen effects may not be changed significantly during the growing process of pre-ovulatory follicles.

3.2.4 Oocyte Maturation

The germinal disc of the oocyte contains the germinal vesicle and numerous cytoplasmic elements, such as elongated membrane-bound vesicles, mitochondria, and glycogen granules in the yellow follicles of quail. The cytoplasmic projections of granulosa cells interdigitate with microvilli on the surface of the oocyte, and form spot desmosomes and gap junctions with the oolemma. By 1 h before ovulation, the oocyte and the granulosa cells are disconnected, and oocyte maturation resumes, including the formation of the second maturation spindle just beneath the surface of the oocyte, and the release of the first polar body in the perivitelline space (Yoshimura et al. 1993b).

This resumption of oocyte maturation is probably caused by LH that increases 6 h before ovulation, because an experimental injection with LH induces the oocyte maturation. At 2 h after injection of LH, the germinal vesicles in the largest follicle, but not in the second largest follicle, begin to break down, and membrane-bound vesicles increase in number and size in the surrounding germinal disc (Fig. 3.5; Yoshimura et al. 1993c). The junctions between the oocyte surface and the granulosa cell projections dissociate with an accumulation of fluids in the perivitelline space. The first maturation spindle is formed 4 h after injection of LH, whereas the first polar body and the second maturation spindle are formed 6 h after LH stimulation, namely just before ovulation.

Mori et al. (1991) investigated the maturation-promoting factor and a homolog of fission yeast cdc2+ gene product ($p34^{cdc2}$) during the final 24 h of maturation of quail oocytes. They found a 15-times increase in kinase activity of $p34^{cdc2}$ in the oocyte germinal disk at maturation, and presence of 32 and 34 kDa immunoreactive protein using anti-$p34^{cdc2}$ in immature oocytes, and the appearance of an additional band close to the 32-kDa protein during maturation. Thus, it is suggested that this factor is involved in the oocyte maturation in birds. More recently, Elis et al. (2008) reported that eight genes (*btg4*, *chkmos*, *wee*, *zpA*, *dazL*, *cvh*, *zar1*, and *ktfn*) were preferentially expressed in the maturing oocyte. Some of these genes may be potentially involved in oocyte maturation.

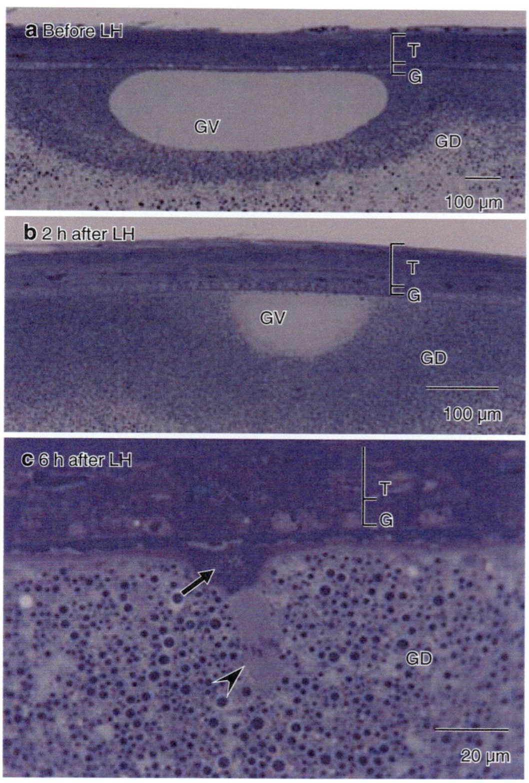

Fig. 3.5 Germinal disc of Japanese quail undergoing break down during oocyte maturation induced by luteinizing hormone injection. **a** Ten hours before the expected time of ovulation and before LH injection. Note the large germinal vesicle (GV) located in the germinal disc (GD). **b** Two hours after LH injection. The germinal vesicle (GV) is breaking down and forming the folds of its membrane. **c** Six hours after LH injection. Note the second maturation spindle (*arrowhead*) and the first polar body (*arrow*). *G* granulosa layer, *T* theca layer (Yoshimura et al. 1993c)

3.2.5 Ovulation

Ovulation occurs through the rupture of the stigma of follicles, where blood vessels except for capillaries are not supplied. In the theca externa of the stigma region, the rough endoplasmic reticulum in the fibroblasts develops conspicuously 30 min before ovulation, suggesting that synthesis of some protein factors may be activated at this phase. Then, a few minutes before ovulation, the theca externa becomes thinner with a conspicuous disintegration of the collagenous fibers into individual fibrils, and the granulosa layer and theca interna disappear completely in the middle part of the stigma. These structural changes may contribute to the fragility of the stigma tissue and induction of the rupture of the follicle (Yoshimura and Koga 1982). The fragility of the stigma tissues may be due to the reduction of connective tissue matrix caused by proteolysis. Degradation of glycosaminoglycans that occurs specifically in the stigma region prior to ovulation may be one of the factors responsible for the fragility of these tissues (Jackson et al. 1991). Neutral and acid proteases as well as collagenase activities increase in the post-ovulatory follicle immediately after ovulation or at some later time after ovulation (Tojo et al. 1982). Matrix metalloproteinases (MMP3 and MMP9) expression levels increase during follicular maturation, and MMP3 reaches maximal expression in the largest follicle, whereas MMP9 levels continue to rise in post-ovulatory follicles (Zhu et al. 2014).

The fragile stigma may be ruptured by the tensile strength of the follicular wall. The theca layer and smooth muscle bands in the superficial layer have contraction activity, which is enhanced by the neurotransmitters and prostaglandins. The increased tensile strength arising from contraction may promote the rupture of the stigma (Yoshimura et al. 1983). The post-ovulatory follicles undergo regression without forming corpus luteum, unlike in mammals, although the largest post-ovulatory follicle may synthesize prostaglandins (Saito et al. 1987). The regression of chicken post-ovulatory follicles seems to be an inflammatory event mediated by cytokines and chemokines similar to luteolysis in the mammalian corpus luteum (Sundaresan et al. 2008).

Ovulation is regulated by complex hormonal interactions, whereas it is well accepted that ovulation is induced by the pre-ovulatory surge of LH, which is caused by the positive feedback of progesterone. Testosterone also participates in the regulation of the ovulatory process. The temporal correlation between the testosterone, progesterone, and LH pre-ovulatory surge, the suppressive effect of testosterone inhibitor on ovulation, the promotion of progesterone secretion by testosterone from granulosa cells with an increased mRNA expression of StAR protein (steroidogenic acute regulatory protein) and P450scc enzymes, and with LH receptor mRNA expression in granulosa cells, are described in a recent review (Rangel and Gutierrez 2014).

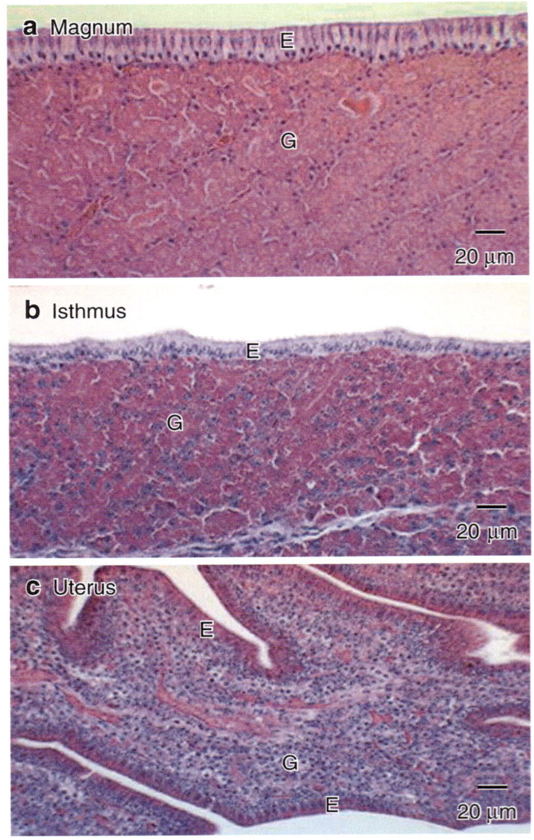

Fig. 3.6 Mucosal tissues of the magnum (**a**), isthmus (**b**) and uterus (**c**) of laying hens. The surface of the mucosa is lined by ciliated pseudostratified epithelium, and the tubular glands are well developed in the lamina propria in all tissues. The tubular gland cells contain amorphous substances and granules in the magnum and isthmus respectively, whereas the glandular cells in the uterus are less stained. *E* surface epithelium, *G* tubular glands. (Reproduced with agreement of Japanese Association of Veterinary Anatomists)

3.3 Oviducts

3.3.1 Structure of the Oviduct and Egg Formation

The oviductal mucosal layer of the laying hens consists of the lamina propria and ciliated pseudostratified epithelium lining the surface of the mucosa. In the magnum, isthmus, and uterus, the tubular glands are well developed in the lamina propria to sequentially synthesize the albumen, eggshell membrane, and eggshell respectively (Fig. 3.6).

3.3.1.1 Infundibulum

The mucosal folds in the funnel and tubular region of the infundibulum are relatively thin but tall. Fertilization occurs in the infundibulum before the outer perivitelline layer, and chalaza are formed on the ovum by the substances secreted by the cells in the tubular region. Sperm are identified at the basal space of the folds after copulation or insemination. A recent study suggests that a component of outer perivitelline layer is dependent on an estrogen-induced gene that is post-transcriptionally regulated by microRNAs (Lee et al. 2015).

3.3.1.2 Magnum

In the magnum, where the egg albumen layer is formed, the tubular glands synthesize ovalbumen, conalbumen, lysozyme, and antimicrobial peptides. The gene expression level of avidin, ovalbumen, ovomucin, and lysozyme have been shown to be up-regulated by steroid hormones (diethylstilbestrol and progesterone) (Jung et al. 2011). The goblet cells in the surface epithelium synthesize not only mucins but also avidin under the stimulation by progesterone and other steroids (Kunnas et al. 1992).

3.3.1.3 Isthmus

The tubular gland cells in the isthmus contain numerous granules in their cytoplasm, which are probably the substrate to be included in the eggshell membrane. The eggshell membrane formed in this segment is a highly cross-linked fibrous meshwork, which is organized into inner and outer membranes, containing types I, V, and X collagens and other substances (Hincke et al. 2000). The gene expression of collagen X, fibrillin-1, and cysteine rich eggshell membrane protein has been identified in the isthmus tissue (Du et al. 2015). At the distal part of the isthmus, mammillary bodies of the eggshell are formed by the penetration of the outer shell membrane fibers into their tips (Hincke et al. 2000).

3.3.1.4 Uterus (Shell Gland)

In the uterus (shell gland), the egg uptakes fluid into albumen and then calcification for shell formation on the eggshell membrane occurs. The acellular uterine fluid secreted from the tubular gland cells contains various ions such as Ca^{2+}, HCO^{3-}, Na^{+}, K^{+}, and organic matrix protein precursors (Marie et al. 2014). Microarrays have shown a list of 57 proteins potentially secreted into the uterine fluid and participating in the mineralization process (Brionne et al. 2014). The eggshell is formed by mineralization of ionized calcium and bicarbonate in the uterine fluid. These ionic precursors for calcium carbonate ($CaCO^{3}$), namely Ca^{2+} and HCO^{3-}, are supplied by the blood via trans-epithelial transport. Although the exact mechanism by which Ca^{2+} is transported from the blood to the uterine fluid has not yet been established, it has been suggested that the calbindin D28K (a calcium-binding protein) expressed in the tubular gland cells may participate in Ca^{2+} transport (Jonchère et al. 2012). Calbindin D28K mRNA is under the direct control of 1,25$(OH)_2$ vitamin D_3, and estradiol and progesterone may play coregulatory roles (Corradino 1993). The effects of estradiol on the synthesis of calbindin in the shell gland are minor, although it affects plasma Ca level (Bar et al. 1996). It is proposed that the HCO^{3-} is mainly prepared from CO_2 by the carbonic anhydrase 2 in the tubular gland cells (Jonchère et al. 2012). Once calcification is completed, the cuticle is deposited on the shell surface. The cuticle is believed to participate in the prevention of microbial invasion into the eggs. In brown-colored eggshell, protoporphyrin, the main eggshell pigment, is also deposited in the cuticle. The pigments are secreted from the ciliated surface epithelial cells in the uterus.

3.3.1.5 Vagina

In the vagina, numerous mucosal folds lined with ciliated pseudostratified epithelium are developed. The vagina may not play any distinct roles in egg formation. The ciliated epithelial cells and immune system in the vagina are well developed, probably to prevent the invasion of microbes from the cloaca.

3.3.2 Oviposition

Oviposition is the process of expulsion of the egg, which is controlled by various neuroendocrine factors including the arginine vasotocin (AVT) and prostaglandins. AVT is released from the posterior pituitary gland, and its plasma levels are related to oviposition (Koike et al. 1988). It binds to its receptors in the shell gland (Takahashi et al. 1994). The mechanism of AVT action may be mediated by local production of prostaglandins (Rzasa 1984). Prostaglandins and AVT are reported to amplify their action, as PGF2α released in the uterine tissue at the time of oviposition can stimulate further contractions that in turn also increase the release of AVT (Shimada et al. 1987; Takahashi et al. 2011). The body temperature of laying hens has been reported to increase at the time of laying, and such an increase may be caused by high levels of prostaglandins (Kadono and Yamade 1985). Age also can affect oviposition intervals with decreasing egg production in aged hens (Lillpers and Wilhelmson 1993).

3.3.3 Sperm Storage

Avian sperm are stored in the sperm storage tubules (SSTs) in the utero-vaginal junction (UVJ) of the oviduct (Fujii and Tamura 1963; Bakst et al. 1994). This unique feature enables the female to lay fertile eggs for a prolonged period, up to 2–3 weeks in chickens and 10 days in Japanese quail, after mating or artificial insemination. The SSTs are simple invaginations of the surface epithelium in UVJ (Fig. 3.7a). The UVJ surface epithelium is of ciliated pseudostratified type, whereas SST epithelium is a single layer of non-ciliated cells with microvilli on their apical surface (Fig. 3.7b). Although the mechanism by which sperm survive in the SSTs has not been fully established, possible factors responsible such as nutrients supply for sperm, regulation of SST luminal environments, and suppression of immunoresponse to resident sperm, have been suggested, as described in Chap. 11.

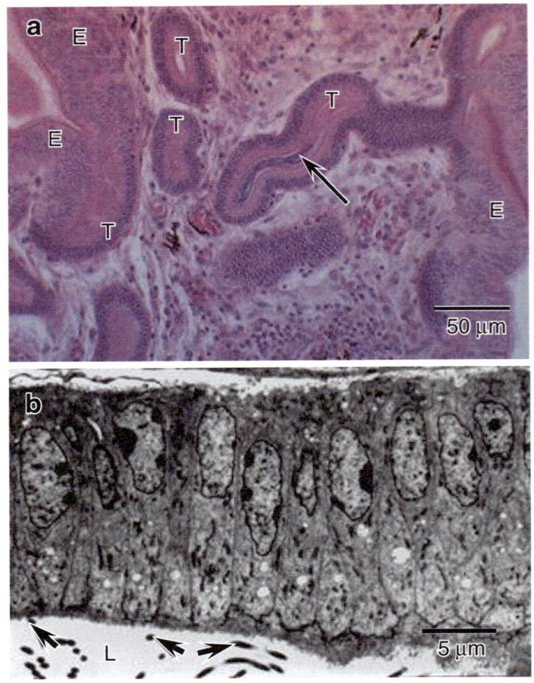

Fig. 3.7 Sperm storage tubules in the mucosal tissues of the utero-vaginal junction of Japanese quail. **a** The sperm storage tubules are formed by the invagination of the mucosal epithelium. *E* mucosal epithelium, *T* sperm storage tubules. **b** The epithelium of the sperm storage tubule consists of a single layer of cells with developed microvilli. *Arrows* show the sperm in the sperm storage tubules. (Reproduced with permission of Japanese Association of Veterinary Anatomists)

3.4 Defense System in the Reproductive Organs

The hen ovary and oviduct could be infected by various bacterial pathogenic microbes such as *Salmonella enteritidis, Salmonella typhimurium*, and *Mycoplasma melagridis* (Takata et al. 2003; De Buck et al. 2004; Wales and Davies 2011). Viral pathogenic infections include avian infectious bronchitis virus and avian influenza viruses (Promkuntod et al. 2006; Chousalkar and Roberts 2007; Nii et al. 2014, 2015). These infections in the ovary and oviduct may cause malformation of eggs and a decline in egg production. Also, contamination in eggs may increase the mortality of embryos. Intestinal microbes such as *Salmonella* organisms may be transported to the ovary and oviduct through the blood stream and consequently colonize these organs. Also, those in the cloaca may ascend the oviduct through the vagina and uterus. The innate and adaptive immune systems in the ovary and oviduct play essential roles in the defense against infection caused by those pathogenic organisms.

The innate immune system is the first line of the defense against infection, followed by activation of the adaptive immune system to remove the microbes. The innate immune response to pathogenic agents is initiated by the recognition of pathogen-associated molecular patterns (PAMPs) by the pattern recognition receptors (PRRs). Toll-like receptors (TLRs) are the primary PRRs that recognize PAMPs (Roach et al. 2005). In chickens, ten TLRs including TLR1 (types 1 and 2), −2 (types 1 and 2), −3, −4, −5, −7, −15, and −21 have been identified (Temperley et al. 2008; Brownlie and Allan 2011). TLR2, which forms a heterodimer with TLR1, recognizes peptidoglycan, lipoteichoic acid and lipoprotein of the Gram-positive bacterial cell wall (Keestra et al. 2007). TLR3 recognizes the double-stranded RNA of infectious viruses (Alexopoulou et al. 2001), whereas TLR4 recognizes lipopolysaccharide (LPS) of Gram-negative bacteria (St. Paul et al. 2013). TLR5 and −7 recognize bacterial flagellin and single-stranded RNA of viruses respectively (St. Paul et al. 2013). TLR15 and −21 are unique TLRs identified in avians. TLR15 recognizes the non-secreted and heat-stable bacterial components, the secreted virulence-associated fungal and bacterial proteases (Nerren et al. 2010; de Zoete et al. 2011). TLR21 recognizes CpG oligodeoxynucleotide of microbes with unmethylated CpG motifs, which is a functional homologue to mammalian TLR9 (Brownlie and Allan 2011). Interaction of TLRs with PAMPs activates the transcription factors such as NFκB and AP-1 to induce pro-inflammatory cytokines, chemokines, and antimicrobial factors.

Various cellular members, such as macrophages and natural killer cells, and antimicrobial substances are involved in the innate immune system. Defensins and cathelicidin are antimicrobial peptides that actively kill Gram-positive and Gram-negative bacteria, protozoa as well as some fungi and enveloped viruses (Xiao et al. 2004; Van Dijk et al. 2008; Cuperus et al. 2013). In addition, they may have a chemotactic function because they can attract T cells, monocytes, and immature dendritic cells while exerting toxic effects on a wide range of normal and malignant cells in mammals (Bulet et al. 2004). The defensins in birds contain six cysteines and belong to the β-defensins group (avian β-defensins: AvBDs). Up to now, 14 AvBDs genes and four cathelicidin genes have been identified in chicken (Xiao et al. 2004; Mageed et al. 2008; Cuperus et al. 2013).

The process of adaptive immune responses is initiated by antigen presentation by the major histocompatibility complex (MHC) expressed in the cells. The antigen-presenting cells (APCs) such as dendritic cells and macrophages incorporate substances in the tissue, and present antigens by MHC class II to helper/inducer T cells (CD4 + T cells). The activated CD4+ T cells, specifically Th2 cells, induce B-cell maturation and antibody synthesis by them, namely IgY, IgM, and IgA. The CD4+ T cells (Th1 cells) may also enhance the phagocytic activity of macrophages and activate CD8+ cytotoxic T cells. The cells intracellularly infected by microorganisms present the antigens by MHC class I, and stimulate cytotoxic/suppressor T cells (CD8+ T cells) to eliminate the infected cells. Chicken premature B cells can be identified by their surface antigen Bu-1, whereas the matured CD4+ T cells and CD8+ T cells both express the surface antigen CD3.

3.4.1 Defense System in the Ovary

3.4.1.1 Innate Immunodefense Systems in the Ovary

Macrophages are localized in the theca layer of pre-ovulatory and postovulatory follicles (Barua et al. 1998a, c). Carbons intravenously injected are incorporated by the fibroblast-like cells in the theca interna (Yoshimura and Okamoto 1998). Thus, it is suggested that these cells phagocytose the foreign substances as the first line of defense when they migrate into the follicles.

The hen follicular tissues express TLRs that induce innate immune response. The mRNAs of TLR1–2, 2–1, 3–5, 7, 15 and 21 are expressed in chicken ovary (Woods et al. 2009; Michailidis et al. 2010). Among them, the expression of TLR2, 4, 5 and 7 in the theca layer and TLR4 and 5 in the granulosa layer was identified in the pre-hierarchal and pre-ovulatory follicles (Subedi et al. 2007). The TLRs in the follicles play roles in recognition of PAMPs to induce the immune process. Also, they may contribute to other follicular functions; namely, the TLR signaling pathway affects steroidogenesis and cell maturation, initiating apoptosis in response to pathogenic stimuli in the granulosa cells (Woods et al. 2009).

In the hen ovary, 11 AvBDs genes among 14 AvBDs are expressed (Michailidis et al. 2012). In the pre-ovulatory follicles, the expression of 6 AvBDs in the theca and 4 AvBDs in the granulosa layer was identified, and the immunoreactive AvBD8, 10 and 12 were localized in the granulosa cells and theca interna cells (Subedi et al.

2007). The expression of AvBDs in the ovary increased with sexual maturation and age of birds (Michailidis et al. 2012). The densities of these AvBD proteins were greater in the preovulatory follicles than in primordial follicles (Fig. 3.8; Subedi et al. 2008; Abdelsalam et al. 2010). *Salmonella enteritidis* inoculation increased the expression of several AvBDs in vivo (Michailidis et al. 2012). Injection of birds with LPS also increased the expression of several AvBDs and pro-inflammatory cytokines in the theca tissue, although no response was observed in the granulosa layer (Subedi et al. 2007; Abdelsalam et al. 2011). Thus, it is likely that LPS upregulates the expression of AvBDs in the ovary in vivo. In contrast, the LPS stimulation of the cultured theca tissue of pre-ovulatory follicles upregulated the expression of pro-inflammatory cytokines, namely IL1β and IL6, but did not affect the expression of AvBD10 and 12 (Abdelsalam et al. 2012). However, the expression of AvBD12 gene and protein was upregulated by IL1β in the cultured theca tissues (Fig. 3.9; Abdelsalam et al. 2012). Thus, it is suggested that recognition of LPS, a Gram-negative bacterial component, by

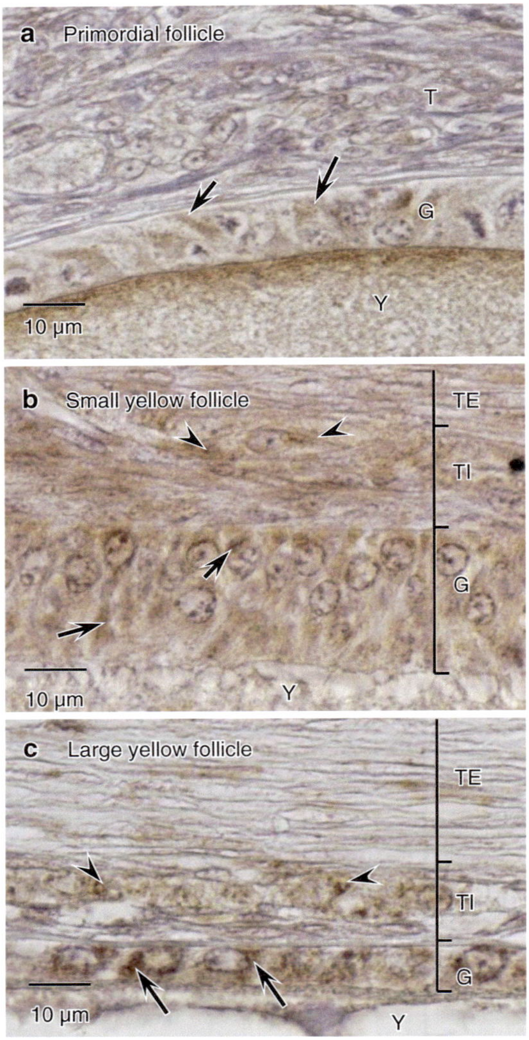

Fig. 3.8 Ovarian follicles immunostained for avian β-defensin 10 (AvBD10). **a** Primordial follicle, **b** Small yellow follicle. **c** Large yellow follicle. Granulosa cells contain immunoreaction products of AvBD10, whereas the density is higher in the large yellow follicle than in primordial and small yellow follicles. Thecal interstitial cells of small and large yellow follicles also contain immunoreaction products (*arrowheads*). *G* granulosa layer, *T* theca layer, *TI* theca interna, *TE* theca externa, *Y* yolk. (Abdelsalam et al. 2010)

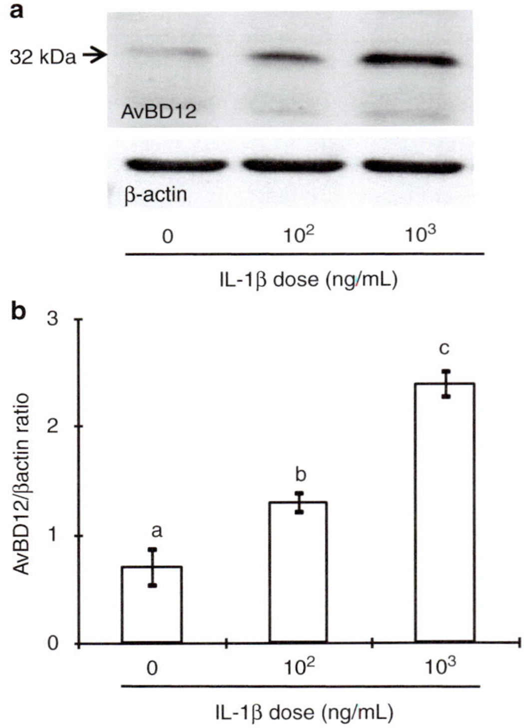

Fig. 3.9 Western blot of avian β-defensin 12 (AvBD12) in the theca layer of ovarian follicles treated with or without IL-1β. **a** Immunoblot of AvBD12 in theca tissue treated with 0 to 10^3 ng/ml of IL-1β for 5 h. **b** The ratio of AvBD12 bands to β-actin band density obtained from immunoblot. Data are the means ± SE ($n = 5$). Bars with different letters (**a–c**) are significantly different ($P < 0.05$). (Abdelsalam et al. 2012)

TLR4 leads to the synthesis of IL1β, which in turn stimulates the AvBD12 expression. Such a complex pathway for the effects of LPS to induce AvBDs may occur also in vivo.

3.4.1.2 Adaptive Immunodefense System in the Oviduct

The immunocompetent cells have been localized in the theca layer of follicular wall. The frequencies of macrophages, MHC class II+ antigen-presenting cells, CD4+ and CD8+ T cells, and B cells in the ovarian theca layer increased with sexual maturation and decreased with aging thereafter (Barua and Yoshimura 1999a; Barua et al. 1998a, b, 2001). The migration of these immunocompetent cells into the follicles could be upregulated by estrogen, suggesting that estrogen may be one of the factors responsible for causing the changes in the immunocompetent cell frequency during sexual maturation and aging (Barua and Yoshimura 1999b; Barua et al. 1998a, b). Infection by *Salmonella enteritidis* also causes the increase of the macrophage and T cell subsets population, namely CD3+, CD4+, and CD8+ T cells, in the ovarian stroma and the follicular tissues, suggesting that the immune response by these cells against *Salmonella* antigens could be induced in the hen ovary (Withanage et al. 2003; Barua and Yoshimura 2004). The influx of T cell subsets and B cells is also enhanced in the chicken ovary containing tumorigenesis, suggesting these lymphocytes play roles in the tumor response in the ovary (Bradaric et al. 2013).

3.4.2 Defense System in the Oviduct

3.4.2.1 Structural Defense System in the Oviduct

The infection by microbes in the oviduct is prevented by the defense system in the mucosa. The mucosal barrier is formed by mucins covering the mucosal surface, epithelial cell junctions, antimicrobial factors, and leukocytes. The movement of cilia on the mucosal surface may also play roles in the excretion of microbes into the oviductal lumen.

The mucin layer and epithelial tight junctions play roles as mucosal barriers that prevent microbial infection in the mucosal tissue (Forster 2008; Linden et al. 2008). The luminal surface of the mucosal tissues is covered by mucin substances that may attach the microbes as decoy substances and prevent the attachment of microorganisms to the cells. The mucin is synthesized by the goblet cells in the mucosal surface epithelium throughout the oviduct. The mucin expression may be stimulated by microbial infection, since its expression in the vagina was upregulated by LPS in which NFκB and activated protein (AP)-1 participate as transcriptional factors (Ariyadi et al. 2014). The potential function to synthesize mucin in the uterine and vaginal tissues was higher in laying hens than in molting hens (Ariyadi et al. 2012), which is probably affected by estrogen since estradiol upregulated mucin expression in association with epithelial development (Ariyadi et al. 2013a). However, estrogen may suppress the response to LPS for mucin induction (Ariyadi et al. 2013a). The tight junctions of epithelial cells play roles in the regulation of permeability of luminal substances including toxic substances or pathogens into the oviductal mucosa tissues. Those junctions are formed between the surface epithelial cells by claudins that assemble in the plasma membrane as the tight junction complex. Expression of claudins in the mucosa of the oviduct was lower in non-laying hens than in laying hens, suggesting that the mucosal barrier affecting the susceptibility may be weakened in the regressed oviduct of molting hens (Ariyadi et al. 2013b). The expression of claudin in the oviduct was also upregulated by estrogen (Ariyadi et al. 2013b).

3.4.2.2 Innate Immunodefense System in the Oviduct

Antimicrobial proteins and peptides are synthesized in the oviductal mucosa and probably play roles in the mucosal barrier. Lysozyme, an enzyme that kills Gram-positive bacteria, is synthesized in the tubular glands of the magnum and incorporated into the albumen, and possibly in the eggshell membrane and eggshell matrix (Short et al. 1996; Hincke et al. 2000).

Lactoperoxidase activity that catalyzes the oxidation of thiocyanate (SCN-) to produce hypothiocyanate (OSCN-), which has a broad-spectrum antimicrobial activity (Shin et al. 2002), was localized at the basal region of the secondary folds of the vagina in laying hens (Yoshimura et al. 2006a).

In the oviduct of sexually mature birds, all types of TLRs apart from TLR1–1 and at least seven AvBDs are expressed (Fig. 3.10; Mageed et al. 2008; Ozoe et al. 2009; Michailidis et al. 2011; Sonoda et al. 2013; Abdel-Mageed et al. 2014). The AvBD mRNA signals and immunoreactive AvBDs were localized in the surface epithelium of the oviduct, suggesting that AvBDs are synthesized by the mucosal epithelial cells (Fig. 3.11; Ohashi et al. 2005; Mageed et al. 2009). The AvBD expression in the oviduct may be increased in association with aging of hens, while decreased in the regressed oviducts of molting hens (Yoshimura et al. 2006b). In addition to AvBDs, gallin (ovodefensin) and cathelicidin are expressed in the oviductal mucosa (Gong et al. 2010; Abdel-Mageed et al. 2016). Gallin is a member of a new defensin-related group that has antimicrobial activity, and is synthesized in the tubular glands of the magnum (Gong et al. 2010).

Salmonella infection as well as LPS stimulation upregulated the expression of several AvBDs in the oviductal tissues (Fig. 3.12; Yoshimura et al. 2006b; Mageed et al. 2008, 2011). The expressions of pro-inflammatory cytokines and chemokines in the oviduct, including IL-1β and IL-6, were also upregulated in response to LPS or avian infectious bronchitis virus in vivo (Nii et al. 2011, 2014). It is also reported that the challenge of cultured vaginal tissue with TLR2, 3 5, 7 and 21 ligands showed a tendency to upregulate IL-1β and IL-6 expression, although their effects on the

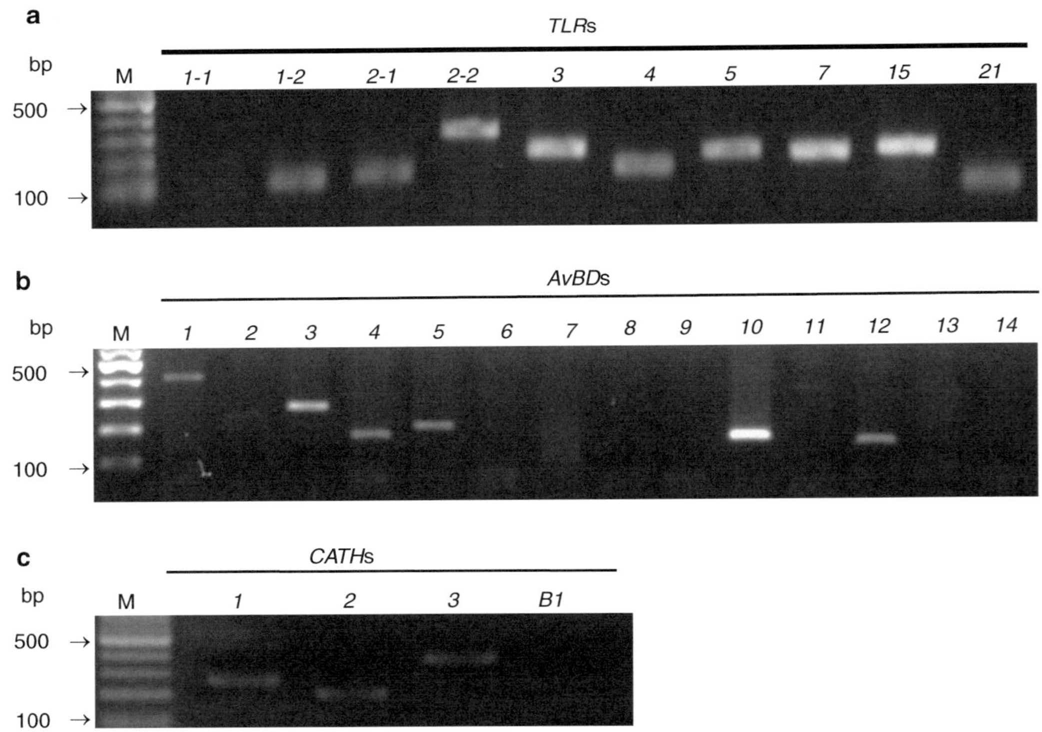

Fig. 3.10 Expression of Toll-like receptors, avian β-defensins and cathlicidins in the vagina of laying hens. **a** Toll-like receptors. **b** Avian β-defensins (Sonoda et al. 2013). **c** Cathlicidins (Abdel-Mageed et al. 2016)

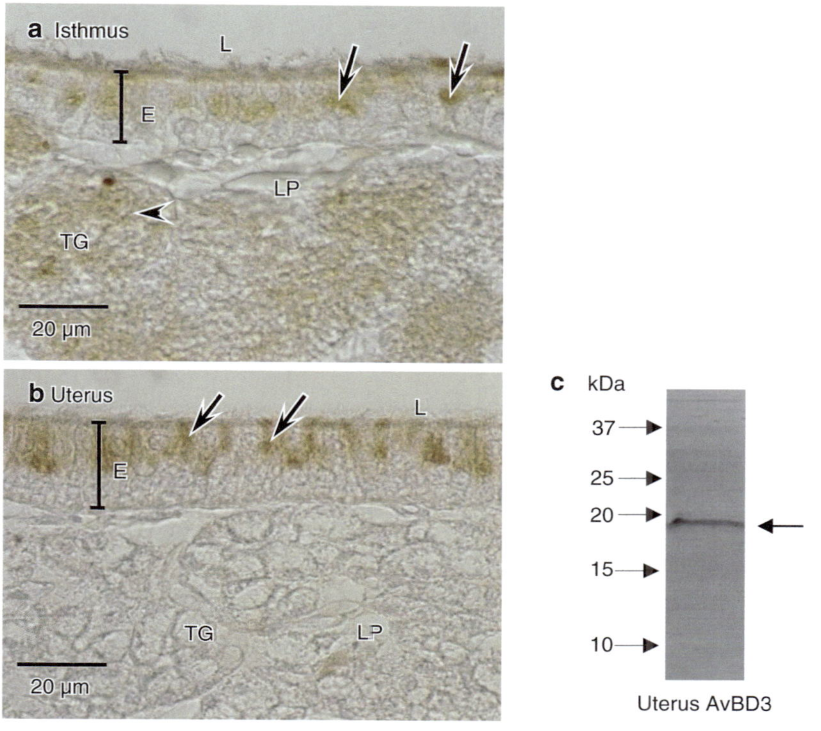

Fig. 3.11 Oviductal tissues of laying hens immunostained for avian β-defensin 3. **a** and **b** Isthmus and uterus. The upper regions of the cytoplasm of the surface epithelial contain immunoreaction products (*long arrows*). The tubular gland cells of the isthmus are weakly stained (*arrowhead*). **c** Western blot in the uterus tissue. *L* lumen, *E* surface epithelium, *LP* lamina propria, *TG* tubular glands. (Mageed et al. 2009)

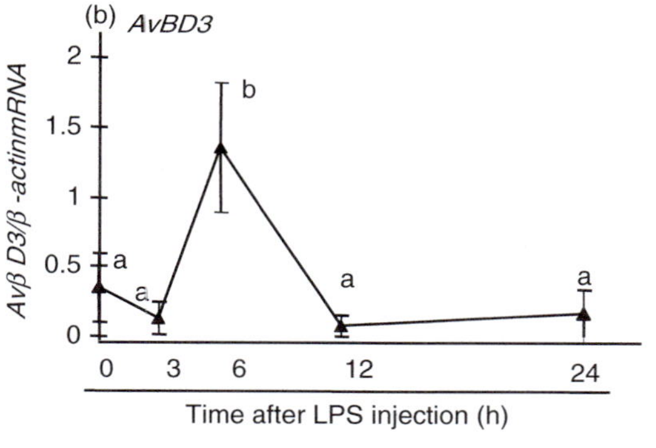

Fig. 3.12 Changes in the expression of avian β-defensin 3 (AvBD3) mRNA in the vagina after i.v. treatment with lipopolysaccharide (LPS). Birds were i.v. injected with 1 mg of LPS/kg of BW, and AvBD3 expression was examined by semiquantitative reverse transcription-PCR at different periods after LPS treatment. Values are mean ± SEM ($n = 4$). **a–c** Values with different letters are significantly different ($P < 0.05$). (Mageed et al. 2008)

AvBD expression were limited (Sonoda et al. 2013; Abdel-Mageed et al. 2014). Furthermore, we have observed that AvBD expression was upregulated by IL1β in the vaginal mucosa (Fig. 3.13; Sonoda et al. 2013). Thus, it is likely that AvBD expression in the oviduct is induced by the direct effects of TLRs ligands and/or by pro-inflammatory cytokines including IL1β induced by the interaction of TLRs and their ligands (Fig. 3.14; Yoshimura 2015). There is also a report suggesting that post-transcriptional regulation of AvBD11 expression may also occur, because microRNA-1615 has been discovered to influence AvBD11 expression in chicken oviduct (Lim et al. 2013).

3.4.2.3 Adaptive Immunodefense System in the Oviduct

Many of the immunocompetent cells are localized in the subepithelial stroma and surface epithelium in the oviduct (Fig. 3.15; Yoshimura et al. 1997; Zheng et al. 1998; Zheng and Yoshimura 1999; Nii et al. 2013). The densities of CD3+ T cells and Bu1+ B cells were greater in the infundibulum and vagina than in the other segments of the oviduct, probably due to the higher propensity to encounter antigens in these segments. These cells increased in density with sexual maturation and decreased during aging (Yoshimura et al. 1997; Zheng et al.

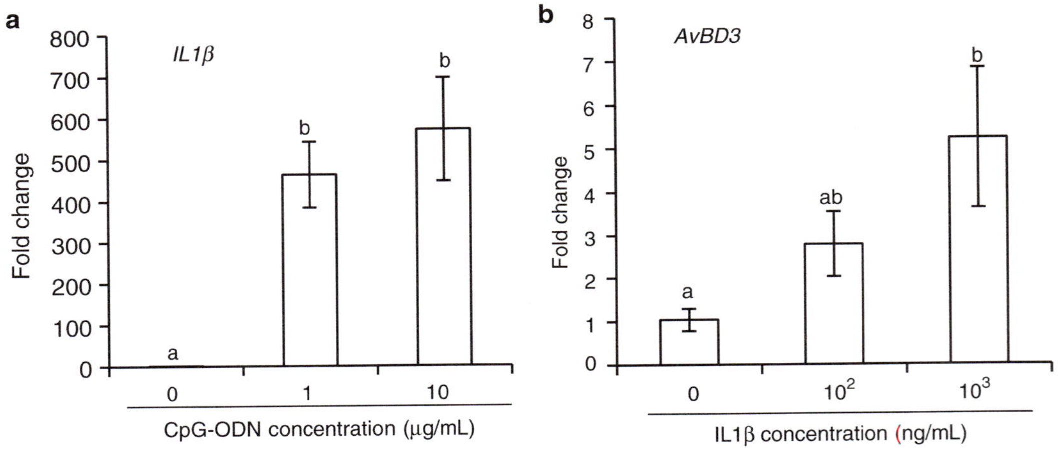

Fig. 3.13 Effects of CpG oligo-DNA on IL1β expression and of IL1β on avian β-defensins 3 expression in the cultured vaginal cells of hens. The cultured vaginal cells were stimulated by 0–10 μg/ml CpG-ODN (**a**), or by 0–10^3 ng/ml IL1β (**b**). Values are mean ± SEM (n = 6). **a, b** The values with different letters are significantly different ($P < 0.05$). (Sonoda et al. 2013)

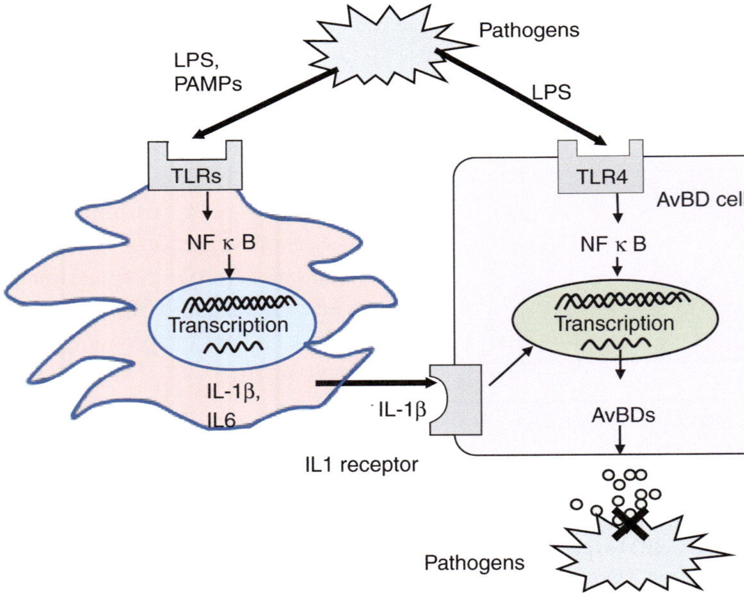

Fig. 3.14 Possible process from infection to induction of avian β-defensins (AvBDs) in the oviductal cells. Lipopolysaccharides of Gram-negative bacteria interact with TLR4, which may then activate transcription factors (NFκB) to induce cytokines such as interleukin (IL)-1β and IL6 (*left side*), as well as AvBDs (*right side*). Other pathogen-associated molecular patterns (PAMPs), such as poly(I:C), flagellin and CpG-ODN, which are ligands of TLR3, 5 and 21, may also induce IL1β and IL6 (*left side*). The synthesized IL1β upregulates the expression of AvBDs. (Yoshimura 2015)

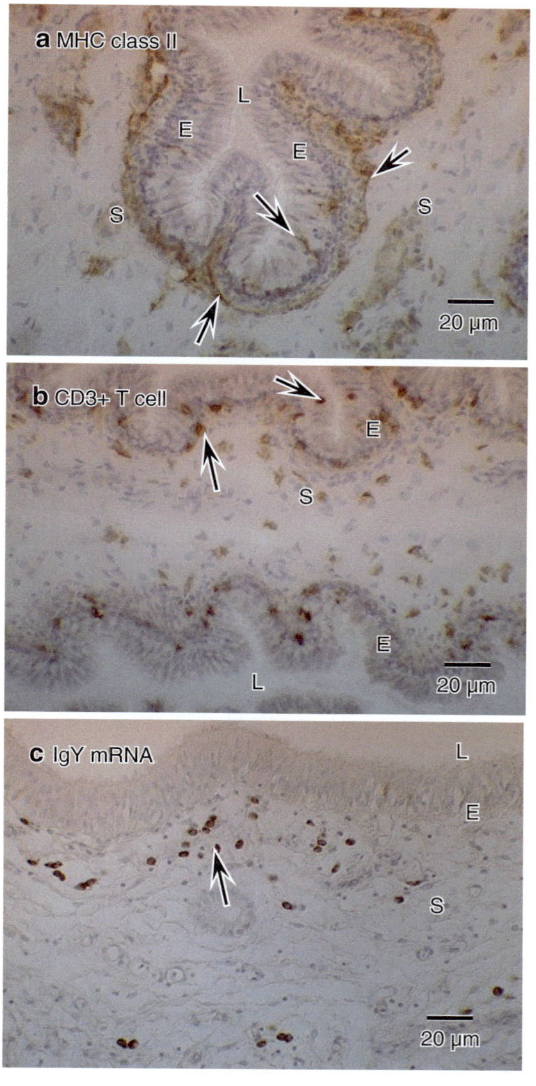

Fig. 3.15 Localization of immunocompetent cells in the oviductal mucosa of laying hens. **a** MHC class II in the vagina, immunohistochemistry. **b** CD3+ T cells in the vagina, immunohistochemistry. **c** IgY mRNA expressing cells in the utero-vaginal junction, in situ hybridization. (Zheng et al. 2000)

1998), and the densities in the epithelial and subepithelial tissues were reduced in molting hens compared with those in laying hens (Yoshimura et al. 1997). Their migration into the oviduct may be affected by estrogen, since injection with diethylstilbestrol (estrogenic compound) increased the densities of immunocompetent cells in the oviduct. Also, the infection of the mucosa may cause the synthesis of cytokines and chemokines that lead to the influx of immunocompetent cells. The challenge of hens with LPS and avian infectious bronchitis virus upregulated the expressions of proinflammatory cytokines and chemokines in association with recruitment of T cells in the oviduct (Nii et al. 2011, 2015).

3.4.3 Transfer of Antimicrobial Substances and Antibodies into Eggs

The eggshell membrane contains AvBD3 and the eggshell matrix contains AvBD3, 10, and 12 (Fig. 3.16; Mageed et al. 2009). The gallin synthesized in the tubular glands of the magnum is incorporated into the albumen (Gong et al. 2010). Other antimicrobial proteins including lysozyme and ovotransferrin have also been identified in the albumen, eggshell membrane, and eggshell (Ahlborn et al. 2006; Gautron et al. 2007; Wellman-Labadie et al. 2008). The presence of antimicrobial activities in the eggshell and its membrane was also demonstrated (Poland and Sheldon 2001; Mine et al. 2003). Thus, the antimicrobial factors synthesized in the oviduct may participate not only in the mucosal barrier in the oviduct but also in the defense system in the eggs.

The egg yolk contains the maternal immunoglobulins that play roles in the protection of neonatal chicks from infection before their adaptive immune system develops. In avian eggs, only IgY is selectively incorporated into the yolk during their growth in the follicles. The possible key amino acid residues of IgY responsible for its specific uptake by receptor-mediated endocytosis of oocyte have been reported (Bae et al. 2010; Murai et al. 2013).

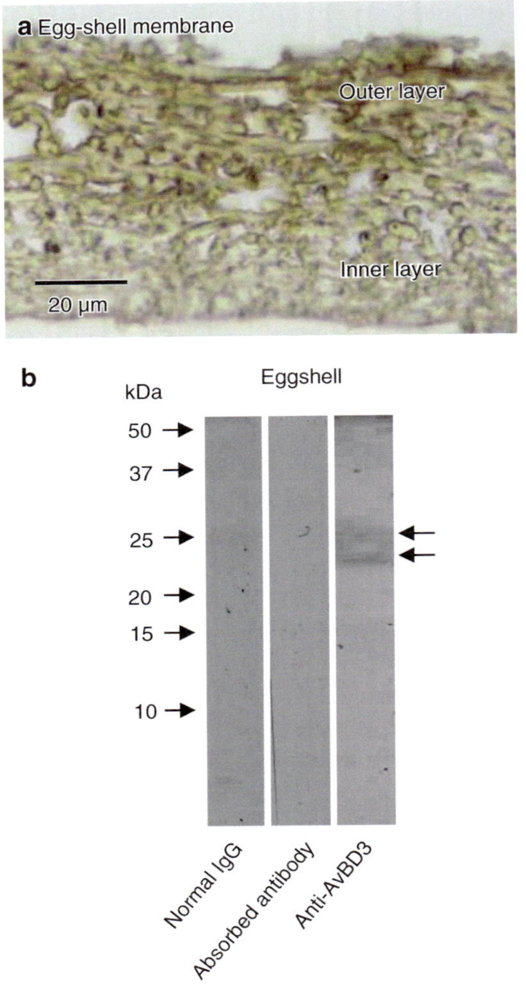

Fig. 3.16 Avian β-defensin 3 (AvBD3) in the eggshell membrane and eggshell. **a** Eggshell membrane immunostained for AvBD3. **b** Western blotting for AvBD-3 in proteins from the extract of eggshell. Lanes 1 and 2 were incubated with normal rabbit IgG and preabsorbed antibodies (control staining), lane 3 was immunostained by anti-AvBD3 antibody. Specific bands are identified at 23 and 25 kDa. (Mageed et al. 2009)

3.5 Chicken as a Model of Spontaneous Ovarian Cancer and Uterine Fibroids

Chickens have long been used to elucidate the etiology and molecular mechanism of pathogenesis, and to determine the efficacy of preventive interventions of many human diseases including leukemia (Taizi et al. 2006), cystic fibrosis (Craig-Schmidt et al. 1986), endometriosis (Nap et al. 2004), multiple sclerosis (Barnes 1986), systemic sclerosis (Beyer et al. 2010), immunological disorders (Davison 2003) which are difficult to study in humans. Ovarian cancer (OVCA), a lethal gynecological malignancy, is responsible for a high case-to-death ratio of women. Early detection of OVCA improves the survival rate of patients. Due to the non-specificity of symptoms at an early stage and the lack of an effective early detection test, OVCA in most cases is detected at a late stage, often accompanied with high rates of recurrence and death (Siegel et al. 2014). Due to these obstacles, access to early-stage OVCA tissue and serum is a significant barrier to the establishment of an early detection test. Thus, an animal model of OVCA that recapitulates many aspects of the disease is urgently needed. The laying hens are the only widely available and easily accessible animals which develop OVCA naturally and spontaneously with a high rate of incidence (Fredrickson 1987; Barua et al. 2009a) (Fig. 3.17a, b). In addition, the histopathology and mode of dissemination of ovarian carcinomas in hens are remarkably similar to OVCA in humans (Barua et al. 2009a). Furthermore, ovarian tumors in hens also express several tumor markers including tumor suppressor *p53,* Kirsten rat sarcoma (K-ras), human epidermal growth factor receptor 2 (*HER-2/neu)* oncogenes (Hakim et al. 2009), selenium-binding protein 1 (SELENBP1) (Stammer et al. 2008), CA-125 (Jackson et al. 2007) which are similarly expressed in human ovarian tumors. Incessant ovulations lead to the frequent rupturing of ovarian surface. Recurrent wear and tear of ovarian surface epithelium due to ovulatory insults increases chances of mutations resulting in uncontrolled cell growth leading to ovarian malignancy (Murdoch 2008). In addition, frequent ovulation also exposes ovarian surface epithelium at the site of rupture and the fimbrial epithelium at the site of receiving of the ovulated egg to the inflammatory cytokines. Longstanding unresolved chronic inflammation has been suggested as a risk factor for malignancy. Thus, frequent ovulation in commercial hens has a risk

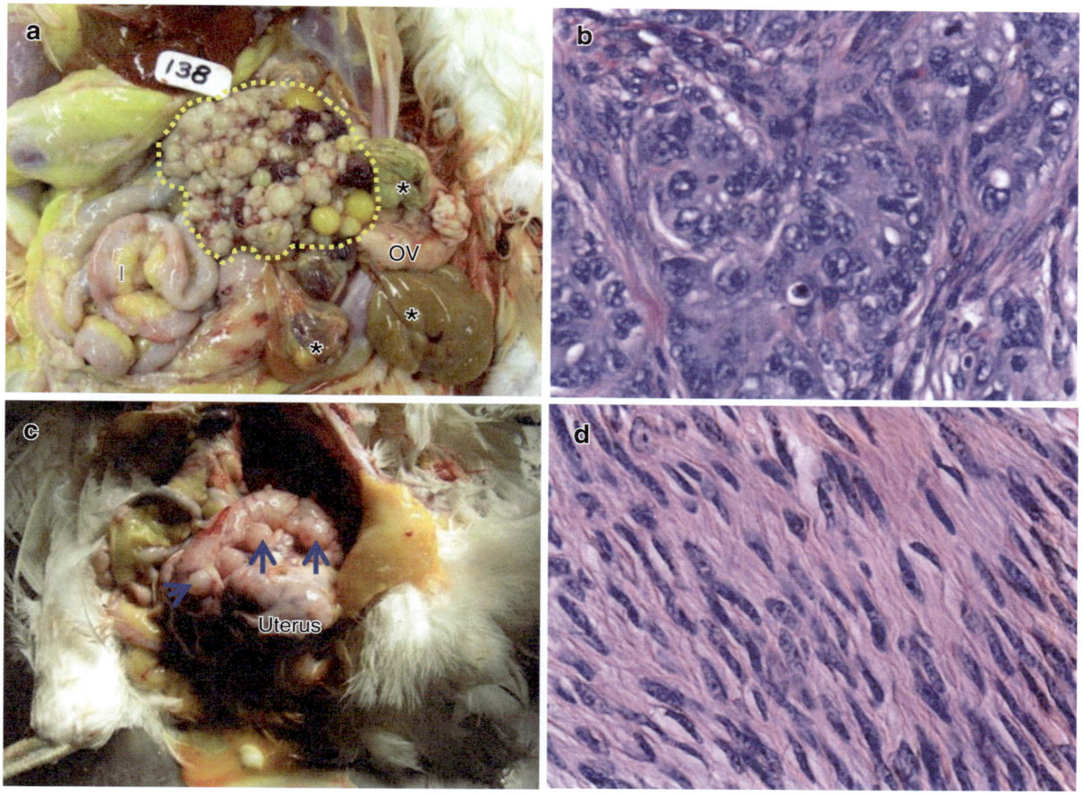

Fig. 3.17 Chicken as a model of human diseases. **a** Gross presentation of spontaneous ovarian cancer in hens. The tumor appears like a mass of cauliflower as observed in women. **b** Section of an ovarian serous tumor in a hen showing tumor cells containing large pleomorphic nuclei. **c** Gross presentation of fibroids protruded from the surface of magnum as observed in uterine fibroids in women. **d** Section of a fibroid. Tumor cells contain elongated, spindle-shaped nuclei arranged in a bundle-shape pattern with different directions (whorled). *I* Intestine; *OV* Oviduct, *Yellow dotted circle* tumor mass in the ovary, * Atretic large hierarchical follicles; *arrows* indicate the examples of fibroids

factor for high incidence of spontaneous OVCA making them a suitable model for human OVCA.

Laying hens are also being used to test molecular-targeted, contrast-enhanced ultrasound imaging agents for the detection of OVCA at early stage (Barua et al. 2014, 2015). Serum biomarkers represent non-invasive methods for screening of early stage OVCA identified by ultrasonography. Ovarian tumors in hens share many similarities with human OVCA including spontaneous incidence, tumor histology, mode of OVCA dissemination (Barua et al. 2009a), and expression of molecular markers mentioned above. Thus, laying hens possess tremendous advantages over other animal models for the establishment of serum biomarkers for early detection of OVCA. Immunotherapy represents a safer and more precise method of targeted treatment compared to traditional, cytotoxic chemotherapies or dangerous radiotherapies. Historically, the hen has had major contributions to the field of immunology, as the delineation of T and B cells was discovered using chickens. Moreover, there are also similarities between human and avian immune functions including anti-tumor immune responses (Barua et al. 2007b, 2009b) as well as tumor-induced immunosuppression (Khan et al. 2012; Barua et al. 2013; Bradaric et al. 2013) and laying hens are being exploited for the development of anti-OVCA immunotherapy (Barua et al. 2013). Also, laying hen represents potential for the development of anti-OVCA vaccines as it expresses several tumor-associated ovarian antigens including mesothelin (Yu et al. 2011), death receptor 6(DR6) (Barua et al. 2016) and selenium-binding protein 1 (Stammer et al. 2008), similarly expressed by ovarian tumors in women. Thus, these antigens can be used for developing vaccines using the laying-hen model.

Uterine fibroids (UF) are benign tumors of the uterus with 25–30% incidence rates in women of reproductive age. UFs are the leading cause of hysterectomy, responsible for significant amount of public health cost. In addition, UFs also affect the reproductive health and rate of fertility in women. Laying hens develop fibroids in the magnum and/or shell glands (uterus) spontaneously, with significant similarities to fibroids in women including the expression of TGF, BCL2, SMA, and desmin (Fig. 3.17c, d; Machado et al. 2012). Thus, the laying hens also offer a suitable model for devising preventive interventions against uterine fibroid formation.

Conclusion

We described the processes of egg formation that are regulated by the complex endocrine system. Not only reproductive organs but also other organs such as the liver and bones participate in its successful process. Maintenance of the health of hens is essential to obtain the maximum reproductive performance. Specifically, production of safe eggs and healthy chicks require the health and normal functioning of reproductive organs. As described in this chapter, the defense system including innate and acquired immune systems are developed in the ovary and oviduct. Studies on the vaccines have contributed to enhance the immunodefense functions against the infection by many pathogenic microbes. However, the strategy to enhance the innate immune functions is not well developed. The elements for innate immune systems, such as the pattern recognition receptors (TLRs), proinflammatory cytokines, and antimicrobial peptides, are expressed and respond to the pathogenic agents in both the ovary and oviduct. It would be an important strategy to enhance the functions of the innate immune system, in addition to more efficient vaccine development, in the reproductive organs. Recent studies suggest that the innate immune system may form immune memory as adaptive immune system, and the trained immunity for innate immune system may include the control of cellular function such as epigenetic reprogramming (Netea et al. 2016). Novel approaches to enhance the innate immune system are expected for maintenance of the health in the reproductive organs in hens.

References

Abdel-Mageed AM, Isobe N, Yoshimura Y. Effects of different TLR ligands on the expression of proinflammatory cytokines and avian beta-defensins in the uterine and vaginal tissues of laying hens. Vet Immunol Immunopathol. 2014;162:132–41.

Abdel-Mageed AM, Isobe N, Yoshimura Y. Effects of viral-associated molecular patterns on the expression of cathelicidins in the hen vagina. J Poult Sci. 2016;53:240–47.

Abdelsalam M, Isobe N, Yoshimura Y. Changes in the localization of immunoreactive avian beta-defensin-8,-10 and-12 in hen ovarian follicles during follicular growth. J Poul Sci. 2010;47:77–84.

Abdelsalam M, Isobe N, Yoshimura Y. Effects of lipopolysaccharide on the expression of proinflammatory cytokines and chemokines and influx of leukocytes in the hen ovary. Poult Sci. 2011;90:2054–62.

Abdelsalam M, Isobe N, Yoshimura Y. Effects of lipopolysaccharide and interleukins on the expression of avian beta-defensins in hen ovarian follicular tissue. Poult Sci. 2012;91:2877–84.

Ahlborn GJ, Clare DA, Sheldon BW, Kelly RW. Identification of eggshell membrane proteins and purification of ovotransferrin and beta-NAGase from hen egg white. Protein J. 2006;25:71–81.

Alexopoulou L, Holt AC, Medzhitov R, Flavell RA. Recognition of double-stranded RNA and activation of NF-kappaB by Toll-like receptor 3. Nature. 2001;413:732–8.

Ariyadi B, Isobe N, Yoshimura Y. Differences in the mucosal surface barrier formed by mucin in the lower oviductal segments between laying and molting hens. Poult Sci. 2012;91:1173–8.

Ariyadi B, Isobe N, Yoshimura Y. Induction of mucin expression by estrogen and lipopolysaccharide in the lower oviductal segments in hens. Poult Sci. 2013a;92:3205–13.

Ariyadi B, Isobe N, Yoshimura Y. Expression of tight junction molecule "claudins" in the lower oviductal segments and their changes with egg-laying phase and gonadal steroid stimulation in hens. Theriogenology. 2013b;79:211–8.

Ariyadi B, Isobe N, Yoshimura Y. Toll-like receptor signaling for the induction of mucin expression by lipopolysaccharide in the hen vagina. Poult Sci. 2014;93:673–9.

Bae HD, Kobayashi M, Horio F, Murai A. Identification of the amino acid residues involved in human IgG transport into egg yolks of Japanese quail (*Coturnix japonica*). Mol Immunol. 2010;47:1404–10.

Bahr JM, Wang SC, Huang MY, Calvo FO. Steroid concentrations in isolated theca and granulosa layers of preovulatory follicles during the ovulatory cycle of the domestic hen. Biol Reprod. 1983;29:326–34.

Bakst M, Wishart G, Brillard J. Oviductal sperm selection, transport, and storage in poultry. Poul Sci Rev. 1994;5:117–43.

Bar A, Vax E, Hunziker W, Halevy O, Striem S. The role of gonadal hormones in gene expression of calbindin (Mr 28,000) in the laying hen. Gen Comp Endocrinol. 1996;103:115–22.

Barnes DM. Bird chimeras may be models for certain neurological diseases. Science. 1986;232:930–2.

Barua A, Edassery SL, McNeal S, Bahr JM, Bitterman P, Basu S, Sharma S, Abramowicz JS. Enhancement of ovarian tumor detection by DR6-targeted ultrasound imaging agents in laying hen model of spontaneous ovarian cancer. International Journal of Gynecological Cancer. 2016;26(8):1375–85.

Barua A, Yoshimura Y. Effects of aging and sex steroids on the localization of T cell subsets in the ovary of chicken, Gallus domesticus. Gen Comp Endocrinol. 1999a;114:28–35.

Barua A, Yoshimura Y. Immunolocalization of MHC-II+ cells in the ovary of immature, young laying and old laying hens Gallus domesticus. J Reprod Fertil. 1999b;116:385–9.

Barua A, Yoshimura Y. Ovarian cell-mediated immune response to Salmonella enteritidis infection in laying hens (Gallus domesticus). Poult Sci. 2004;83:997–1002.

Barua A, Yoshimura Y, Tamura T. The effects of age and sex steroids on the macrophage population in the ovary of the chicken, Gallus domesticus. J Reprod Fertil. 1998a;114:253–8.

Barua A, Yoshimura Y, Tamura T. Effects of ageing and oestrogen on the localization of immunoglobulin-containing cells in the chicken ovary. J Reprod Fertil. 1998b;114:11–6.

Barua A, Yoshimura Y, Tamura T. Localization of macrophages in the ovarian follicles during the follicular growth and postovulatory regression in chicken, Gallus domesticus. Poult Sci. 1998c;77:1417–21.

Barua A, Michiue H, Yoshimura Y. Changes in the localization of MHC class II positive cells in hen ovarian follicles during the processes of follicular growth, postovulatory regression and atresia. Reproduction. 2001;121:953–7.

Barua A, Abramowicz JS, Bahr JM, Bitterman P, Dirks A, Holub KA, Sheiner E, Bradaric MJ, Edassery SL, Luborsky JL. Detection of ovarian tumors in chicken by sonography: a step toward early diagnosis in humans? J Ultrasound Med. 2007a;26:909–19.

Barua A, Bradaric MJ, Kebede T, Espionosa S, Edassery SL, Bitterman P, Rotmensch J, Luborsky JL. Anti-tumor and anti-ovarian autoantibodies in women with ovarian cancer. Am J Reprod Immunol. 2007b;57:243–9.

Barua A, Bitterman P, Abramowicz JS, Dirks AL, Bahr JM, Hales DB, Bradaric MJ, Edassery SL, Rotmensch J, Luborsky JL. Histopathology of ovarian tumors in laying hens: a preclinical model of human ovarian cancer. Int J Gynecol Cancer. 2009a;19:531–9.

Barua A Edassery SL, Bitterman P, Abramowicz JS, Dirks AL, Bahr JM, Hales DB, Bradaric MJ, Luborsky JL. Prevalence of antitumor antibodies in laying hen model of human ovarian cancer. Int J Gynecol Cancer. 2009b;19:500–7.

Barua A Bradaric MJ, Bitterman P, Abramowicz JS, Sharma S, Basu S, Lopez H, Bahr JM. Dietary supplementation of Ashwagandha (*Withania somnifera*, Dunal) enhances NK cell function in ovarian tumors in the laying hen model of spontaneous ovarian cancer. Am J Reprod Immunol. 2013;70:538–50.

Barua A, Yellapa A, Bahr JM, Machado SA, Bitterman P, Basu S, Sharma S, Abramowicz JS. Enhancement of ovarian tumor detection with $\alpha v\beta 3$ integrin-targeted ultrasound molecular imaging agent in laying hens: a preclinical model of spontaneous ovarian cancer. Int J Gynecol Cancer. 2014;24:19–28.

Barua A, Yellapa A, Bahr JM, Machado SA, Bitterman P, Basu S, Sharma S, Abramowicz JS. VEGFR2-targeted ultrasound imaging agent enhances the detection of ovarian tumors at early stage in laying hens, a preclinical model of spontaneous ovarian cancer. Ultrason Imaging. 2015;37:224–37.

Beyer C, Schett G, Distler O, Distler JH. Animal models of systemic sclerosis: prospects and limitations. Arthritis Rheum. 2010;62:2831–44.

Bradaric MJ, Penumatsa K, Barua A, Edassery SL, Yu Y, Abramowicz JS, Bahr JM, Luborsky JL. Immune cells in the normal ovary and spontaneous ovarian tumors in the laying hen (Gallus domesticus) model of human ovarian cancer. PLoS One. 2013;8(9):e74147.

Brionne A, Nys Y, Hennequet-Antier C, Gautron J. Hen uterine gene expression profiling during eggshell formation reveals putative proteins involved in the supply of minerals or in the shell mineralization process. BMC Genomics. 2014;15:220. doi:10.1186/1471-2164-15-220.

Brownlie R, Allan B. Avian toll-like receptors. Cell Tissue Res. 2011;343:121–30.

Bulet P, Stöcklin R, Menin L. Anti-microbial peptides: from invertebrates to vertebrates. Immunol Rev. 2004;198:169–84.

Calvo FO, Bahr JM. Adenylyl cyclase system of the small preovulatory follicles of the domestic hen: responsiveness to follicle-stimulating hormone and luteinizing hormone. Biol Reprod. 1983;29:542–7.

Chousalkar KK, Roberts JR. Ultrastructural study of infectious bronchitis virus infection in infundibulum and magnum of commercial laying hens. Vet Microbiol. 2007;122:223–36.

Corradino RA. Calbindin D28K regulation in precociously matured chick egg shell gland in vitro. Gen Comp Endocrinol. 1993;91:158–66.

Craig-Schmidt MC, Faircloth SA, Teer PA, Weete JD, Wu CY. The essential fatty acid deficient chicken as a model for cystic fibrosis. Am J Clin Nutr. 1986;44:816–24.

Cuperus T, Coorens M, Van Dijk A, Haagsman HP. Avian host defense peptides. Dev Comp Immunol. 2013;41:352–69.

Davison TF. The immunologists' debt to the chicken. Br Poult Sci. 2003;44:6–21.

De Buck J, Van Immerseel F, Haesebrouck F, Ducatelle R. Colonization of the chicken reproductive tract and egg contamination by Salmonella. J Appl Microbiol. 2004;97:233–45.

De Zoete MR, Bouwman LI, Keestra AM, Van Putten JPM. Cleavage and activation of a Toll-like receptor by microbial proteases. Proc Natl Acad Sci U S A. 2011;108:4968–73.

Du JW, Hincke MT, Rose-Martel M, Hennequet-Antier C, Brionne A, Cogburn LA, Nys Y, Gautron J. Identifying specific proteins involved in eggshell membrane formation using gene expression analysis and bioinformatics. BMC Genomics. 2015;16:792.

Elis S, Batellier F, Couty I, Balzergue S, Martin-Magniette ML, Monget P, Blesbois E, Govoroun MS. Search for the genes involved in oocyte maturation and early embryo development in the hen. BMC Genomics. 2008;9:110.

Forster C. Tight junctions and the modulation of barrier function in disease. Histochem Cell Biol. 2008;130:55–70.

Fredrickson TN. Ovarian tumors of the hen. Environ Health Perspect. 1987;73:35–51.

Fujii S, Tamura T. Location of sperms in the oviduct of the domestic fowl with special reference to storage of sperms in the vaginal gland. J Fac Fish Anim Husb, Hiroshima Univ. 1963;5:145–63.

Gautron J, Murayama E, Vignal A, Morisson M, Mckee MD, Rehault S, Labas V, Belghazi M, Vidal ML, Nys Y, Hincke MT. Cloning of ovocalyxin-36, a novel chicken eggshell protein related to lipopolysaccharide-binding proteins, bactericidal permeability-increasing proteins, and plunc family proteins. J Biol Chem. 2007;282:5273–86.

Gong DQ, Wilson PW, Bain MM, Mcdade K, Kalina J, Herve-Grepinet V, Nys Y, Dunn IC. Gallin; an antimicrobial peptide member of a new avian defensin family, the ovodefensins, has been subject to recent gene duplication. BMC Immunol. 2010;11:12.

Hakim AA, Barry CP, Barnes HJ, Anderson KE, Petitte J, Whitaker R, Lancaster JM, Wenham RM, Carver DK, Turbov J, Berchuck A, Kopelovich L, Rodriguez GC. Ovarian adenocarcinomas in the laying hen and women share similar alterations in p53, ras, and HER-2/neu. Cancer Prevention Research. 2009;2(2):114–21.

Hincke MT, Gautron J, Panheleux M, Garcia-Ruiz J, Mckee MD, Nys Y. Identification and localization of lysozyme as a component of eggshell membranes and eggshell matrix. Matrix Biol. 2000;19:443–53.

Jackson JA, Friberg AC, Bahr JM. Preovulatory changes in glycosaminoglycans and collagen content in the stigma region of the follicle of the domestic hen. Biol Reprod. 1991;45:301–7.

Jackson E, Anderson K, Ashwell C, Petitte J, Mozdziak PE. CA125 expression in spontaneous ovarian adenocarcinomas from laying hens. Gynecol Oncol. 2007;104:192–8.

Johnson AL. Reproduction in the female. In: Whittow GG, editor. Sturkie's avian physiology. 5th ed. San Diego: Academic Press; 2000. p. 569–96.

Johnson AL. Ovarian follicle selection and granulosa cell differentiation. Poult Sci. 2015;94:781–5.

Jonchère V, Brionne A, Gautron J, Nys Y. Identification of uterine ion transporters for mineralisation precursors of the avian eggshell. BMC Physiol. 2012;12:10.

Jung JG, Park TS, Kim JN, Han BK, Lee SD, Song G, Han JY. Characterization and application of oviductal epithelial cells in vitro in Gallus domesticus. Biol Reprod. 2011;85:798–807.

Kadono H, Yamade T. Changes of body temperature related to oviposition and ovulation induced by LH in the domestic hen. Nihon Juigaku Zasshi. 1985;47:55–61.

Keestra AM, De Zoete MR, Van Aubel RA, Van Putten JP. The central leucine-rich repeat region of chicken TLR16 dictates unique ligand specificity and species-specific interaction with TLR2. J Immunol. 2007;178:7110–9.

Khan MF, Bahr JM, Yellapa A, Bitterman P, Abramowicz JS, Edassery SL, Basu S, Rotmensch J, Barua A. Expression of leukocyte inhibitory immunoglobulin-like transcript 3 receptors by ovarian tumors in laying hen model of spontaneous ovarian cancer. Transl Oncol. 2012;5:85–91.

Koike TI, Shimada K, Cornett LE. Plasma levels of immunoreactive mesotocin and vasotocin during oviposition in chickens: relationship to oxytocic action of the peptides in vitro and peptide interaction with myometrial membrane binding sites. Gen Comp Endocrinol. 1988;70:119–26.

Kunnas TA, Joensuu TK, Viitala KK, Sopanen P, Tuohimaa P, Kulomaa MS. Induction of avidin messenger-ribonucleic-acid in the chick oviduct by progesterone and other steroids. Endocrinology. 1992;130:3421–6.

Lee SI, Ji MR, Jang YJ, Jeon MH, Kim JS, Park JK, Jeon IS, Byun SJ. Characterization and miRNA-mediated posttranscriptional regulation of vitelline membrane outer layer protein I in the adult chicken oviduct. In Vitro Cell Dev Biol Anim. 2015;51:222–9.

Lillpers K, Wilhelmson M. Age-dependent changes in oviposition pattern and egg production traits in the domestic hen. Poult Sci. 1993;72:2005–11.

Lim W, Jeong W, Kim J, Yoshimura Y, Bazer FW, Han JY, Song G. Expression and regulation of β-defensin 11 in the oviduct in response to estrogen and in ovarian tumors of chickens. Mol Cell Endocrinol. 2013;366:1–8.

Linden SK, Sutton P, Karlsson NG, Korolik V, McGuckin MA. Mucin in the mucosal barrier to infection. Mucosal Immunol. 2008;1:183–97.

Machado SA, Bahr JM, Hales DB, Braundmeier AG, Quade BJ, Nowak RA. Validation of the aging hen

(*Gallus gallus* domesticus) as an animal model for uterine leiomyomas. Biol Reprod. 2012;87:86.

Mageed AMA, Isobe N, Yoshimura Y. Expression of avian beta-defensins in the oviduct and effects of lipopolysaccharide on their expression in the vagina of hens. Poult Sci. 2008;87:979–84.

Mageed AMA, Isobe N, Yoshimura Y. Immunolocalization of avian beta-defensins in the hen oviduct and their changes in the uterus during eggshell formation. Reproduction. 2009;138:971–8.

Mageed AMA, Isobe N, Yoshimura Y. Changes in the density of immunoreactive avian β-Defensin-3 and-11 in the hen uterus in response to lipopolysaccharide inoculation. Poult Sci. 2011;48:73–7.

Marie P, Labas V, Brionne A, Harichaux G, Hennequet-Antier C, Nys Y, Gautron J. Data set for the proteomic inventory and quantitative analysis of chicken uterine fluid during eggshell biomineralization. Data Brief. 2014;1:65–9.

Michailidis G, Theodoridis A, Avdi M. Transcriptional profiling of Toll-like receptors in chicken embryos and in the ovary during sexual maturation and in response to Salmonella enteritidis infection. Anim Reprod Sci. 2010;122:294–302.

Michailidis G, Theodoridis A, Avdi M. Effects of sexual maturation and Salmonella infection on the expression of Toll-like receptors in the chicken vagina. Anim Reprod Sci. 2011;123:234–41.

Michailidis G, Avdi M, Argiriou A. Transcriptional profiling of antimicrobial peptides avian beta-defensins in the chicken ovary during sexual maturation and in response to Salmonella enteritidis infection. Res Vet Sci. 2012;92:60–5.

Mine Y, Oberle C, Kassify Z. Eggshell matrix proteins as defense mechanism of avian eggs. J Agric Food Chem. 2003;51:249–53.

Mori M, Yamashita M, Yoshikuni M, Fukada S, Nagahama Y. Maturation-promoting factor and P34cdc2 kinase during oocyte maturation of the Japanese quail. Dev Biol. 1991;146:246–9.

Murai A, Murota R, Doi K, Yoshida T, Aoyama H, Kobayashi M, Horio F. Avian IgY is selectively incorporated into the egg yolks of oocytes by discriminating Fc amino acid residues located on the C upsilon 3/C upsilon 4 interface. Dev Comp Immunol. 2013;39:378–87.

Murdoch WJ. Ovulatory factor in ovarian carcinogenesis. Adv Exp Med Biol. 2008;622:119–28.

Nap AW, Dunselman GA, de Goeij AF, Evers JL, Groothuis PG. Inhibiting MMP activity prevents the development of endometriosis in the chicken chorioallantoic membrane model. Hum Reprod. 2004;19:2180–7.

Nerren JR, He HQ, Genovese K, Kogut MH. Expression of the avian-specific toll-like receptor 15 in chicken heterophils is mediated by Gram-negative and Gram-positive bacteria, but not TLR agonists. Vet Immunol Immunopathol. 2010;136:151–6.

Netea MG, Joosten LA, Latz E, Mills KH, Natoli G, Stunnenberg HG, O'Neill LA, Xavier RJ. Trained immunity: a program of innate immune memory in health and disease. Science. 2016;352:aaf1098.

Nii T, Sonoda Y, Isobe N, Yoshimura Y. Effects of lipopolysaccharide on the expression of proinflammatory cytokines and chemokines and the subsequent recruitment of immunocompetent cells in the oviduct of laying and molting hens. Poult Sci. 2011;90:2332–41.

Nii T, Isobe N, Yoshimura Y. Effects of repeated lipopolysaccharide stimulation on the development of antigen-presenting cells and T cells pool in hen vagina. Poult Sci. 2013;50:83–9.

Nii T, Isobe N, Yoshimura Y. Effects of avian infectious bronchitis virus antigen on eggshell formation and immunoreaction in hen oviduct. Theriogenology. 2014;81:1129–38.

Nii T, Isobe N, Yoshimura Y. The effect of estrogen on the early cytotoxic response to IB virus infection in hen oviduct. Vet Immunol Immunopathol. 2015;164:56–66.

Nishio S, Kohno Y, Iwata Y, Arai M, Okumura H, Oshima K, Nadano D, Matsuda T. Glycosylated chicken ZP2 accumulates in the egg coat of immature oocytes and remains localized to the germinal disc region of mature eggs. Biol Reprod. 2014;91:107.

Nitta H, Osawa Y, Bahr JM. Multiple steroidogenic cell populations in the thecal layer of preovulatory follicles of the chicken ovary. Endocrinology. 1991;129:2033–340.

Ohashi H, Subedi K, Nishibori M, Isobe N, Yoshimura Y. Expressions of antimicrobial peptide gallinacin-1,-2 and −3 mRNAs in the oviduct of laying hens. Poult Sci. 2005;42:337–45.

Ozoe A, Isobe N, Yoshimura Y. Expression of Toll-like receptors (TLRs) and TLR4 response to lipopolysaccharide in hen oviduct. Vet Immunol Immunopathol. 2009;127:259–68.

Poland AL, Sheldon BW. Altering the thermal resistance of foodborne bacterial pathogens with an eggshell membrane waste by-product. J Food Prot. 2001;64:486–92.

Promkuntod N, Antarasena C, Prommuang P, Prommuang P. Isolation of avian influenza virus A subtype H5N1 from internal contents (albumen and allantoic fluid) of Japanese quail (*Coturnix Coturnix japonica*) eggs and oviduct during a natural outbreak. Ann N Y Acad Sci. 2006;1081:171–3.

Qin N, Fan XC, Zhang YY, Xu XX, Tyasi TL, Jing Y, Mu F, Wei ML, Xu RF. New insights into implication of the SLIT/ROBO pathway in the prehierarchical follicle development of hen ovary. Poult Sci. 2015;94:2235–346.

Rangel PL, Gutierrez CG. Reproduction in hens: is testosterone necessary for the ovulatory process? Gen Comp Endocrinol. 2014;203:250–61.

Roach JC, Glusman G, Rowen L, Kaur A, Purcell MK, Smith KD, Hood LE, Aderem A. The evolution of vertebrate Toll-like receptors. Proc Natl Acad Sci U S A. 2005;102:9577–82.

Rzasa J. The effect of arginine vasotocin on prostaglandin production of the hen uterus. Gen Comp Endocrinol. 1984;53:260–3.

Saito N, Sato K, Shimada K. Prostaglandin levels in peripheral and follicular plasma, the isolated theca and granulosa layers of pre- and postovulatory follicles, and the myometrium and mucosa of the shell gland (uterus) during a midsequence-oviposition of the hen (Gallus domesticus). Biol Reprod. 1987;36:89–96.

Schneider WJ. Vitellogenin receptors: oocyte-specific members of the low-density lipoprotein receptor supergene family. Int Rev Cytol. 1996;166:103–37.

Shimada K, Saito N, Itogawa K, Koike TI. Changes in plasma concentrations of arginine vasotocin after intrauterine injections of prostaglandin F-2 alpha and acetylcholine at various times during the oviposition cycle of the domestic hen (Gallus domesticus). J Reprod Fertil. 1987;80:143–50.

Shin K, Yamauchi K, Teraguchi S, Hayasawa H, Imoto I. Susceptibility of Helicobacter pylori and its urease activity to the peroxidase–hydrogen peroxide–thiocyanate antimicrobial system. J Med Microbiol. 2002;51:231–7.

Short ML, Nickel J, Schmitz A, Renkawitz R. Lysozyme gene expression and regulation. Experientia. 1996;75:243–57.

Siegel R, Ma J, Zou Z, Jemal A. Cancer statistics, 2014. CA Cancer J Clin. 2014;64:9–29.

Sonoda Y, Mageed AMA, Isobe N, Yoshimura Y. Induction of avian beta-defensins by CpG oligodeoxynucleotides and proinflammatory cytokines in hen vaginal cells in vitro. Reproduction. 2013;145:621–31.

St Paul M, Brisbin JT, Abdul-Careem MF, Sharif S. Immunostimulatory properties of Toll-like receptor ligands in chickens. Vet Immunol Immunopathol. 2013;152:191–9.

Stammer K, Edassery SL, Barua A, Bitterman P, Bahr JM, Hales DB, Luborsky JL. Selenium-binding protein 1 expression in ovaries and ovarian tumors in the laying hen, a spontaneous model of human ovarian cancer. Gynecol Oncol. 2008;109:115–21.

Subedi K, Isobe N, Nishibori M, Yoshimura Y. Changes in the expression of toll-like receptor mRNAs during follicular growth and in response to lipopolysaccharide in the ovarian follicles of laying hens. J Reprod Dev. 2007;53:1227–35.

Subedi K, Isobe N, Yoshimura Y. Changes in the localization of immunoreactive avian beta-defensin-12 in ovarian follicles during follicular growth and in response to lipopolysaccharide. Poult Sci. 2008;45:210–4.

Sundaresan NR, Saxena VK, Sastry KV, Nagarajan K, Jain P, Singh R, Anish D, Ravindra PV, Saxena M, Ahmed KA. Cytokines and chemokines in postovulatory follicle regression of domestic chicken (*Gallus gallus* domesticus). Dev Comp Immunol. 2008;32:253–64.

Taizi M, Deutsch VR, Leitner A, Ohana A, Goldstein RS. A novel and rapid in vivo system for testing therapeutics on human leukemias. Exp Hematol. 2006;34:1698–708.

Takahashi T, Kawashima M, Kamiyoshi M, Tanaka K. Arginine vasotocin receptor-binding in the hen uterus (shell gland) before and after oviposition. Eur J Endocrinol. 1994;130:366–72.

Takahashi T, Tajima H, Nakagawa-Mizuyachi K, Nakayama H, Kawashima M. Changes in prostaglandin F2alpha receptor bindings in the hen oviduct uterus before and after oviposition. Poult Sci. 2011;90:1767–73.

Takata T, Liang J, Nakano H, Yoshimura Y. Invasion of Salmonella enteritidis in the tissues of reproductive organs in laying Japanese quail: an immunocytochemical study. Poult Sci. 2003;82:1170–3.

Temperley ND, Berlin S, Paton IR, Griffin DK, Burt DW. Evolution of the chicken Toll-like receptor gene family: a story of gene gain and gene loss. BMC Genomics. 2008;9:62.

Tojo H, Fujii M, Ogawa K. Proteolytic-enzymes and gonadal-hormones of the ovarian follicle wall during ovulation in the domestic-fowl (Gallus-Domesticus). J Reprod Fertil. 1982;64:73–7.

Van Dijk A, Veldhuizen EJ, Haagsman HP. Avian defensins. Vet Immunol Immunopathol. 2008;124:1–18.

Wales AD, Davies RH. A critical review of Salmonella Typhimurium infection in laying hens. Avian Pathol. 2011;40:429–36.

Wellman-Labadie O, Picman J, Hincke MT. Antimicrobial activity of cuticle and outer eggshell protein extracts from three species of domestic birds. Br Poult Sci. 2008;49:133–43.

Withanage GS, Sasai K, Fukata T, Miyamoto T, Lillehoj HS, BABA E. Increased lymphocyte subpopulations and macrophages in the ovaries and oviducts of laying hens infected with Salmonella enterica serovar Enteritidis. Avian Pathol. 2003;32:583–90.

Woods DC, Schorey JS, Johnson AL. Toll-like receptor signaling in hen ovarian granulosa cells is dependent on stage of follicle maturation. Reproduction. 2009;137:987–96.

Xiao YJ, Hughes AL, Ando J, Matsuda Y, Cheng JF, Skinner-Noble D, Zhang GL. A genome-wide screen identifies a single beta-defensin gene cluster in the chicken: implications for the origin and evolution of mammalian defensins. BMC Genomics. 2004;5(1):56.

Yoshimura Y. Avian β-defensins expression for the innate immune system in hen reproductive organs. Poult Sci. 2015;94:804–9.

Yoshimura Y, Bahr JM. Localization of progesterone receptors in pre- and postovulatory follicles of the domestic hen. Endocrinology. 1991;128:323–30.

Yoshimura Y, Koga O. Ultrastructural changes of the stigma of the follicle during the process of ovulation in the hen. Cell Tissue Res. 1982;224:349–59.

Yoshimura Y, Okamoto T. Phagocytosis of carbon particles by theca interna fibroblasts in hen ovary. Japanese Poult Sci. 1998;35:314–8.

Yoshimura Y, Tanaka K, Koga O. Studies on the contractility of follicular wall with special reference to the mechanism of ovulation in hens. Br Poult Sci. 1983;24:213–8.

Yoshimura Y, Chang C, Okamoto T, Tamura T. Immunolocalization of androgen receptor in the small, preovulatory, and postovulatory follicles of laying hens. Gen Comp Endocrinol. 1993a;91:81–9.

Yoshimura Y, Okamoto T, Tamura T. Ultrastructural changes of oocyte and follicular wall during oocyte maturation in the Japanese quail (*Coturnix Coturnix japonica*). J Reprod Fertil. 1993b;97:189–96.

Yoshimura Y, Okamoto T, Tamura T. Electron microscope observations on LH-induced oocyte maturation in Japanese quail (*Coturnix Coturnix japonica*). J Reprod Fertil. 1993*c*;98:401–7.

Yoshimura Y, Okamoto T, Tamura T. Changes in localization of ovarian immunoreactive estrogen receptor during follicular development in hens. Gen Comp Endocrinol. 1995;100:368–74.

Yoshimura Y, Okamoto T, Tamura T. Localisation of MHC class II, lymphocytes and immunoglobulins in the oviduct of laying and moulting hens. Br Poult Sci. 1997;38:590–6.

Yoshimura Y, Mizuguchi M, Fujii M, Isobe N. Enzyme histochemical localization of peroxidase activity in the oviduct of immature laying and non-laying hens. Poult Sci. 2006a;43:401–7.

Yoshimura Y, Ohashi H, Subedi K, Nishibori M, Isobe N. Effects of age, egg-laying activity, and Salmonella inoculation on the expressions of gallinacin mRNA in the vagina of the hen oviduct. J Reprod Dev. 2006b;52:211–8.

Yu Y, Edassery SL, Barua A, Abramowicz JS, Bahr JM, Hellstrom I, Luborsky JL. The hen model of human ovarian cancer develops anti-mesothelin autoantibodies in response to mesothelin-expressing tumors. J Ovarian Res. 2011;4:12.

Zheng WM, Yoshimura Y. Localization of macrophages in the chicken oviduct: effects of age and gonadal steroids. Poult Sci. 1999;78:1014–8.

Zheng WM, Yoshimura Y, Tamura T. Effects of age and gonadal steroids on the localization of antigen-presenting cells, and T and B cells in the chicken oviduct. J Reprod Fertil. 1998;114:45–54.

Zheng WM, Izaki J, Furusawa S, Yoshimura Y. Localization of immunoglobulin G gamma-chain mRNA-expressing cells in the oviduct of laying and diethylstilbestrol-treated immature hens. Gen Comp Endocrinol. 2000;120:345–52.

Zhu GY, Kang L, Wei QQ, Cui XX, Wang SZ, Chen YX, Jiang YL. Expression and regulation of MMP1, MMP3, and MMP9 in the chicken ovary in response to gonadotropins, sex hormones, and TGFB1. Biol Reprod. 2014;90(3):57.

Development and Preservation of Avian Sperm

4

Atsushi Asano and Atsushi Tajima

Abstract

Terminally differentiated avian sperm consist of a head which male genetic material locates and flagellum that provides the motive force to propel them towards the fertilization site. The apical end of the sperm head accommodates a secretory vesicle, called an acrosome, that undergoes acrosome reaction releasing proteolytic content to penetrate the perivitelline membrane of an egg. Transcriptionally and translationally inactive, sperm need to rely on these distinct compartments in which different functions are preassembled, in order to achieve the goal of "fertilization". How are these complex structures with high functionality formed? Spermatogenesis is divided into an early stage in which diploid spermatogonia is proliferated into round spermatids thorough mitotic and meiotic divisions, and a late stage in which round spermatids are transformed into sperm though nuclear condensation and elongation of the sperm head, and formation of accessory structures. Recently, it was reported in aves that morphologically differentiated sperm undergo post-testicular maturation during passage through the male genital tract, suggesting that a similar system to mammals might be involved in the acquisition of fertilizing ability in avian sperm. Investigation for mechanisms underlying how sperm regulate their functions which are necessary to achieve fertilization is important for developing reproductive biotechnology in aves, because cryopreservation of poultry sperm is still not reliable for use in commercial production or for the preservation of genetic resources. In this review, we firstly provide an update on avian spermatogenesis, and then discuss the uniqueness of structure and functions of avian sperm, highlighting

A. Asano (✉) • A. Tajima (✉)
Faculty of Life and Environmental Sciences,
University of Tsukuba, 1-1-1 Tennoudai, Tsukuba,
Ibaraki 305-8572, Japan
e-mail: asano.atsushi.ft@u.tsukuba.ac.jp;
tajima.atsushi.gb@u.tsukuba.ac.jp

T. Sasanami (ed.), *Avian Reproduction*, Advances in Experimental Medicine and Biology 1001,
DOI 10.1007/978-981-10-3975-1_4

differences from mammalian sperm. Lastly, we discuss the molecular mechanism and current techniques of cryopreservation for avian sperm.

Keywords
Acrosome reaction • Birds • Epididymis • Fertilization • Flagellar motility • Post-testicular maturation • Signaling pathways • Spermatogenesis • Sperm cryopreservation

4.1 Spermatogenesis in Aves

Spermatogenesis is a complicated process including mitotic proliferation of spermatogonia, meiotic division of primary spermatocyte into round spermatid, and morphological differentiation of round spermatid into spermatozoa. Although spermatogenesis is extensively studied in mammals, there are only limited reports detailing spermatogenesis in aves. It has been shown in mammals that spermatogonia is categorized into three main patterns based on cellular appearances: type A, intermediate, and type B. These patterns are now subdivided in some species of mammals. For example, type A spermatogonia of rats is divided into generations of A_0, A_1, A_2, A_3, and A_4, and only A_0 spermatogonia contains stem cell activity for sperm production (Clermont and Bustos-Obregon 1968). However, a different pattern of spermatogonia has been identified in primates, showing the presence of only dark type A (A_d), pale type A (A_p), and type B (Clermont 1963; Clermont and Antar 1973; Chowdhury and Steinberger 1976; Schulze 1979; Hadley and Dym 1983), suggesting species-dependent differences in the cytological pattern of spermatogonia. On the other hand, the number of studies on avian spermatogonia is very small. Although it is not fair to make comparisons in a classification of spermatogonia between studies because of different experimental conditions used, it appears that there is species-difference between birds (Jones and Lin 1993).

After proliferation of spermatogonia, they commence two different meiotic phases to be divided into round spermatids with haploid set of chromosomes. Primary spermatocyte is pre-meiotic cells that are characterized by a conspicuous nuclear structure. Lin et al. (1990) showed in Japanese quail that primary spermatozoa are classified into eight different phases based on the structure of their chromosomes: preleptone, leptone, zygotene, pachytene, diplotene, diakinesis, metaphase, and anaphase. This suggests that the kinetics of multiploid spermatocyte in the quail is similar to that in mammals. Secondary spermatocytes are produced by first meiotic division, and they have a relatively short life compared to primary spermatocyte (0.15 days vs 3.86 days) because another meiotic division occurs immediately to produce round spermatid (Lin et al. 1990). This characteristic is shared with mammals.

Spermiogenesis is the later phase in the process of spermatogenesis, and during this period, the haploid, round spermatids undergo morphological differentiation to become sperm. Gunawardana and Scott (1977) classified the spermatid into four phases based on the nuclear changes; (1) spermatids with round nuclei, (2) spermatids with irregular nuclei, (3) spermatids with long nuclei containing coarse chromatin granules, and (4) spermatids with long homogeneously dense nuclei. The transformation of round spermatids into sperm occurs when they move toward the adluminal side of seminiferous tubules, and the dynamic feature within the tubules is generated by Sertoli cells, which are the only non-germinal, somatic cells present in seminiferous epithelium. Seminiferous epithelium is structured by stratified cellular arrangements, which shows spermatids at various steps of spermiogenesis in close apposition. Development of germ cells is synchronous within each generation and between successive generations produced from the same spermatogonia, which results in a special association of

Stages of the spermatogenic cycles

I	II	III	IV	V	VI	VII	VIII	IX	X
Ad	Ad	Ad	Ad	Ad	Ad	Ad	Ad	Ad	Ad
Ad	Ad	Ad	Ad	Ad	Ad	Ad	Ad	Ad	AP1
AP1	AP1	AP2	AP2	AP2	AP2	B	B	B	B
B	B	B	L	Z	P	P	P	P	P
P	P	P	P	P	P	DP	DK	M-An	II
1	2	3	4	5	6	7	8	9	10
11	11	11	12	12					

Fig. 4.1 The cycles of the seminiferous epithelium in the Japanese quail, showing ten associations of germ cells (Lin and Jones 1992). *Ad* dark type A spermatogonia, *AP1* pale type A1 spermatogonia, *AP2* pale type A2 spermatogonia, *B* type B spermatogonia, *L* leptone primary spermatocyte, *Z* zygotene primary spermatocyte, *P* pachytene primary spermatocyte, *DP* diplotene primary spermatocyte, *DK* diakinesis primary spermatocyte, *M-An* methapase to anaphase primary spermatocyte, *II* secondary spermatocyte, *1–12* step1–step12 spermatids

germ cells in the seminiferous epithelium. The associations of successive generations are referred to as cellular associations of the seminiferous epithelium, representing different combinations of germ cells. In Japanese quail, 10 cellular associations have been reported to be present in a seminiferous tubule (Fig. 4.1) (Lin and Jones 1992). The successive cellular associations are spatially arranged in sequential order along with seminiferous tubule forming a wave of spermatogenesis.

Although a fundamental characteristic of spermatogenesis is shared between aves and mammals, avian spermiogenesis is divided into 12 steps based on biogenesis of the sperm accessory structures and the progression of nuclear condensation (Fig. 4.2) (Aire 2014), unlike mammals (ex. 19 steps in rat, 16 steps in mouse, 14 steps in ungulates and primates (Sutovsky and Manandhar 2007). Based on the progression of acrosome biogenesis, these steps can be classified into four different phases as follows: Golgi phase, cap phase, acrosome phase, and maturation phase. In the Golgi phase (step 1–3), spermatids have a spherical nucleus with scattered chromatin aggregations and chromatin condensation started later in this phase. The Golgi complex contains numerous proacrosomal granules which are later budded from the trans-Golgi network and coalesce to form large acrosomal vesicle. The acrosomal vesicle is juxtaposed close to or making contact with the nucleus. In the contact point between the acrosomal vesicle and the nucleus, nuclear membrane begins to be thickened. Although the acrosomal vesicle shares characteristics in shape between mammals and aves, there could be difference in its contents. For example, the acrosomal vesicle is a hollow and round structure which is occupied with electron-dense materials surrounded by electron-lucent space (Burgos and Fawcett 1955; Leblond and Clermont 1952). However, the acrosomal vesicle in aves is a membrane vesicle that is filled with homogeneous and moderately electron-dense materials (Okamura and Nishiyama 1976; Nagano 1962; Aire 2003). The centriolar complex comprised of proximal and distal centrioles emerges close to the developing acrosome, nearly making contact with the nucleus, and becomes a fledgling flagellum. In the cap phase (step 4–6), spherical nuclei of spermatids continue to be condensed and are elongated to become pear-shaped. The acrosome becomes elongated, and its rostal end makes contact with spermatid plasma membrane adjacent to plasma membrane of Sertoli cells. The thickened nuclear membrane at the contact point between the acrosome vesicle and nucleus invaginates into the nucleoplasm, forming the endonuclear canal where the primordium of the filamentous structure in acrosome, the perforatorium, is later formed. Formation and composition of the perforatorium are extensively studied in rodents (Oko et al. 1990; Oko and Clermont

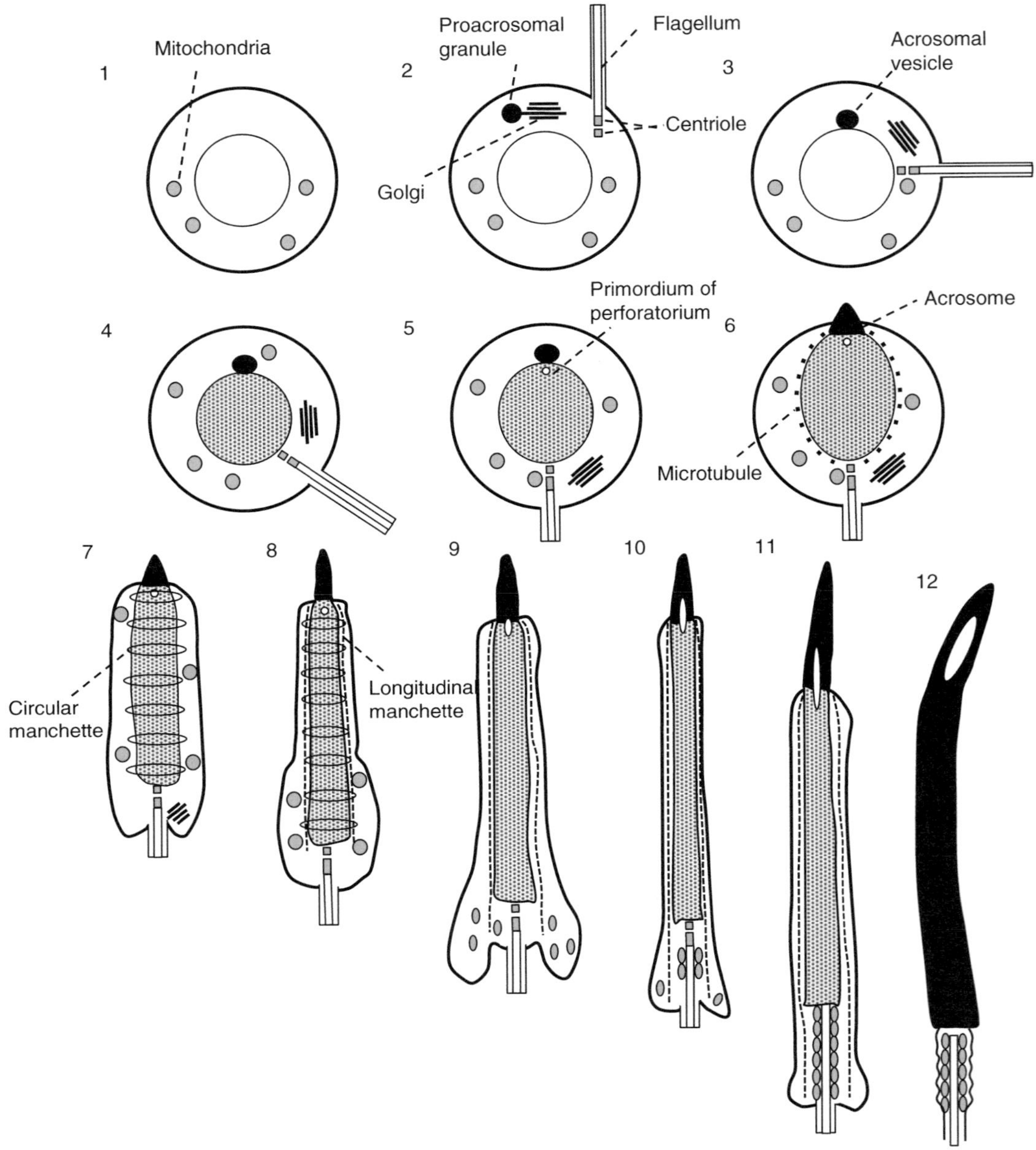

Fig. 4.2 A schematic diagram showing spermiogenesis of non-passerine birds which was described in previous review (Aire 2014). Spermiogenesis of aves consists of 12 steps that are categorized into four phases: Golgi phase (1–3), cap phase (4–6), acrosome phase (7–9), maturation phase (10–12)

1988; Selvaraj et al. 2010). Previously, the perforatorium was believed to be involved in the structural arrangement of acrosome (Oko and Morales 1994). However, a recent study concerning the mutant loss of a major perforatorium component demonstrated no difference in structure of the perforatorium (Selvaraj et al. 2010), leaving the role of the perforatorium unclear. The manchette is a transient microtubular structure which is comprised of circular manchette (CM) and longitudinal manchette (LM). CM appears to be responsible for an initial transformation in shape of the spermatid nucleus (Soley 1997), consistent with results showing that disruption of the manchette caused deformation of spermatid nucleus (Russell et al. 1991). Both manchettes are formed in close apposition to the nucleus from the region of the acrosome–nuclear junction to the position where the centriolar complex is located (Soley 1997; Aire 2003). In the acrosome phase (step 7–9), spermatid nucleus become slimmer and elongated. CM and LM are well established along with external surface of nucleus. The flagellum is enclosed in fibrous sheath. Mitochondria becomes elongated to be later aligned along with flagellum. In the maturation phase (step 10–12), the spermatid nucleus continues to elongate, with gentle curvature of the head region. The elongating acrosome forms a cap-like structure over the end of nucleus and accommodates the perforatorium that extends from the base of endonuclear canal to the apex of the acrosome. The elongated mitochondria is aligned with the midpiece and forms a helical sheath. Dissolution and disassembly of manchette occurs, which appears to make way for the elongated mitochondria to be apposed to proximal axons. Matured spermatid is eventually externalized to luminal surface of seminiferous tubules, being held with a slip of Sertoli cell cytoplasm. In general, the released spermatozoa, upon spermiation, do not have a cytoplasmic droplet in most of aves other than emu (du Plessis and Soley 2014) and domestic duck (Simoes et al. 2012). Residual bodies formed at spermiation are phagocytised by Sertoli cells, as in mammals. Testicular spermatozoa, characterized by low motility, can fertilize eggs when the semen was inseminated in the magnum. However, no fertilized eggs were obtained when the testicular semen was utilized in intravaginal insemination (Howarth 1983), suggesting importance of flagellar motility in fertilizing ability of avian sperm.

4.2 Post-testicular Sperm Maturation in Aves

After spermiation, matured sperm make the transition into epididymis, which is an accessory reproductive tract. In mammals, it is well known that sperm undergo functional maturation during epididymal transit, so-called epididymal maturation (Cornwall and Hann 1995). This process involves structural modification, changes in lipid and protein compositions of the plasma membranes (Jones 1998), and development of flagellar motility (Morton et al. 1978) and is known to be a prerequisite for mammalian sperm to acquire fertilizing ability. Although a complex mechanism behind epididymal maturation is still under investigation, it appears to be necessary for this functional maturation that sperm incorporate functional proteins secreted from epithelium into the plasma membrane during epididymal transit (Rankin et al. 1992; Sullivan 2015). On the other hand, it was a long-standing dogma that there is no maturational process during epididymal transit in avian sperm due to retrogression of the epididymis. However, recent studies suggest that fowl sperm gradually acquire abilities necessary for fertilization during passage of male genital tract prior to ejaculation (Ahammad et al. 2011a, b). Another study showed in Japanese quail that sperm motility is greatly increased at their entry into epididymis (Nixon et al. 2014). This phenomenon was also reported in fowl sperm collected in different regions of male genital tract (Howarth 1983). Although what mechanisms are involved in these functional changes are currently unknown, previous study showed that subset of secreted proteins in epididymis binds to the sperm surface and remains adherent even after prolonged incubation in female reproductive tract (Esponda and Bedford 1985; Morris et al. 1987), suggesting that incorporation of epididymal secreted proteins to sperm surface might provide sperm functions necessary for fertilization. Together with these, it is conceivable that epididymal maturation might be required to become fully able to fertilize an egg in avian sperm. Post-testicular maturation also appears to occur in sperm structure. Tingari (1973) found that the acrosomal cap is loosely juxtaposed to the sperm nucleus before spermiation, but in the excurrent duct it becomes in close apposition. In addition to this, the plasma membrane overlying the acrosomal cap becomes more tortuous during transit of sperm from testicular lumen to epididymal tubule. The role of this structural change in sperm during transit of the male genital tract remains unknown.

It is well known that mammalian sperm, after ejaculation, cannot fertilize an egg unless spending time in the female reproductive tract (Chang 1951; Austin 1951). This process is known as "capacitation", which involves modification of lipid composition (Davis 1976) and permeability and fluidity of the sperm plasma membrane (Abou-haila and Tulsiani 2009), the activation of complex signaling pathways (Visconti et al. 1995a, b), and enhancement of acrosomal responsiveness (Cross 1999; Iborra et al. 2000). On the other hand, a process like capacitation in mammals has not been detected in avian sperm (Howarth 1970). Another study showed that fowl sperm undergo acrosome reaction immediately after stimulation with inner peri-vitelline membrane (Lemoine et al. 2008), confirming that extra maturation in female reproductive tract is not needed for avian sperm to fertilize an egg. Interestingly, several signaling molecules involved in pathways to initiate capacitation in mammals are now reported to play a role in induction of acrosome reaction in avian sperm (Ashizawa et al. 2004, 2006; Lemoine et al. 2008). This suggests that sperm might accommodate inherent signaling pathways to changes taking place during evolution processes, perhaps resulting in divergent functional consequences of each signaling pathway between mammalian and avian sperm.

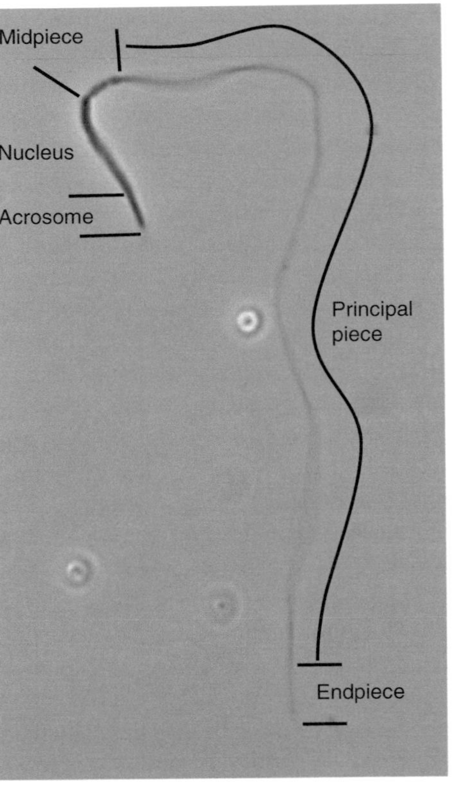

Fig. 4.3 Compartmentalization of chicken sperm. Sperm consist of three sub-cellular compartments: acrosome, nucleus, and flagellum. Flagellum is divided into midpiece where unique mitochondria are lined with flagellum, principal piece, and endpiece. The sub-cellular compartmentalization is believed to be important for sperm to efficiently utilize preassembled cellular pathways to fertilize an egg

4.3 Sperm Structure and Physiology

Terminally differentiated, sperm are extremely streamlined cells for achieving a single goal: "fertilization". Although sperm morphology greatly differs between aves and mammals, it is a shared characteristic that sperm consist of three compartments: acrosome, nucleus, and flagellum (Fig. 4.3). These compartments bear specific and distinct functions to be exerted when needed. In general, the sperm head of a non-passeridan bird is a cylindrical structure, so that there is difficulty to distinguish neck region forming connection between head and midpiece in most aves unless there is cytoplasmic residue at base of the nucleus, as seen in emu (du Plessis and Soley 2014). On the other hand, the sperm head of various passeridan birds displays a helical shape absent in perforatorium (Aire 2014), suggesting a species-dependent difference in sperm-head morphology. The acrosome is a secretory vesicle positioned close to the plasma membrane at tip of the sperm head, and surrounds the perforatorium that extends from the endonuclear canal (Korn et al. 2000; Aire 2014). In mammals, it is well known that acrosomal plasma membrane undergoes membrane fusion with outer acrosomal membrane, leading to an exocytotic event called acrosome reaction that is a prerequisite for both mammalian and avian sperm to penetrate into the egg's extra-cellular matrix. However, a recent study showed in fowl sperm that acrosome reaction occurs by shedding acrosome cap in intact form, suggesting the difference in mechanism underlying acrosome reaction between groups (Ahammad et al. 2013). This view is consistent with the fact that avian sperm are able to undergo acrosome reaction soon after ejaculation in response to physiological stimuli (Lemoine et al. 2008), while mammalian sperm need to remain in the female reproductive tract for a while to acquire the ability to induce an acrosome reaction (Chang 1951). Flagellum is divided into midpiece, principal piece, and end piece, which are all dedicated to propelling sperm toward fertilization site in oviduct. In mammals, energy production for maintenance of motility relies on adenosine triphosphate (ATP) produced in the flagellum by glycolysis and mitochondrial oxidation (Amaral et al. 2013). Several studies in mammalian sperm have suggested that mitochondrial oxidation taking place within the midpiece is the dominant source of ATP for sperm motility, although glycolysis can produce ATP along with the flagellum (du Plessis et al. 2015). Although there are only limited studies on ATP production in avian sperm, a previous study using domestic fowl showed that mitochondrial ATP synthesis plays an important role in maintenance of sperm motility rather than glycolysis (Froman and Feltmann 1998). However, the metabolic pathway for ATP production is poorly understood compared to mammals.

With relevance to sperm motility, fowl sperm, unlike mammalian sperm, display reversible inhibition when temperature is raised from 30 °C to 40 °C (equivalent to avian body temperature). This inhibition is rescued by decreasing the temperature or addition of Ca^{2+} (Ashizawa et al. 1994b). This suggests that Ca^{2+} plays a key role in stimulation of fowl sperm motility in vivo. The CatSper family are sperm-specific ion channels that are crucial for Ca^{2+} influx to maintain flagellar motility in mammalian sperm (Tamburrino et al. 2014; Ren et al. 2001). Recent study of evolutional genomics demonstrated that genomic sequences of CatSper were degenerated in chicken (Cai and Clapham 2008), suggesting that there is a difference in the mechanism for Ca^{2+} influx between mammals and aves. Previous study demonstrated that Ca^{2+} ATPase plays an important role in regulation of temperature-dependent motility in fowl sperm (Thomson and Wishart 1991), suggesting a possibility that an increase in intercellular Ca^{2+} content could depend on efflux rather than influx in avian sperm. Several signaling pathways appear to be involved in calcium-regulated flagellar motility in fowl sperm. For example, pharmacological experiments have shown that a specific inhibitor of phosphatidylinositol-3 kinase inhibited restoration of flagellar movement when sperm were incubated with Ca^{2+} at 40 °C (Ashizawa et al. 2009). Furthermore, activation of protein kinase C was reported to inhibit flagellar movement enhanced by Ca^{2+} addition at 40 °C (Ashizawa et al. 1994a). In addition to these, using demembranated fowl sperm, specific inhibitors of protein phosphatase type 1 and type 2A restored motility reduced by incubation at 40 °C (Ashizawa et al. 1994c). These studies suggest that status of protein phosphorylation is critical for flagellar movement dependent on temperature in fowl sperm. This is confirmed by previous study showing that 43 KDa axonemal protein undergoes phosphorylation independently of cAMP when the incubating temperature is changed into 30 °C from avian body temperature (Ashizawa et al. 1992). A study with dog sperm showed that regulation of flagellar motility includes a mechanism relevant to the control of phosphorylation of sperm proteins present in axonemal components (Tash and Means 1982), suggesting similarity between mammals and aves. Calmodulin is a Ca^{2+}-binding protein that regulates cell motility and membrane fusion by its ability to activate key regulatory enzymes (Means et al. 1982). It has been reported previously that calmodulin plays an important role as the signal inducer in Ca^{2+}-regulated motility of fowl sperm (Ashizawa et al. 1994b). Recently, it was demonstrated that some of the calmodulin-dependent protein kinases are present mostly in flagellum of

Table 4.1 Signaling molecules associated with induction of acrosome reaction in avian sperm

Molecules	Known functions	Species	References
Protein phosphatase type 2B	Serine/threonine protein phosphatase	Chicken	Ashizawa et al. (2004 #1823)
Protein phosphatase type 1 and type 2A	Serine/threonine protein phosphatase	Chicken	Ashizawa et al. (2006 #1793)
Protein kinase C (PKC)	Serine/threonine protein kinase	Chicken	Ashizawa et al. (2006 #1793)
Protein kinase A (PKA)	Serine/threonine protein kinase	Chicken	Lemoine et al. (2009 #1794)
Phosphoinositide 3-kinase (PI3K)	Phosphorylation of phosphatidyl inositol	Chicken	Lemoine et al. (2009 #1794)
Mitogen-activated protein kinase 1 (MAPK1)	Serine/threonine protein kinase	Chicken	Lemoine et al. (2009 #1794)
5`-AMP-activated protein kinase (AMPK)	Serine/threonine protein kinase	Chicken	Nguyen et al. (2014 #1791)
Ca^{2+}/calmodulin-dependent protein kinase kinases (CaMKKs)	Serine/threonine protein kinase	Chicken	Nguyen et al. (2016 #1792)

chicken sperm, and stimulates phosphorylation of 5`-AMP-activated protein kinase (AMPK) that was found to enhance sperm motility, acrosomal responsiveness, and production of lactate that is used for ATP production in chicken sperm (Nguyen et al. 2014, 2016), together suggesting that activation of calmodulin is linked to stimulation of motility via activation of AMPK. Further study is needed to determine whether this signaling cascade is involved in phosphorylation of axonemal proteins in sperm flagellum in order to stimulate motility.

To date, in aves, several signaling cascades have been reported to be involved in induction of acrosomal reaction which is a prerequisite for penetrating peri-vitelline membrane (PVM), analogous to the zona pellucida, extracellular matrix of mammalian oocytes (Table 4.1). For example, specific inhibitors for protein-phosphatase-type 1 (PP1) and -type 2A (PP2A) stimulated acrosome reaction in response to binding to PVM under absence of Ca^{2+} in fowl sperm (Ashizawa et al. 2006). Furthermore, a specific inhibitor for protein kinase A (PKA) inhibited acrosome reaction induced by binding to PVM under presence of Ca^{2+} in chicken sperm (Lemoine et al. 2009). In mammalian sperm, activation of PKA plays a key role for capacitation priming to acrosome reaction. Accumulated literature has demonstrated in mammalian sperm that activation of PKA is induced by inactivation of PP2A, leading to enhancement of acrosome responsiveness (Visconti et al. 2011). This suggests that there might be a similar signaling cascade in avian sperm, although the capacitation process does not appear to be present in avian sperm to induce acrosome reaction. Overall, signaling cascades associated with induction of acrosomal reaction are poorly understood in avian compared to mammalian sperm. A recent study demonstrated in chicken sperm that AMPK is activated via its phosphorylation, and contributes to maintenance of flagellar motility as well as induction of acrosome reaction (Nguyen et al. 2014). Although the relationship between PKA cascade and AMPK phosphorylation has not been understood in avian sperm, a previous study in mammalian sperm demonstrated that AMPK phosphorylation is mediated via activation of PKA (Hurtado de Llera et al. 2014), suggesting that AMPK is involved in downstream of PKA. This finding could provide a potential investigation to determine PKA-involved signaling cascade in avian sperm. What mechanism is behind the regulation of PKA activity in avian sperm is not known yet. Accumulated studies performed with an in-vitro culture system demonstrated that sterol efflux from plasma membrane is an initial trigger to stimulate PKA activation (Visconti et al. 1999a, b), suggesting

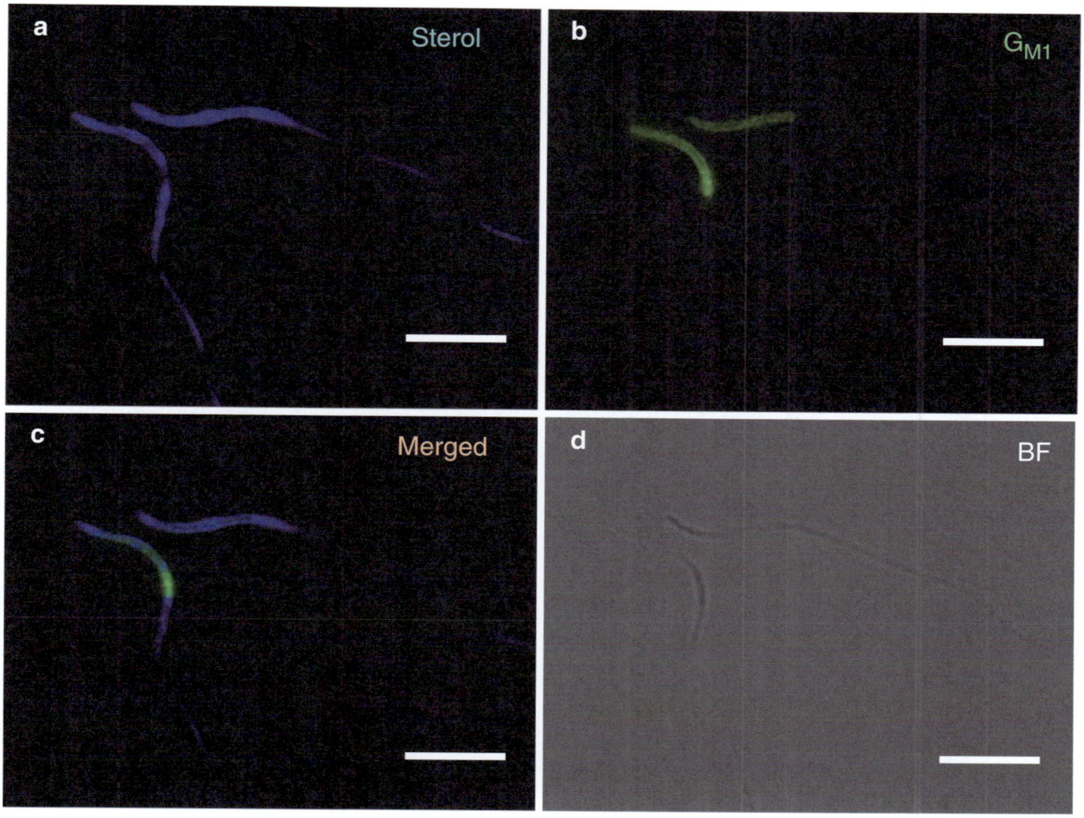

Fig. 4.4 Localization of membrane rafts in chicken sperm. The plasma membranes overlaying the sperm head region are reported to possess enormous membrane domain enriched in specific lipids. Membrane rafts are dynamic membrane domains enriched in both sterols and ganglioside G_{M1} (G_{M1}), and play an important role in regulation of diverse cellular functions in several cell types

that membrane change is upstream of PKA activation. Membrane rafts are specific regions enriched in sterols and functional proteins, and play an important role as signaling platforms in diverse cellular processes in different cell types (Simons and Toomre 2000). Recently, studies showed that membrane rafts are involved in capacitation and acrosome reaction in mammalian sperm (Thaler et al. 2006; Asano et al. 2013). It should be noted that membrane rafts have also been found to be present in the plasma membrane overlying the head region of chicken sperm (Fig. 4.4) (Asano et al. 2016). Consistent with this, recent microscopic study of sperm from White Pekin duck demonstrated that an enormous membrane domain enriched in sterols is present in sperm head region (Majhi et al. 2016), suggesting the presence of membrane rafts in sperm across avian species. Functional roles of membrane rafts in avian sperm have not been understood yet. Recently, we demonstrated in chicken sperm that sterol efflux from the plasma membranes induced an early apoptotic change characterized by translocation of phosphatidylserine (PS) onto the external surface (Ushiyama et al. 2016). PS translocation is known as a signal to attract various phagocytes for engulfment of apoptotic cells. In immune cells, membrane rafts are involved in PS translocation taking place in response to physiological stimulation (Ishii et al. 2005). Considering the regulatory roles of sterols in the organization and functionality of membrane rafts in mammalian sperm (Asano et al. 2013; Shadan et al. 2004; Khalil et al. 2006), it is conceivable that membrane rafts of chicken sperm regulate apoptotic change leading to engulfment by phagocytes in the female genital tract. Further study will be required to stimulate our understanding of the functional roles of membrane rafts in avian sperm.

4.4 Sperm Production Rate

Because of the economic importance and the prevalence of artificial insemination in domestic animals, two different concepts with regard to sperm production rate were proposed by Amann (1970). One of them was daily sperm output (DSO), which is defined as a total number of ejaculated spermatozoa over a period of time expressed on a per-day basis. The other was daily sperm production (DSP), which is defined as a total number of sperm produced per day by the two testes. DSO is apparently correlated with DSP. Raviers et al. (1981) reported in rooster that DSP is about 80 to 120 × 10^6 per gram of testis tissue. They also reported that the DSP, the testis weight, and the total number of Sertoli cells per testis are highly intercorrelated. Photoperiod appears to influence the DSP and testis weight at sexual maturity. Raviers (1980) proposed at least three parameters regarding the action of the day length upon reproductive performance: (1) the day length itself, (2) its variations before and after sexual maturity, and (3) its deviation into one or several photoperiods during a 24-h cycle. This is consistent with a previous report that the age of sexual maturity can be modified to a large extent by the use of certain photoschedules in male turkeys (Noirault et al. 2006). Control of the photoperiod would be useful to improve reproductive performance in aves.

4.5 Sperm Cryopreservation

The importance of securing poultry genetic resources, such as local breeds, research resources, and grandparent stocks from unexpected accidents or disease outbreaks can not be overemphasized, particularly after the recent worldwide epidemic of avian influenza. In many mammalian species, the genetic resource conservation has been achieved by integrating several key technologies, i.e., artificial insemination, in-vitro fertilization, embryo transfer, and cryopreservation of sperm and ova.

In oviparous species including aves, however, it is not possible to simply apply the strategy utilized for genetic resource conservation of mammals, primarily due to the technical inability of freezing large yolk-laden eggs. An alternative strategy, therefore, had to be developed to reduce the risk of losing valuable poultry genetic resources. It has been suggested that the poultry genetic resources can be most efficiently retrieved when sperm from a genetically valuable poultry genetic resource are cryopreserved beforehand and used to inseminate female germline chimeras (Tajima 2002). In this respect, semen cryopreservation technology plays a key role in the efficient retrieval of poultry genetic resources through germline chimeras.

Historically, Shaffner et al. (1941) reported that when fowl semen was supplemented with levulose, 30% of the sperm could be revived after freezing to −79 °C. Some sperm retained fertilizing ability even under this condition, but all of the fertilized eggs died within 10 to 15 h after fertilization (Shaffner 1942). Cryoprotective evidence of glycerol for sperm of higher vertebrate was first shown by Polge et al. (1949). They diluted chicken semen with Ringer's solution containing 40% glycerol and observed full recovery of motility after thawing. However, when the glycerolated semen was inseminated into more than 2000 hens, none of the eggs were fertile regardless of whether they were inseminated with fresh or frozen semen (Smith and Polge 1950). In the meantime, Sloviter (1951), who was freezing red blood cells, found that frozen–thawed red blood cells survived without hemolysis if the glycerol was removed by dialysis prior to transfusion. Polge (1951), therefore, dialyzed chicken semen to reduce the glycerol concentration, and 35 live chicks were successfully produced. Since then, a number of studies have been performed to freeze chicken semen (Sexton et al. 1978; Lake et al. 1981; Donoghuea and Wishart 2000; Long 2006; Blesbois 2007).

Three cryoprotective agents, i.e., glycerol, dimethylsulfoxide (DMSO), and dimethylacetamide (DMA), have since been found to be effective in protecting domestic chicken sperm to varying degrees of freezing damage (Tajima et al. 1990). Among these three cryoprotective agents, glycerol has been most widely used for freezing chicken semen because it has the highest post-thaw fertility rates (Tajima et al. 1990). However, it has been reported that the glycerol concentration must be decreased to below 0.163 M prior to insemination to obtain fertile eggs, due to the contraceptive effect of glycerol (Lake et al. 1981). In addition to this, the fertilizing ability of avian sperm is highly sensitive to freezing process. For example, comparison of fertility between chicken fresh and frozen sperm demonstrated that only 1–6% of the frozen sperm kept their fertilizing ability in chicken (Wishart 1985). In line with this, frozen–thawed chicken sperm only retained 19.7% of the relative fertilizing ability to fresh sperm (Tajima et al. 1989). It was not fully understood why the fertilizing ability of chicken sperm declined following freezing. However, a recent cell biological study demonstrated in chicken sperm that the frozen–thawed procedure causes an early apoptotic change, characterized by external translocation of phosphatidylserine (PS) to the sperm surface (Ushiyama et al. 2016). Considering that PS translocation is an initial signal to attract phagocytes for ingestion of apoptotic cells, the decline of fertilizing ability in frozen–thawed chicken sperm might involve the apoptotic response in the female genital tract.

On the other hand, post-thaw fertility tends to be lower in both DMSO and dimethylacetamide (DMA) methods compared to the glycerol method although, unlike glycerol, there is no need to remove these two cryoprotective agents prior to artificial insemination (Tajima et al. 1990). Further investigation, therefore, will be required to develop a cryopreservation method of chicken semen with high post-thaw fertility by direct insemination with semen that contains the cryoprotective agent. Recently, a highly practical method for cryopreservation of sperm collected from genetically valuable chicken breeds was developed by using N-methylacetamide as a new cryoprotective agent, showing that cryopreserved sperm retained high fertility despite the presence of the agent in semen at insemination (Sasaki et al. 2010).

Despite continuous and intensive efforts to cryopreserve turkey sperm to meet with the industrial necessity, technical progress of turkey sperm cryopreservation is slow compared with cryopreservation of chicken sperm (Oderkirk and Buckland 1977; Graham et al. 1982a, b; Iaffaldano et al. 2011; Long et al. 2014). Apparently, there are species-specific differences in sperm survival after cryopreservation. It has been reported that post-thaw survivability can be enhanced by adding species-based osmoprotectant/ATP combinations in a cryodiluent where DMA is included (Blanco et al. 2011). In addition, it has been reported that cryopreserved chicken sperm with glycerol and DMA differs in the ability to undergo acrosome reaction (Mocé et al. 2010). These studies indicate that customization of freezing protocol is necessary for sperm of each avian species.

Alternative approaches to improve post-thaw fertility are to reduce stresses applied to sperm during cryopreservation. For example, Vitamin E and catalase were shown to be beneficial for membrane integrity, motility and tolerance for oxidative stress in cryopreserved rooster semen (Moghbeli et al. 2016). Addition of antioxidants such as N-acetyl-L-cysteine (NAC) and superoxide dismutase (SOD) is beneficial for the function of frozen–thawed chicken sperm by preventing the reduction in motility, viability, and mitochondrial membrane potential, as well as protecting from apoptotic changes in sperm (Partyka et al. 2013).

To date, most of the efforts to improve post-thaw fertility have focused on empirical approaches, such as types of cryoprotectant, and freezing protocol of cryopreservation. However, it is still not understood why avian sperm lose fertilizing ability after cryopreservation. Therefore, the conventional approaches combined with comprehensive investigations for molecular and cellular basis of cryopreservation would contribute to generating a technical breakthrough necessary for the improvement of efficacy of sperm cryopreservation in aves.

References

Abou-haila A, Tulsiani DRP. Signal transduction pathways that regulate sperm capacitation and the acrosome reaction. Arch Biochem Biophys. 2009;485(1):72–81. doi:10.1016/j.abb.2009.02.003.

Ahammad MU, Nishino C, Tatemoto H, Okura N, Kawamoto Y, Okamoto S, Nakada T. Maturational changes in motility, acrosomal proteolytic activity, and penetrability of the inner perivitelline layer of fowl sperm, during their passage through the male genital tract. Theriogenology. 2011a;76(6):1100–9. doi:10.1016/j.theriogenology.2011.05.017.

Ahammad MU, Nishino C, Tatemoto H, Okura N, Kawamoto Y, Okamoto S, Nakada T. Maturational changes in the survivability and fertility of fowl sperm during their passage through the male reproductive tract. Anim Reprod Sci. 2011b;128(1–4):129–36. doi:10.1016/j.anireprosci.2011.09.010.

Ahammad MU, Nishino C, Tatemoto H, Okura N, Okamoto S, Kawamoto Y, Nakada T. Acrosome reaction of fowl sperm: evidence for shedding of the acrosomal cap in intact form to release acrosomal enzyme. Poult Sci. 2013;92(3):798–803. doi:10.3382/ps.2012-02523.

Aire TA. Ultrastructural study of spermiogenesis in the turkey, *Meleagris gallopavo*. Br Poult Sci. 2003;44(5):674–82. doi:10.1080/00071660310001643651.

Aire TA. Spermiogenesis in birds. Spermatogenesis. 2014;4(3):e959392. doi:10.4161/21565554.2014.959392.

Amann R. Sperm production rates. In: Johnson AD, Gomes WR, Vandemark NL, editors. The testis, vol. 1. New York: Academic Press; 1970. p. 433–82.

Amaral A, Lourenco B, Marques M, Ramalho-Santos J. Mitochondria functionality and sperm quality. Reproduction. 2013;146(5):R163–74. doi:10.1530/REP-13-0178.

Asano A, Nelson-Harrington JL, Travis AJ. Phospholipase B is activated in response to sterol removal and stimulates acrosome exocytosis in murine sperm. J Biol Chem. 2013;288(39):28104–15. doi:10.1074/jbc.M113.450981.

Asano A, Kanbe H, Ushiyama A, Tajima A. Organization of membrane rafts in chicken sperm. J Poult Sci. 2016;53(3):233–9.

Ashizawa K, Katayama S, Tsuzuki Y. Regulation of flagellar motility by temperature-dependent phosphorylation of a 43 kDa axonemal protein in fowl spermatozoa. Biochem Biophys Res Commun. 1992;185(2):740–5. doi:10.1016/0006-291X(92)91688-M.

Ashizawa K, Katayama S, Kobayashi T, Tsuzuki Y. Possible role of protein kinase C in regulation of flagellar motility and intracellular free Ca^{2+} concentration of fowl spermatozoa. J Reprod Fertil. 1994a;101(3):511–7.

Ashizawa K, Tomonaga H, Tsuzuki Y. Regulation of flagellar motility of fowl spermatozoa: evidence for the involvement of intracellular free Ca^{2+} and calmodulin. J Reprod Fertil. 1994b;101(2):265–72.

Ashizawa K, Wishart GJ, Tomonaga H, Nishinakama K, Tsuzuki Y. Presence of protein phosphatase type 1 and its involvement in temperature-dependent flagellar movement of fowl spermatozoa. FEBS Lett. 1994c;350(1):130–4.

Ashizawa K, Wishart GJ, Ranasinghe ARAH, Katayama S, Tsuzuki Y. Protein phosphatase-type 2B is involved in the regulation of the acrosome reaction but not in the temperature-dependent flagellar movement of fowl spermatozoa. Reproduction. 2004;128(6):783–7. doi:10.1530/rep.1.00327.

Ashizawa K, Wishart GJ, Katayama S, Takano D, Ranasinghe AR, Narumi K, Tsuzuki Y. Regulation of acrosome reaction of fowl spermatozoa: evidence for the involvement of protein kinase C and protein phosphatase-type 1 and/or -type 2A. Reproduction. 2006;131(6):1017–24. doi:10.1530/rep.1.01069.

Ashizawa K, Omura Y, Katayama S, Tatemoto H, Narumi K, Tsuzuki Y. Intracellular signal transduction pathways in the regulation of fowl sperm motility: evidence for the involvement of phosphatidylinositol 3-kinase (PI3-K) cascade. Mol Reprod Dev. 2009;76(7):603–10. doi:10.1002/mrd.20995.

Austin CR. Observations on the penetration of the sperm in the mammalian egg. Aust J Sci Res B. 1951;4:581–96.

Blanco JM, Long JA, Gee G, Wildt DE, Donoghue AM. Comparative cryopreservation of avian spermatozoa: benefits of non-permeating osmoprotectants and ATP on turkey and crane sperm cryosurvival. Anim Reprod Sci. 2011;123(3–4):242–8. doi:10.1016/j.anireprosci.2010.12.005.

Blesbois E. Current status in avian semen cryopreservation. Worlds Poult Sci J. 2007;63:213–22.

Burgos MH, Fawcett DW. Studies on the fine structure of the mammalian testis. I. Differentiation of the spermatids in the cat (Felis domestica). J Biophys Biochem Cytol. 1955;1(4):287–300.

Cai X, Clapham DE. Evolutionary genomics reveals lineage-specific gene loss and rapid evolution of a sperm-specific ion channel complex: catspers and catsper? PLoS One. 2008;3(10):e3569. doi:10.1371/journal.pone.0003569.

Chang MC. Fertilizing capacity of spermatozoa deposited into the fallopian tubes. Nature. 1951;168(4277):697–8.

Chowdhury AK, Steinberger E. A study of germ cell morphology and duration of spermatogenic cycle in the baboon, Papio anubis. Anat Rec. 1976;185(2):155–69. doi:10.1002/ar.1091850204.

Clermont Y. The cycle of the seminiferous epithelium in man. Am J Anat. 1963;112:35–51. doi:10.1002/aja.1001120103.

Clermont Y, Antar M. Duration of the cycle of the seminiferous epithelium and the spermatogonial renewal in the monkey *Macaca arctoides*. Am J Anat. 1973;136(2):153–65. doi:10.1002/aja.1001360204.

Clermont Y, Bustos-Obregon E. Re-examination of spermatogonial renewal in the rat by means of seminiferous tubules mounted "in toto". Am J Anat. 1968;122(2):237–47. doi:10.1002/aja.1001220205.

Cornwall GA, Hann SR. Specialized gene expression in the epididymis. J Androl. 1995;16(5):379–83.

Cross NL. Effect of methyl-beta-cyclodextrin on the acrosomal responsiveness of human sperm. Mol Reprod Dev. 1999;53(1):92–8. doi:10.1002/(SICI)1098-2795(199905)53:1<92::AID-MRD11>3.0.CO;2-Q.

Davis BK. Influence of serum albumin on the fertilizing ability in vitro of rat spermatozoa. Proc Soc Exp Biol Med. 1976;151(2):240–3.

Donoghuea AM, Wishart GJ. Storage of poultry semen. Anim Reprod Sci. 2000;62(1–3):213–32.

Esponda P, Bedford JM. Surface of the rooster spermatozoon changes in passing through the Wolffian duct. J Exp Zool. 1985;234(3):441–9. doi:10.1002/jez.1402340311.

Froman DP, Feltmann AJ. Sperm mobility: a quantitative trait of the domestic fowl (Gallus domesticus). Biol Reprod. 1998;58(2):379–84.

Graham EF, Nelson DS, Schmehl MK. Development of extender and techniques for frozen turkey semen. 1. Development. Poult Sci. 1982a;61(3):550–7.

Graham EF, Nelson DS, Schmehl MK. Development of extender and techniques for frozen turkey semen. 2. Fertility trials. Poult Sci. 1982b;61(3):558–63.

Gunawardana VK, Scott MG. Ultrastructural studies on the differentiation of spermatids in the domestic fowl. J Anat. 1977;124(Pt 3):741–55.

Hadley MA, Dym M. Spermatogenesis in the vasectomized monkey: quantitative analysis. Anat Rec. 1983;205(4):381–6. doi:10.1002/ar.1092050403.

Howarth B Jr. An examination for sperm capacitation in the fowl. Biol Reprod. 1970;3(3):338–41.

Howarth B Jr. Fertilizing ability of cock spermatozoa from the testis epididymis and vas deferens following intramagnal insemination. Biol Reprod. 1983;28(3):586–90.

Hurtado de Llera A, Martin-Hidalgo D, Gil MC, Garcia-Marin LJ, Bragado MJ. The calcium/CaMKKalpha/beta and the cAMP/PKA pathways are essential upstream regulators of AMPK activity in boar spermatozoa. Biol Reprod. 2014;90(2):29. doi:10.1095/biolreprod.113.112797.

Iaffaldano N, Romagnoli L, Manchisi A, Rosato MP. Cryopreservation of turkey semen by the pellet method: effects of variables such as the extender, cryoprotectant concentration, cooling time and warming temperature on sperm quality determined through principal components analysis. Theriogenology. 2011;76(5):794–801. doi:10.1016/j.theriogenology.2011.04.012.

Iborra A, Companyo M, Martinez P, Morros A. Cholesterol efflux promotes acrosome reaction in goat spermatozoa. Biol Reprod. 2000;62(2):378–83.

Ishii H, Mori T, Shiratsuchi A, Nakai Y, Shimada Y, Ohno-Iwashita Y, Nakanishi Y. Distinct localization of lipid rafts and externalized phosphatidylserine at the surface of apoptotic cells. Biochem Biophys Res Commun. 2005;327(1):94–9. doi:10.1016/j.bbrc.2004.11.135.

Jones R. Plasma membrane structure and remodelling during sperm maturation in the epididymis. J Reprod Fertil Suppl. 1998;53:73–84.

Jones RC, Lin M. Spermatogenesis in birds. Oxf Rev Reprod Biol. 1993;15:233–64.

Khalil MB, Chakrabandhu K, Xu H, Weerachatyanukul W, Buhr M, Berger T, Carmona E, Vuong N, Kumarathasan P, Wong PTT, Carrier D, Tanphaichitr N. Sperm capacitation induces an increase in lipid rafts having zona pellucida binding ability and containing sulfogalactosylglycerolipid. Dev Biol. 2006;290(1):220–35. doi:10.1016/j.ydbio.2005.11.030.

Korn N, Thurston RJ, Pooser BP, Scott TR. Ultrastructure of spermatozoa from Japanese quail. Poult Sci. 2000;79(3):407–14.

Lake PE, Ravie O, McAdam J. Preservation of fowl semen in liquid nitrogen: application to breeding programmes. Br Poult Sci. 1981;22(1):71–7. doi:10.1080/00071688108447865.

Leblond CP, Clermont Y. Spermiogenesis of rat, mouse, hamster and guinea pig as revealed by the periodic acid–fuchsin sulfurous acid technique. Am J Anat. 1952;90(2):167–215. doi:10.1002/aja.1000900202.

Lemoine M, Grasseau I, Brillard JP, Blesbois E. A reappraisal of the factors involved in in-vitro initiation of the acrosome reaction in chicken spermatozoa. Reproduction. 2008;136(4):391–9. doi:10.1530/REP-08-0094.

Lemoine M, Dupont J, Guillory V, Tesseraud S, Blesbois E. Potential involvement of several signaling pathways in initiation of the chicken acrosome reaction. Biol Reprod. 2009;81(4):657–65. doi:10.1095/biolreprod.108.072660.

Lin M, Jones RC. Renewal and proliferation of spermatogonia during spermatogenesis in the Japanese quail, *Coturnix coturnix japonica*. Cell Tissue Res. 1992;267(3):591–601.

Lin M, Jones RC, Blackshaw AW. The cycle of the seminiferous epithelium in the Japanese quail (*Coturnix coturnix japonica*) and estimation of its duration. J Reprod Fertil. 1990;88(2):481–90.

Long JA. Avian semen cryopreservation: what are the biological challenges? Poult Sci. 2006;85(2):232–6.

Long JA, Purdy PH, Zuidberg K, Hiemstra SJ, Velleman SG, Woelders H. Cryopreservation of turkey semen: effect of breeding line and freezing method on post-thaw sperm quality, fertilization, and hatching. Cryobiology. 2014;68(3):371–8. doi:10.1016/j.cryobiol.2014.04.003.

Majhi RK, Kumar A, Yadav M, Kumar P, Maity A, Giri SC, Goswami C. Light and electron microscopic study of mature spermatozoa from White Pekin duck (*Anas platyrhynchos*): an ultrastructural and molecular analysis. Andrology. 2016;4(2):232–44. doi:10.1111/andr.12130.

Means AR, Tash JS, Chafouleas JG. Physiological implications of the presence, distribution, and regulation of calmodulin in eukaryotic cells. Physiol Rev. 1982;62(1):1–39.

Mocé E, Grasseau I, Blesbois E. Cryoprotectant and freezing-process alter the ability of chicken sperm to acrosome react. Anim Reprod Sci. 2010;122(3–4):359–66. doi:10.1016/j.anireprosci.2010.10.010.

Moghbeli M, Kohram H, Zare-Shahaneh A, Zhandi M, Sharafi M, Nabi MM, Zahedi V, Sharideh H. Are the optimum levels of the catalase and vitamin E in rooster semen extender after freezing–thawing influenced by sperm concentration? Cryobiology. 2016; doi:10.1016/j.cryobiol.2016.03.008.

Morris SA, Howarth B Jr, Crim JW, Rodriguez de Cordoba A, Esponda P, Bedford JM. Specificity of sperm-binding Wolffian duct proteins in the rooster and their persistence on spermatozoa in the female host glands. J Exp Zool. 1987;242(2):189–98. doi:10.1002/jez.1402420210.

Morton BE, Sagadraca R, Fraser C. Sperm motility within the mammalian epididymis: species variation and correlation with free calcium levels in epididymal plasma. Fertil Steril. 1978;29(6):695–8.

Nagano T. Observations on the fine structure of the developing spermatid in the domestic chicken. J Cell Biol. 1962;14:193–205.

Nguyen TM, Alves S, Grasseau I, Metayer-Coustard S, Praud C, Froment P, Blesbois E. Central role of 5′-AMP-activated protein kinase in chicken sperm functions. Biol Reprod. 2014;91(5):121. doi:10.1095/biolreprod.114.121855.

Nguyen TM, Combarnous Y, Praud C, Duittoz A, Blesbois E. Ca^{2+}/calmodulin-dependent protein kinase kinases (CaMKKs) effects on AMP-activated protein kinase (AMPK) regulation of chicken sperm functions. PLoS One. 2016;11(1):e0147559. doi:10.1371/journal.pone.0147559.

Nixon B, Ewen KA, Krivanek KM, Clulow J, Kidd G, Ecroyd H, Jones RC. Post-testicular sperm maturation and identification of an epididymal protein in the Japanese quail (*Coturnix coturnix japonica*). Reproduction. 2014;147(3):265–77. doi:10.1530/REP-13-0566.

Noirault J, Brillard J-P, Bakst MR. Effect of various photoperiods on testicular weight, weekly sperm output and plasma levels of LH and testosterone over the reproductive season in male turkeys. Theriogenology. 2006;66(4):851–9. doi:10.1016/j.theriogenology.2005.11.025.

Oderkirk AHF, Buckland RB. A comparison of diluents and cryopreservatives for freezing Turkey semen. Poult Sci. 1977;56(6):1861–7. doi:10.3382/ps.0561861.

Okamura F, Nishiyama H. The early development of the tail and the transformation of the shape of the nucleus of the spermatid of the domestic fowl, *Gallus gallus*. Cell Tissue Res. 1976;169(3):345–59. doi:10.1007/BF00219607.

Oko R, Clermont Y. Isolation, structure and protein composition of the perforatorium of rat spermatozoa. Biol Reprod. 1988;39(3):673–87.

Oko R, Morales CR. A novel testicular protein, with sequence similarities to a family of lipid binding proteins, is a major component of the rat sperm perinuclear theca. Dev Biol. 1994;166(1):235–45. doi:10.1006/dbio.1994.1310.

Oko R, Moussakova L, Clermont Y. Regional differences in composition of the perforatorium and outer periacrosomal layer of the rat spermatozoon as revealed by immunocytochemistry. Am J Anat. 1990;188(1):64–73. doi:10.1002/aja.1001880108.

Partyka A, Nizanski W, Bajzert J, Lukaszewicz E, Ochota M. The effect of cysteine and superoxide dismutase on the quality of post-thawed chicken sperm. Cryobiology. 2013;67(2):132–6. doi:10.1016/j.cryobiol.2013.06.002.

du Plessis L, Soley JT. A re-evaluation of sperm ultrastructure in the emu, *Dromaius novaehollandiae*. Theriogenology. 2014;81(8):1073–84. doi:10.1016/j.theriogenology.2014.01.034.

du Plessis SS, Agarwal A, Mohanty G, van der Linde M. Oxidative phosphorylation versus glycolysis: what fuel do spermatozoa use? Asian J Androl. 2015;17(2):230–5. doi:10.4103/1008-682X.135123.

Polge C. Functional survival of fowl spermatozoa after freezing at −79 degrees C. Nature. 1951;167(4258):949–50.

Polge C, Smith AU, Parkes AS. Revival of spermatozoa after vitrification and dehydration at low temperatures. Nature. 1949;164(4172):666.

Rankin TL, Tsuruta KJ, Holland MK, Griswold MD, Orgebin-Crist MC. Isolation, immunolocalization, and sperm-association of three proteins of 18, 25, and 29 kilodaltons secreted by the mouse epididymis. Biol Reprod. 1992;46(5):747–66.

Raviers MD. Photoperiodism, testis development and sperm production in the fowl. 9th Int Cong Anim Reprod Artif Insem. 1980;II:519–26.

Raviers MD, Williams J, Brillard J. Predicting the adult daily sperm output after the first ejaculates in cocks raised under different photoschedules. Reprod Nutr Dev. 1981;21:1113–24.

Ren D, Navarro B, Perez G, Jackson AC, Hsu S, Shi Q, Tilly JL, Clapham DE. A sperm ion channel required for sperm motility and male fertility. Nature. 2001;413(6856):603–9. doi:10.1038/35098027.

Russell LD, Russell JA, MacGregor GR, Meistrich ML. Linkage of manchette microtubules to the nuclear envelope and observations of the role of the manchette in nuclear shaping during spermiogenesis in rodents. Am J Anat. 1991;192(2):97–120. doi:10.1002/aja.1001920202.

Sasaki K, Tatsumi T, Tsutsui M, Niinomi T, Imai T, Naito M, Tajima A, Nishi Y. Method for cryopreserving semen from Yakido roosters using N-Methylacetamide as a cryoprotective agent. J Poult Sci. 2010;47:297–301.

Schulze C. Morphological characteristics of the spermatogonial stem cells in man. Cell Tissue Res. 1979;198(2):191–9.

Selvaraj V, Asano A, Page JL, Nelson JL, Kothapalli KS, Foster JA, Brenna JT, Weiss RS, Travis AJ. Mice lacking FABP9/PERF15 develop sperm head abnormalities but are fertile. Dev Biol. 2010;348(2):177–89. doi:10.1016/j.ydbio.2010.09.019.

Sexton TJ, Buckland RB, Lopez R. Comparison of two procedures for freezing semen from cocks of high and low fertility with frozen semen. Poult Sci. 1978;57(2):550–2.

Shadan S, James PS, Howes EA, Jones R. Cholesterol efflux alters lipid raft stability and distribution during capacitation of boar spermatozoa. Biol Reprod. 2004;71(1):253–65.

Shaffner C. Longitivity of fowl spermatozoa in frozen condition. Science. 1942;96:337.

Shaffner C, Henderson E, Card C. Viability of spermatozoa of the chicken under various environmental conditions. Poult Sci. 1941;20:259–65.

Simoes K, Orsi AM, Artoni SM. Ultrastructure of the spermatozoa of the domestic duck (*Anas platyrhynchos* sp.). Anat Histol Embryol. 2012;41(3):202–8. doi:10.1111/j.1439-0264.2011.01124.x.

Simons K, Toomre D. Lipid rafts and signal transduction. Nature Rev Mol Cell Biol. 2000;1(1):31–9.

Sloviter H. In vitro survival of rabbits' red cells recovered after freezing. Lancet. 1951;1:1350–1.

Smith A, Polge C. Survival of spermatozoa at low temperatures. Nature. 1950;166:668–9.

Soley JT. Nuclear morphogenesis and the role of the manchette during spermiogenesis in the ostrich (*Struthio camelus*). J Anat. 1997;190(Pt 4):563–76.

Sullivan R. Epididymosomes: a heterogeneous population of microvesicles with multiple functions in sperm maturation and storage. Asian J Androl. 2015;17(5):726–9. doi:10.4103/1008-682X.155255.

Sutovsky P, Manandhar G. The sperm cell. In: Jonge CJD, editor. The sperm cell. Cambridge: Cambridge University Press; 2007. p. 1–30.

Tajima A. Production of germ-line chimeras and their application in domestic chicken. Avian Poult Biol Rev. 2002;13:15–30.

Tajima A, Graham EF, Hawkins DM. Estimation of the relative fertilizing ability of frozen chicken spermatozoa using a heterospermic competition method. J Reprod Fertil. 1989;85(1):1–5. doi:10.1530/jrf.0.0850001.

Tajima A, Graham EF, Shoffner RN, Otis JS, Hawkins DM. Cryopreservation of semen from unique lines of chicken germ plasm. Poult Sci. 1990;69(6):999–1002.

Tamburrino L, Marchiani S, Minetti F, Forti G, Muratori M, Baldi E. The CatSper calcium channel in human sperm: relation with motility and involvement in progesterone-induced acrosome reaction. Hum Reprod. 2014;29(3):418–28. doi:10.1093/humrep/det454.

Tash JS, Means AR. Regulation of protein phosphorylation and motility of sperm by cyclic adenosine monophosphate and calcium. Biol Reprod. 1982;26(4):745–63.

Thaler CD, Thomas M, Ramalie JR. Reorganization of mouse sperm lipid rafts by capacitation. Mol Reprod Dev. 2006;73(12):1541–9.

Thomson MF, Wishart GJ. Temperature-mediated regulation of calcium flux and motility in fowl spermatozoa. J Reprod Fertil. 1991;93(2):385–91.

Tingari MD. Observations on the fine structure of spermatozoa in the testis and excurrent ducts of the male fowl, Gallus domesticus. J Reprod Fertil. 1973;34(2):255–65.

Ushiyama A, Ishikawa N, Tajima A, Asano A. Comparison of membrane characteristics between freshly ejaculated and cryopreserved sperm in the chicken. J Poult Sci. 2016;53:305–12.

Visconti PE, Bailey JL, Moore GD, Pan D, Olds-Clarke P, Kopf GS. Capacitation of mouse spermatozoa. I. Correlation between the capacitation state and protein tyrosine phosphorylation. Development. 1995a;121(4):1129–37.

Visconti PE, Moore GD, Bailey JL, Leclerc P, Connors SA, Pan D, Olds-Clarke P, Kopf GS. Capacitation of mouse spermatozoa. II. Protein tyrosine phosphorylation and capacitation are regulated by a cAMP-dependent pathway. Development. 1995b;121(4):1139–50.

Visconti PE, Galantino-Homer H, Ning X, Moore GD, Valenzuela JP, Jorgez CJ, Alvarez JG, Kopf GS. Cholesterol efflux-mediated signal transduction in mammalian sperm. Beta-cyclodextrins initiate transmembrane signaling leading to an increase in protein tyrosine phosphorylation and capacitation. J Biol Chem. 1999a;274(5):3235–42.

Visconti PE, Ning X, Fornes MW, Alvarez JG, Stein P, Connors SA, Kopf GS. Cholesterol efflux-mediated signal transduction in mammalian sperm: cholesterol release signals an increase in protein tyrosine phosphorylation during mouse sperm capacitation. Dev Biol. 1999b;214(2):429–43.

Visconti PE, Krapf D, de la Vega-Beltran JL, Acevedo JJ, Darszon A. Ion channels, phosphorylation and mammalian sperm capacitation. Asian J Androl. 2011;13(3):395–405. doi:10.1038/aja.2010.69.

Wishart GJ. Quantitation of the fertilising ability of fresh compared with frozen and thawed fowl spermatozoa. Br Poult Sci. 1985;26(3):375–80. doi:10.1080/00071668508416825.

Avian Egg and Egg Coat

5

Hiroki Okumura

Abstract

An ovulated egg of vertebrates is surrounded by unique extracellular matrix, the egg coat or zona pellucida, playing important roles in fertilization and early development. The vertebrate egg coat is composed of two to six zona pellucida (ZP) glycoproteins that are characterized by the evolutionarily conserved ZP-domain module and classified into six subfamilies based on phylogenetic analyses. Interestingly, investigations of biochemical and functional features of the ZP glycoproteins show that the roles of each ZP-glycoprotein family member in the egg-coat formation and the egg–sperm interactions seemingly vary across vertebrates. This might be one reason why comprehensive understandings of the molecular basis of either architecture or physiological functions of egg coat still remain elusive despite more than 3 decades of intensive investigations. In this chapter, an overview of avian egg focusing on the oogenesis are provided in the first section, and unique features of avian egg coat, i.e., perivitelline layer, including the morphology, biogenesis pathway, and physiological functions are discussed mainly on chicken and quail in terms of the characteristics of ZP glycoproteins in the following sections. In addition, these features of avian egg coat are compared to mammalian zona pellucida, from the viewpoint that the structural and functional varieties of ZP glycoproteins might be associated with the evolutionary adaptation to their reproductive strategies. By comparing the egg coat of birds and mammals whose reproductive strategies are largely different, new insights into the molecular mechanisms of vertebrate egg–sperm interactions might be provided.

H. Okumura
Department of Applied Biological Chemistry,
Faculty of Agriculture, Meijo University, Nagoya,
Japan
e-mail: hokumura@meijo-u.ac.jp

T. Sasanami (ed.), *Avian Reproduction*, Advances in Experimental Medicine and Biology 1001,
DOI 10.1007/978-981-10-3975-1_5

Keywords
Egg coat • Zona pellucida • Sperm–zona interaction • ZP glycoprotein • Extracellular matrix • Evolution

5.1 Introduction to Avian Egg

The egg is a highly specialized cell that is equipped for generating a new individual. Before ovulation, the egg precursor cells, i.e., oocytes, undergo cytoplasmic maturation in the ovarian follicle to obtain the capacities to support male pronucleus formation, block to polyspermy and early embryonic development (Tanghe et al. 2002). During the oocyte maturation in oviparous species including birds, all the materials or nutrients that are required for early development of the embryo are accumulated in the cytoplasm. Especially in birds, the egg cytoplasm contains an extremely large amount of egg yolk as a result of the unique, and well-established oogenesis or follicular development (Wyburn et al. 1965; Johnson 2015). The oogenesis is initiated in the primordial follicles, and the follicles carrying the primary oocytes grow during sexual maturation to become the primary follicles with ~1 to 2 mm in diameter, in which the granulosa cell layer surrounding the oocyte starts to develop. In chicken, the primary follicles incorporate the protein-rich yolk until 8 to 9 days before ovulation to be the white follicles with ~6 to 8 mm in diameter. Subsequently, one of the white follicles is selected to start rapid uptake of a yellow yolk containing large amount of lipid and xanthophyll, and increase in size up to ~40 mm in diameter and ~16 g in weight until ovulation. The preovulatory hierarchy of maturing yellow follicles (Fig. 5.1) that is associated with the ovulation cycle of chicken is due to the daily or near-daily rhythms in the selection of white follicles to grow further (Johnson 2015). It is suggested that the granulosa cell layer surrounding the oocyte plays key roles in regulation of the follicle selection via several signaling systems that is mediated by G protein-coupled receptors (Johnson 2015). In addition, a unique extracellular matrix surrounding egg cell, i.e., the egg coat being designated as the perivitelline layer in birds and corresponding to mammalian zona pellucida, already start to appear between the oocyte plasma membrane and granulosa cell layer at the beginning of the rapid growth phase of the yellow follicle (Wyburn et al. 1965). During the rapid growing of yellow follicle, the perivitelline layer stretches and extends until ovulation, and plays important roles in fertilization and early development (see following sections).

The egg yolk in matured chicken oocyte contains an abundance of lipid (more than 30% of the weight) that mainly comes from very low-density lipoprotein (VLDL) and vitellogenin (VTG) being synthesized in and secreted from the liver into the bloodstream, depending on estrogen stimulation (Schneider 2007; Li et al. 2014). It is suggested that either VLDL or VTG that is transported in the bloodstream to the ovary is incorporated into the developing oocyte by endocytosis via the single

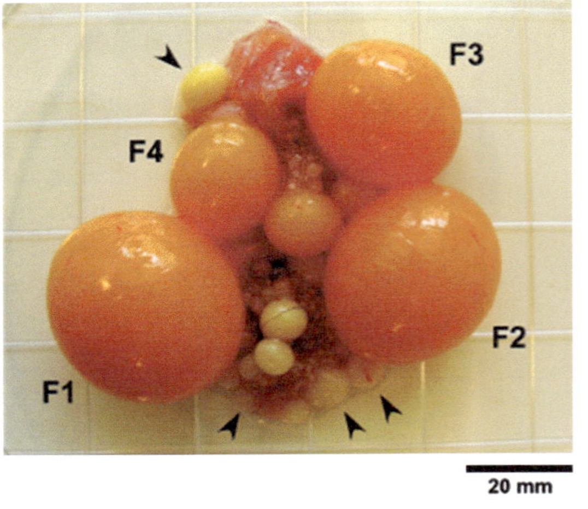

Fig. 5.1 A photograph of chicken ovary excised from a laying hen. The excised chicken ovary was placed on a culture dish with 20 × 20 mm^2 grid and photographed. The four largest yellow follicles are indicated by *F1–F4*. Several white follicles are indicated by *arrowhead*. The mature oocyte in the largest yellow follicle (*F1*) will be ovulated in the living hen. *Bar*: 20 mm

receptor molecule, named as LR8, on the plasma membrane to provide nutrients for the embryo (Schneider 2007; Stifani et al. 1990). On the other hand, there are some other egg-yolk proteins which are characteristic of birds, for example the immunoglobulin IgY, a counterpart of mammalian IgG. IgY is synthesized in the B cells and secreted continuously into the blood (~5 to 7 mg/ml of serum), transported to the ovary and incorporated into the growing oocyte by endocytosis via the Fc receptor on the plasma membrane (Dias da Silva and Tambourgi 2010; Mann and Mann 2008). Chicken IgY can be obtained from a single egg in very large amounts (~100 mg/egg), and furthermore, does not activate or interact with the mammalian complement system, rheumatoid factors, or Fc receptors. For this reason, IgY has the potential to be an excellent tool for immunodiagnostics and immunotherapy for humans (Dias da Silva and Tambourgi 2010).

During follicular development, avian oocyte is arrested in the prophase of the first meiosis similarly to mammalian one. The oocyte resumes the first meiosis a few hours before the ovulation, and is arrested again in metaphase of the second meiosis until ovulation (Hughes 1963; Nakamura et al. 2013; Perry 1987). In avian oocyte having characteristics of telolecithal egg, the female pronucleus is contained in the germinal disc region, which is apparently distinguished as a white plaque (~2 to 3 mm in diameter in matured chicken oocyte) on the surface of the animal pole. In the germinal disc region, many smooth membranous structures, glycogen granules and some organelles including mitochondria and Golgi apparatus are embedded in the dense cytoplasmic matrix surrounding the germinal vesicle (oocyte nucleus), but do not contain the above-mentioned yellow yolk (Perry et al. 1978; Bakst and Howarth 1977a). Birds have a female-heterogametic (ZW female) sex determination system, in which spermatozoa carry the Z sex chromosomes and oocytes carry the Z or W sex chromosomes, and therefore, sex is determined by which sex chromosomes are retained in the oocyte in the first meiosis. It has been reported that concentrations of steroid hormones being supplied from maternal ovary into the egg yolk influence sex determination in peafowl and zebra finch, that is, the concentrations of androstenedione and testosterone are higher in the male egg, and those of 5-α-dihydrotestosterone and 17-β-estradiol are higher in the female egg (Petrie et al. 2001). Furthermore, it has been reported that sex ratio of offspring is biased toward females in chicken and quail, with the administration of progesterone and corticosterone respectively (Correa et al. 2005; Pike and Petrie 2006). Although it is suggested that steroid hormones affect the meiosis of birds to bias sex ratio, the underlying mechanisms remain to be elucidated (Adkins-Regan et al. 2013).

In birds, the ovulated oocyte is enveloped only with the egg coat, perivitelline layer (Bramwell and Howarth 1992), although the remnants of a small numbers of granulosa cells from the follicle are trapped on the surface (Perry 1987). After ovulation, the oocyte is captured by the infundibulum, the anterior region of the oviduct where the fertilization occurs, in which several spermatozoa penetrate the perivitelline layer and fuse with the oocyte at the above-mentioned germinal disc region to form male pronuclei (Perry 1987; Nakamura et al. 2013). In ~1–6 h after the capture by the infundibulum, oocyte resume the second meiosis, and the resultant female pronucleus fuses with one of the male pronuclei to form the zygote nucleus in the next ~3 h (Perry 1987). While birds exhibit the physiological polyspermy that multiple spermatozoa enter into single oocyte at fertilization (Birkhead et al. 1994), the number of spermatozoa that bind to and penetrate through the perivitelline layer covering the germinal disc region is significantly higher than that covering other regions of egg surface (Bramwell and Howarth 1992), although the mechanisms by which spermatozoa is attracted to the germinal disc region remain unclear (Nishio et al. 2014).

After ovulation and capture by the infundibulum, the oocyte or fertilized egg with the surrounding perivitelline layer is further enveloped with several matrices, namely the continuous membrane, outer perivitelline layer, chalazae, albumen, peri-albumen, shell membrane and outer crust, in this order as the egg moves through the oviduct to be spawned (Menkhorst and Selwood 2008). It is noteworthy that the formation of the outer perivitelline membrane prevents spermatozoa from excess penetration through the perivitelline layer and fusion with the oocyte (Bakst and Howarth 1977b). These post-ovulatory extracellular coats are composed of materials that are synthesized in and secreted by the oviduct epithelial cells (Menkhorst and Selwood 2008).

In the avian egg, the above-mentioned egg coat, i.e., perivitelline layer in birds, is one of the extracellular structures that are evolutionarily conserved among all the vertebrates. Although the egg coat is composed of few classes of orthologous ZP glycoproteins, their structural and physiological features considerably vary among vertebrates. In the following sections, unique features of the avian egg coat including the morphology, biogenesis pathway, and roles in egg–sperm interactions are discussed mainly in chicken and quail, being compared to that of mammalian zona pellucida, in terms of the characteristics of ZP-glycoprotein components.

5.2 Morphology and Physiology of Vertebrate Egg Coat

The egg coat of vertebrates is a multifunctional extracellular matrix surrounding the oocyte and early embryo. Although all the vertebrate egg coat is a filamentous meshwork composed of the zona pellucida (ZP) glycoproteins (Bleil and Wassarman 1980), there are large differences in the matrix morphology. For example in mouse, the egg coat, referred to as zona pellucida in mammals, with 7 μm thickness, consists of a fine meshwork of thin filaments and surrounds a relatively small oocyte (~85 μm in diameter) (Greve and Wassarman 1985). In contrast, chicken has a thinner egg coat with 2 to 4 μm thickness, consisting of a relatively rough meshwork of thick filaments in spite of the huge oocyte (~40 mm in diameter) (Wyburn et al. 1965; Bakst and Howarth 1977a). Unlike in eutherian mammals, avian oocytes markedly increase their size; with the egg-yolk accumulation during follicular development from the immature white follicles to the mature yellow follicles, and especially during the final 8 to 9 days before ovulation, the oocyte diameter quickly increases from ~7 to 40 mm (Wyburn et al. 1965). The avian egg coat must, therefore, stretch and extend quickly throughout this rapid growth of the oocyte. A photograph of a chicken ovary that is excised from a laying hen, and a confocal immunofluorescent image of chicken egg-coat matrix using an antibody against one of the ZP glycoproteins (ZP1) are shown in Figs. 5.1 and 5.2 respectively.

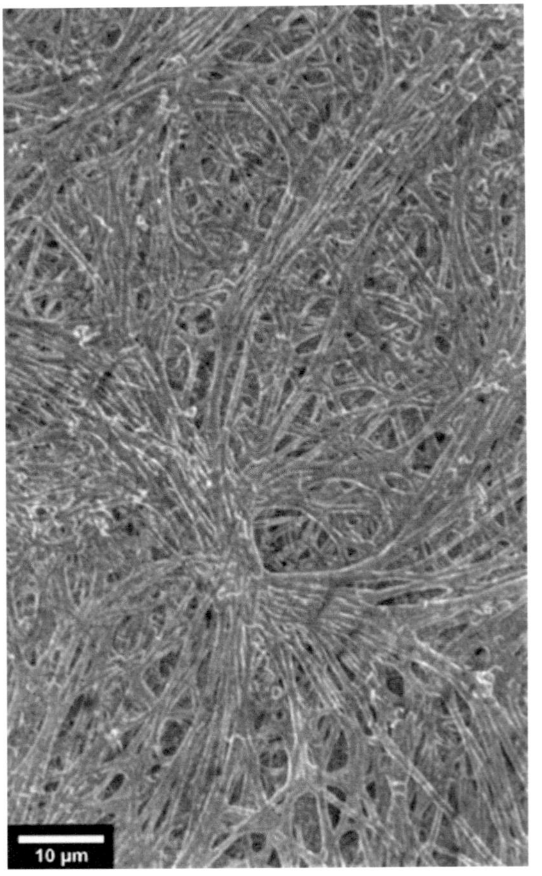

Fig. 5.2 An immunofluorescent image of chicken egg coat. Chicken egg coat was observed under a confocal immunofluorescent microscopy using an antibody against one of the chicken ZP glycoproteins, ZP1. The surface of filamentous egg-coat matrix with a diameter of approximately 0.8–1.4 μm was significantly labeled with the antibody. Detailed procedures for the immunofluorescent staining and confocal laser-scanning microscopy of chicken egg coat were based on our previous study (Okumura et al. 2015). The stacks of 16 images taken at 0.4 μm intervals along the Z axis were merged. *Bar*: 10 μm

After ovulation, the egg coat plays important roles in gamete recognition, induction of sperm acrosomal exocytosis, i.e., acrosome reaction, blocking polyspermy at the egg-coat level, and protecting the early embryo until hatching (Rankin and Dean 2000). The earliest works on the physiological functions of mammalian egg coat, zona pellucida, demonstrated that the capacitated and acrosome-intact spermatozoa bind to the egg coat in a species-specific manner (Hartmann and Gwatkin 1971; Hanada and Chang 1972; Hartmann et al. 1972) and subsequently undergo the acrosome reaction, followed by their penetration through the egg coat (Saling and Storey 1979; Saling et al. 1979). Interestingly, however, recent studies on mammalian fertilization with new experimental strategies have raised questions about whether the acrosome reaction is induced only by the sperm binding to the egg coat (Jin et al. 2011) and is responsible for the sperm penetration or passing through the egg coat (Inoue et al. 2011) under physiological conditions. Therefore, even though it is clear that the egg coat has an ability to induce or accelerate the acrosome reaction (Florman and Storey 1982; Koyanagi et al. 1988; Uto et al. 1988), it is necessary to reconsider the physiological roles of the egg coat in egg–sperm interactions (Yanagimachi 2011).

The earliest works also demonstrated in mammals that the first fusion of spermatozoa to the egg cell triggers exocytosis of the egg cortical granule, and the secreted trypsin-like protease cleaves sperm receptors on the zona pellucida to block further sperm penetration (Gwatkin et al. 1973; Gwatkin and Williams 1977), resulting in a block to polyspermy (Hoodbhoy and Talbot 1994; Burkart et al. 2012). Interestingly, some non-mammalian vertebrates, including birds, exhibit the physiological polyspermy in which several spermatozoa penetrate the egg coat and enter into the egg cell, although only one of them fuse with the egg pronucleus (Birkhead et al. 1994; Perry 1987). These findings indicate that the roles of egg coat in blocking polyspermy are evolutionarily variable depending on the reproductive strategies of the vertebrates, and further investigations of the egg coat of non-mammalian vertebrates are needed for comprehensive understanding of the physiology of egg coat.

5.3 Egg-Coat Components, "Zona Pellucida (ZP) Glycoproteins"

The aforementioned sperm–egg coat (or sperm–zona) interactions are mediated by the egg-coat components, i.e., zona pellucida (ZP) glycoproteins, and the sperm proteins such as zonadhesin (Gao and Garbers 1998; Hardy and Garbers 1995), SED1 (Ensslin and Shur 2003), ADAM3 (Kim et al. 2005; Yamaguchi et al. 2006), CRISP1 (Maldera et al. 2014), acrosin/proacrosin (Jones 1987; Baba et al. 1994), or the egg cortical-granule protease, ovastacin (Burkart et al. 2012). However, physiological contributions of the sperm–zona interactions to normal fertilization remain controversial as above (Inoue et al. 2011; Yanagimachi 2011). Therefore, the structure and function of the egg-coat ZP glycoproteins are one of the keys to understanding how the egg coat is involved in fertilization.

Structural characteristics of the ZP glycoproteins have been well defined by recent advances in genomic studies and structural biology, and summarized in some articles including our previous paper (Okumura et al. 2015) as follows. The ZP glycoproteins are characterized by a conserved ~260-aa-long ZP-domain module (Bork and Sander 1992) containing two unique immunoglobulin-like folds, named ZP-N and ZP-C domains respectively, connected by a linker region (Jovine et al. 2004; Han et al. 2010). ZP glycoproteins are classified into six subfamilies in total, designated ZP1, ZP2/ZPA, ZP3/ZPC, ZP4/ZPB, ZPD, and ZPAX, based on phylogenetic analyses (Spargo and Hope 2003; Goudet et al. 2008), and these subfamilies can be distinguished by their domain architectures in the N-terminal flanking region of the ZP-N domain and structural diversities of the ZP-C domain. On the N terminus of ZP-N domain either in the ZP1 or ZP4 homologue, there is an additional ZP-N domain (named N-terminal domain) followed by a trefoil domain. The N-terminal domain of ZP1 but not ZP4 possesses a free Cys residue that is thought to form intermolecular disulfide bonds (Callebaut et al. 2007), and there is a Pro/Thr/Leu-rich region between the N-terminal and trefoil domains in ZP1 but not in ZP4 (Bausek et al. 2000; Hoodbhoy et al. 2005; Izquierdo-Rico et al. 2009). Furthermore, in the N-terminal flanking region of ZP-N domain, there are three additional ZP-N repeats (named ZP-N1, ZP-N2, and ZP-N3 respectively) in the ZP2 homologue (Callebaut et al. 2007), a variable region (named N-terminal peptide) in the ZP3 homologue (Pan et al. 2000; Waclawek et al. 1998; Han et al. 2010), and a calcium-binding EGF-like domain in avian ZPD homologue (Okumura et al. 2004; Sato et al. 2009). In addition, the ZP3 homologue possesses the type I ZP-C domain containing eight Cys residues and carrying the ZP-C subdomain at the C terminus (Han et al. 2010), whereas the other homologues possess the type II ZP-C domain containing six Cys residues (Monné et al. 2006; Jovine et al. 2005), except that amphibian ZPD homologues possess an unusual type-II-like ZP-C domain with four Cys residues (Lindsay et al. 2002; Okumura et al. 2004; Sato et al. 2009). On the other hand, the ZP-C domain of all the ZP-glycoprotein subfamilies is followed by a C-terminal propeptide containing the external hydrophobic patch (EHP) and, with some exceptions including the known avian ZP1, a transmembrane domain (TM) and a short cytoplasmic tail (Monné et al. 2006). The C-terminal propeptide is cleaved off from the mature protein by proteolytic processing at a consensus furin cleavage site (CSCF) located between ZP-N domain and the EHP (Han et al. 2010). Schematic diagrams of domain architectures are drawn in the ZP1, ZP2, ZP3, ZP4, avian ZP1 (possessing a long repetitive Pro/Thr/Leu-rich repeat but not the TM domain), and avian ZPD homologues and shown in Fig. 5.3, using the examples of mouse (ZP1 and ZP2), human

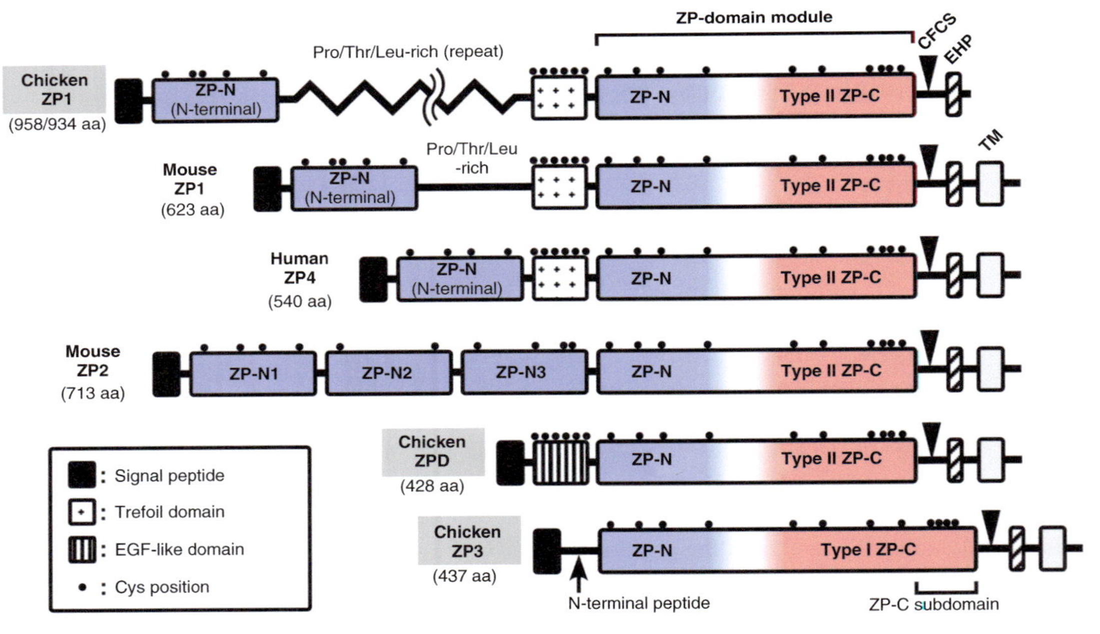

Fig. 5.3 Schematic diagrams of domain architectures of ZP glycoproteins composing mammalian and avian mature egg coat. The domain architectures of ZP1, ZP2, ZP3, ZP4, avian ZP1 (possessing a long repetitive Pro/Thr/Leu-rich repeat domain but not the TM domain), and avian ZPD (possessing a calcium-binding EGF-like domain) homologues are illustrated using the examples of mouse (ZP1 and ZP2), human (ZP4) and chicken (ZP3, avian ZP1, and ZPD) with the domain or region names that are described in the text. The schemes of these ZP glycoproteins are arranged according to their structural relationships. The relative positions of cysteine (Cys) residues are indicated by *dots*. Chicken ZP1, ZP3, and ZPD are marked with *shaded texts* and *asterisks*

(ZP4), and chicken (ZP3, avian ZP1, and ZPD) ZP glycoproteins. Molecular mechanisms of chicken ZP3 in the secretion from ZP3-expressing cell, the egg-coat formation, and the interaction with sperm molecule(s) are illustrated together with the determined crystal structure in Fig. 5.4.

Interestingly, it is suggested that some of the genes encoding ZP glycoproteins have been independently lost or have become pseudogenes during evolution (Goudet et al. 2008), and consequently, the mature egg coat is constituted of ZP3 and ZP4 in teleost fish (Darie et al. 2004; Sano et al. 2010, 2013), ZP2, ZP3, ZP4, ZPD, and ZPAX in amphibians (*Xenopus*) (Lindsay et al. 2001, 2002), and ZP1, ZP3, ZPD, and under detectable or very low levels of ZP2, ZP4, and ZPAX in birds (chicken, quail and turkey) (Okumura et al. 2004; Sato et al. 2009; Mann 2008; Nishio et al. 2014; Benson et al. 2009). It is noteworthy that the mammalian egg coat is composed of ZP1, ZP2, ZP3, and ZP4 in some classes including hamster, rat, cat, and human (Hoodbhoy et al. 2005; Izquierdo-Rico et al. 2009; Lefiévre et al. 2004; Stetson et al. 2015), ZP2, ZP3, and ZP4 in other ones including pig, cow, and dog (Kudo et al. 1998; Kanai et al. 2007; Blackmore et al. 2004), and ZP1, ZP2, and ZP3 only in mouse (Goudet et al. 2008; Lefiévre et al. 2004; Bleil and Wassarman 1980). Moreover, the expression patterns of ZP glycoproteins might have shifted through evolution, and consequently, all mouse ZP glycoproteins are synthesized in the growing oocyte (Ringuette et al. 1986; Liang et al. 1990; Epifano et al. 1995), whereas in birds (chicken, quail, and turkey), ZP1 is synthesized in the liver and secreted into bloodstream to be transported to the ovary (Bausek et al. 2000; Sasanami et al. 2003b; Benson et al. 2009), and both ZP3 and ZPD are synthesized in the granulosa cells surrounding preovulatory-growing oocyte (Okumura et al. 2004; Takeuchi et al. 1999; Sato et al. 2009; Sasanami et al. 2002; Benson et al. 2009). It has been reported in chicken and/or quail that the expression of ZP1 and ZP3 are stimulated by estrogen and testosterone, respectively (Bausek et al. 2000; Hanafy et al. 2007; Sasanami et al. 2003b; Pan et al. 2001). The secretion mechanisms of avian ZP3 have been well defined in quail, in which the C-terminal propeptide is proteolytically cleaved at the above-mentioned furin cleavage site and the *N*-linked carbohydrates might be involved in the cellular transport systems (Sasanami et al. 2002, 2003a, c, 2005). In contrast to ZP3 and ZPD, ZP2 is expressed exclusively in the granulosa cells (chicken) or oocyte

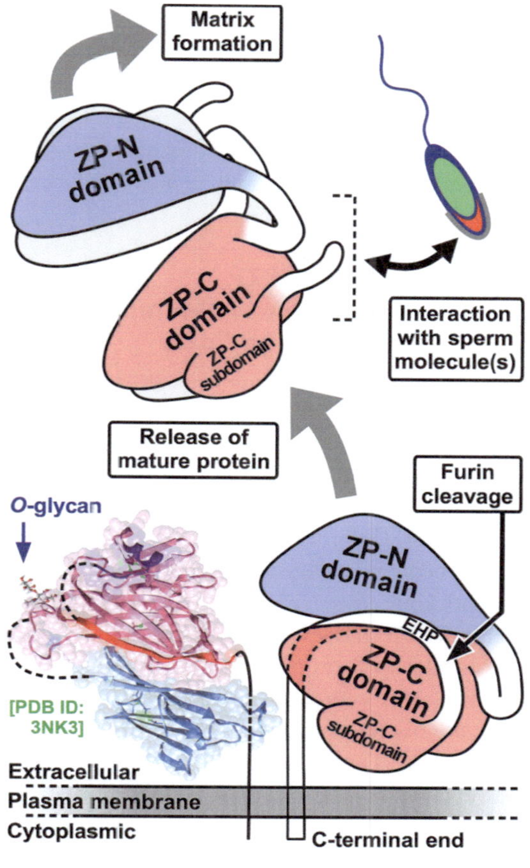

Fig. 5.4 Molecular mechanisms of chicken ZP3 in the secretion from the expressing cell and interaction with sperm molecule(s). Chicken ZP3 is expressed on the surface of the ovarian granulosa cells as a non-covalently linked homodimer. The C-terminal propeptide containing the external hydrophobic patch (EHP) and transmembrane domain is cleaved by furin, and the mature ZP3 is released to form egg-coat matrix and interact with unidentified sperm molecule(s). The determined crystal structure of chicken ZP3 (PDB ID: 3NK3) is shown in one of the ZP3 molecules in the homodimer. The location of *O*-glycan involved in the interaction with spermatozoa is indicated in the structure. (Modified from http://first.lifesciencedb.jp/archives/1505)

(quail) of immature white follicle together with the ZP4 (Kinoshita et al. 2010; Nishio et al. 2014; Serizawa et al. 2011; Rodler et al. 2012) and localized around the germinal-disc region of the yellow follicle to be one of the minor components of the mature egg coat (Nishio et al. 2014).

5.4 Assembly of ZP Glycoproteins in Mammalian and Avian Egg Coats

The first model for the arrangement of ZP glycoproteins in mouse egg-coat filaments was proposed in the 1980s based on biochemical and electron microscopic studies (Greve and Wassarman 1985; Wassarman 1988), and was ascertained and refined by later studies using ZP-knockout mice (Rankin et al. 1996, 1999, 2001). In the model, ZP3 associates with ZP1 and/or ZP2 to form filaments, ZP1 and ZP2 do not associate with each other, and ZP1 cross-links the filaments by forming a disulfide-linked homodimer (Dean 2004; Monné et al. 2006). However, whereas the mouse model demonstrates interconnections among ZP glycoproteins in vertebrate egg coat, the molecular details of the intermolecular contacts and the matrix formation mechanisms remain unexplained.

From 2008 onwards, a series of the first crystal structures has been determined for ZP-N domain of mouse ZP3 (Monné et al. 2008), the type II ZP-C domain of mouse ZP2 (Bokhove et al. 2016), full-length chicken ZP3 containing both ZP-N and the type I ZP-C domains (Han et al. 2010), and for ZP-N and/or the type II ZP-C domains of non-egg-coat ZP-domain proteins including mouse and rat homologues of transforming growth factor (TGF)-beta receptor type 3 (TGFR-3) (Diestel et al. 2013; Lin et al. 2011) and a protease-resistant fragment of human uromodulin (UMOD) containing both ZP-N and ZP-C domains (Bokhove et al. 2016). In our previous study, we utilized the determined configurations of ZP-N and ZP-C domains in the ZP-domain module and the above-mentioned domain architectures of ZP glycoproteins (see Figs. 5.3 and 5.4) to design recombinant domains of chicken ZP1, ZP3, and ZPD (Okumura et al. 2015). In this study, we mapped the intermolecular contacts among ZP glycoproteins involved in the egg-coat matrix formation by using the individual recombinant domains of chicken ZP glycoproteins and the antibodies being raised against them as probes for immunohistochemical and biochemical analyses. Based on the results, together with our previous findings on chicken ZP glycoproteins including the crystal structure of ZP3 (Han et al. 2010), the formation of ZP1–ZP3 complexes (Okumura et al. 2007a, b), and the assembly of ZPD on the egg coat (Okumura et al. 2004), we proposed a model for the architecture of chicken egg-coat matrix as follows. ZP1 and ZP3 assemble into probably helical fibrils through strong interactions between their ZP-C domains in a manner that the ZP-C domains are deeply embedded in the core, the ZP-N domains are located in the medial layer, and that the N-terminal, repeat, trefoil domains of ZP1, the N-terminal peptides of ZP3, and probably the glycan chains on the ZP-N/ZP-C linker regions are exposed on the surface. The resulting ZP1–ZP3 fibrils further assemble into bundles through the intermolecular disulfide bridges between the N-terminal domains of ZP1 and through non-covalent associations between the repeat domains of ZP1 to form the ZP1–ZP3 matrix. In addition, ZPD forms homopolymers, presumably in head-to-head and tail-to-tail arrangements, and attaches to the ZP1–ZP3 fibrils located on the surface of the matrix to form the ZPD aggregates. A model for the molecular architecture of chicken egg-coat matrix (Okumura et al. 2015) is shown in Fig. 5.5. This chicken model is compatible with the above-mentioned mouse model and might be applicable to the egg coat of all other vertebrates by replacing the ZP-glycoprotein components, whereas detailed domain configurations of the ZP glycoproteins, involvements of the non-covalent interactions between ZP3 molecules being observed in the crystal structure (Han et al. 2010), participation of the previously-reported isoforms of ZP3 (Okumura et al. 2012), and amyloid-forming features of the A and G

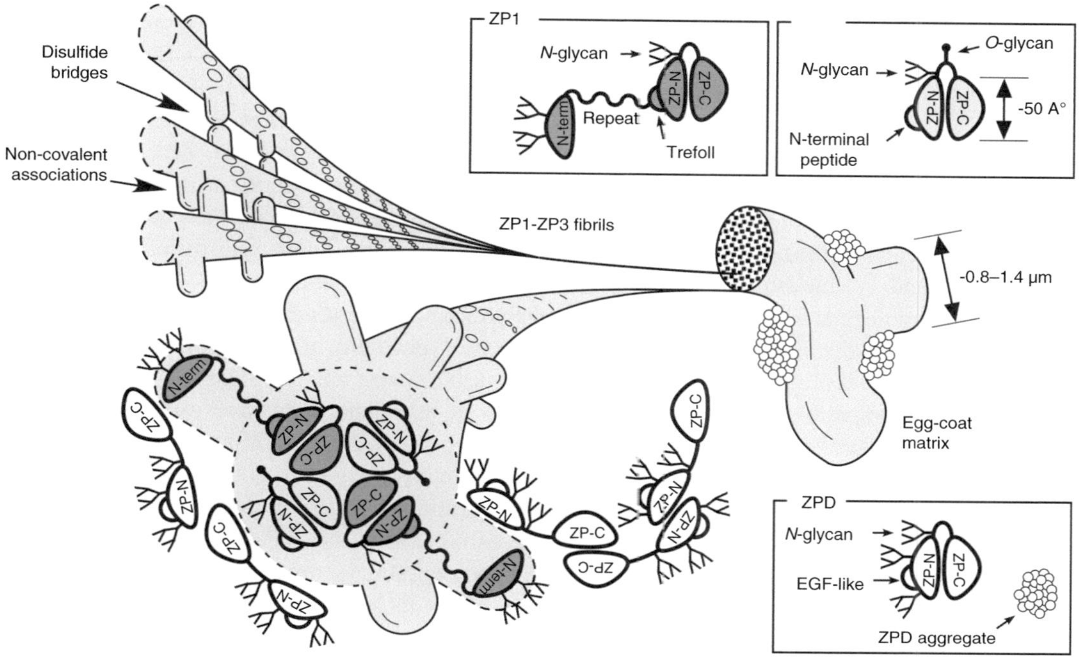

Fig. 5.5 A model for the molecular architecture of chicken egg-coat matrix. A schematic diagram of the chicken egg-coat matrix architecture was drawn based on the results from our previous studies. ZP1 containing the N-terminal (N-term), repeat, trefoil, ZP-N, and ZP-C domains assembles with ZP3 containing the N-terminal peptide, ZP-N, and ZP-C domains, through the interactions between their ZP-C domains, to form the helical ZP1–ZP3 fibrils. The N-terminal, repeat, trefoil domains of ZP1, and the N-terminal peptide of ZP3 are exposed on the surface of the ZP1–ZP3 fibrils to overlie the ZP-N domains. The resulting ZP1–ZP3 fibrils assemble into bundles through both intermolecular disulfide bridges between the N-terminal domains of ZP1, and non-covalent associations between the repeat domains of ZP1, to form the fibrous ZP1–ZP3 matrix. In addition, ZPD containing the calcium-binding EGF-like (EGF-like), ZP-N, and ZP-C domains forms homopolymers in head-to-head and tail-to-tail arrangements, and attaches to the ZP1–ZP3 fibrils located on the surface of the interstices of ZP1–ZP3 matrix to form aggregates. Schematic diagrams of ZP1, ZP3, and ZPD are drawn based on the domain architecture of chicken ZP1 and the crystal structure of full-length chicken ZP3 (PDB ID: 3NK3). The locations of *O*-glycan on the ZP-N/ZP-C linker region of ZP3 are shown, and all the putative *N*-glycosylation sites are occupied in this scheme. [Partially modified from Okumura et al. (2015)]

strands in the ZP-N domains (Louros et al. 2016) in the formation of the ZP1–ZP3 complex remain to be elucidated.

In contrast to mouse zona pellucida, which all the ZP-glycoprotein components are expressed in the oocyte (Ringuette et al. 1986; Liang et al. 1990; Epifano et al. 1995), the core filaments of mature chicken and quail egg coat are assembled from ZP1 and ZP3 that are expressed in distinct tissues, i.e., the liver (Bausek et al. 2000; Sasanami et al. 2003b) and ovarian granulosa cells (Takeuchi et al. 1999; Pan et al. 2001) respectively (Ohtsuki et al. 2004), as described in Sect. 5.3. Therefore, the chicken and quail ZP1–ZP3 complexes could be an excellent model system for investigating the molecular mechanisms of the vertebrate egg-coat formation. In fact, in our previous studies, it is suggested that chicken ZP1 circulating in the bloodstream interacts with the ZP3 being secreted from the granulosa cells on the surface of the oocyte to form soluble ZP1–ZP3 complexes, and subsequently assemble into ZP1–ZP3 fibrils in which approximately half of the ZP1 molecules form the disulfide-linked homodimer to a similar extent as in the mature egg coat (Okumura et al. 2007a, b). The configurations of some domains in the ZP1–ZP3 complexes are probably changed during the assembly into fibrils, for example the ZP-C domain of ZP1 exposed on the surface of the ZP1–ZP3 complexes might become embedded during the egg-coat formation (Okumura et al. 2015). Interestingly, it has been reported in quail that ZP1 purified from the serum but not ZP1 from the egg coat is incorporated into the egg coat after injection into the female bird (Kinoshita et al. 2008), and in chicken that ZP1 does not form an intermolecular disulfide bond in the serum but forms the disulfide-linked homodimer, probably with some conformational changes during the interaction with ZP3 (Okumura et al. 2004, 2007a, b). Further studies are needed to reveal the molecular mechanisms of avian ZP1 in the serum to be incorporated into the egg-coat matrix and of the soluble ZP1–ZP3 complexes to form insoluble filaments.

5.5 Molecular Bases for Egg-Coat Functions in Mammals and Birds

Whereas the physiological roles of the egg coat in vertebrate fertilization remain controversial as above (Yanagimachi 2011), many studies suggest that sugar chains on the ZP glycoproteins play crucial roles in sperm–zona interactions in cooperation with the polypeptide backbone (Han et al. 2010; Florman and Wassarman 1985; Visconti and Florman 2010; Clark 2013). In most cases, it has been shown that ZP glycoproteins individually have abilities to bind to intact or acrosome-reacted spermatozoa. For example, in mouse and human, the C-terminal region of ZP3 is involved in the binding of acrosome-intact spermatozoa in concert with the *O*-linked oligosaccharides on the ZP-N/ZP-C linker region to induce the acrosome reaction (Williams et al. 2006; Bansal et al. 2009; Leyton and Saling 1989; Wassarman 1988; Han et al. 2010) (see Fig. 5.4). It is also suggested in mouse and human that the N-terminal region of ZP2 is involved in taxon-specific binding of the acrosome-intact spermatozoa and the secondary binding of acrosome-reacted spermatozoa during the penetration of egg coat (Avella et al. 2014; Bleil et al. 1988). The cleavage of ZP2 within the ZP-N2 domain by the cortical-granule protease, ovastacin (see Fig. 5.3) is thought to contribute to the block to polyspermy in mammals as above (Burkart et al. 2012). In chicken and quail, on the other hand, ZP1 and ZPD but not ZP3 induce sperm acrosome reaction, although either ZP1 or ZP3 has abilities to bind to the acrosome-intact spermatozoa (Sasanami et al. 2007; Okumura et al. 2004; Han et al. 2010). Furthermore, it has been reported in birds that species specificity of the egg–sperm interactions, or of the sperm binding to ZP3 is not so strict as it is in mammals (Stewart et al. 2004). In addition, chicken ZP2, one of the minor components in the mature egg coat, is thought to be involved in the sperm binding on the germinal-disc region (Nishio et al. 2014). Interestingly, it is shown in some vertebrates that ZP glycoproteins have activities to interact with the spermatozoa only when they form complexes; for example in cow, ZP3/ZP4 heterodimer but not any individual ZP glycoproteins bind to spermatozoa (Kanai et al. 2007), and in chicken, the dimeric form of ZP1 has much higher activity to induce acrosome reaction than the monomeric form (Okumura et al. 2004). These findings imply that the interactions between ZP glycoproteins and sperm molecule(s) are enhanced or regulated by the specific orientations of ZP glycoproteins in the egg-coat matrix, although the roles of individual ZP glycoproteins in the sperm–zona interactions have been diversified during evolution. Further studies will be required to determine the consequences of diverse molecular mechanisms of the sperm–zona interactions on reproductive strategies of vertebrates.

ZPD is a ZP glycoprotein that might play specialized roles in non-mammalian egg coat. The avian ZPD is a unique ZP glycoprotein possessing a calcium-binding EGF-like domain and being deposited on the interstices of ZP1–ZP3 matrix in the egg coat as above (Okumura et al. 2004, 2015; Sato et al. 2009) (see Fig. 5.5). Surprisingly, avian ZPD and one of the non-egg-coat ZP-domain protein, uromodulin, share the calcium-binding EGF-like and the ZP domains, although human uromodulin possesses three additional EGF-like domains and the uncharacterized D8C domain

(Santambrogio et al. 2008; Bokhove et al. 2016). Furthermore, a database homology search using NCBI BLAST (Altschul et al. 1997) against chicken ZPD (GenBank accession number: BAD13713) and human uromodulin (GenBank accession number: NP_001008390) showed that C58–I362 of the chicken ZPD shares a significant sequence homology with C306–I604 of the human uromodulin (29% identities and 48% positives). These data and the recent reports that uromodulin plays important roles in the protection from bacterial infection and the regulation of immunosystems in the urinary tract (Santambrogio et al. 2008; Säemann et al. 2005) imply that avian ZPD might be a close relative of uromodulin and play roles in the defense of oocyte or early embryo in the egg coat. Further investigations might reveal characteristic features of ZPD in the early development of birds.

Conclusions

Birds have evolved a huge egg cell with a large amount of egg yolk, probably associated with their unique reproductive strategies, and therefore, some features of the egg coat surrounding the egg cell are divergent from those of mammals in spite of the fact that all vertebrate egg coat is composed of ZP glycoproteins sharing an evolutionarily conserved ZP-domain module. Considering that the avian oocytes markedly increase their surface area in association with the rapid egg-yolk accumulation during the development of ovarian follicle, it might be beneficial for the accelerated formation of the avian egg coat that the core of the egg-coat matrix is assembled from ZP1 circulating in the bloodstream and ZP3 secreted from the granulosa cells surrounding the oocyte. Furthermore, the repeat domain of avian ZP1 (a long repetitive Pro/Thr/Leu-rich region) might contribute to the elasticity of avian egg-coat matrix during oocyte maturation, because it is suggested that the long repeat domain, probably possessing structural flexibility, plays the role of a linker between the ZP1–ZP3 fibrils. It is noteworthy that there is remarkable species diversity in the repeat domain of avian ZP1 according to the GenBank database, and more interestingly, there are at least two isoforms of chicken ZP1 (GenBank accession numbers: NP_990014 and BAU61867 respectively) that have 24-aa-long differences corresponding with one repeat unit in the repeat domain. Comparative studies of these ZP1 isoforms might reveal functional roles of the highly specialized Pro/Thr/Leu-rich region of avian ZP1 in sperm–zona interactions and egg-coat formation mechanisms. In addition, ZPD found in avian but not in mammalian egg coat shares high sequence homology with mammalian uromodulin, and might therefore be expected to play a role in the defense of the increased surface of avian egg cell. On the other hand, there are some common features of avian and mammalian egg coat, including the molecular mechanisms of the egg-coat assembly and the ZP3 binding to the spermatozoa, and probably the involvement of ZP2 in sperm binding, although avian ZP2 is a minor component in the mature egg coat and localized in the germinal-disc region. Considering that birds and mammals evolved their own reproductive systems independently, these features of ZP glycoproteins that are evolutionarily conserved between birds and mammals could be fundamental in all the vertebrates.

Collectively, further studies of avian egg coat could have implications for understanding not only the characteristic features of avian ZP glycoproteins associated with their unique reproductive strategies, but also the evolutionarily conserved mechanisms underlying the formation of egg coat and sperm–zona interactions involving ZP glycoproteins, by comparing them with mammalian egg coat. Comprehensive understanding of vertebrate egg coat will provide new insights into developing novel approaches to control fertility for human and veterinary medicine, including therapies for infertility being associated with zona-pellucida dysmorphism and/or mutations in the ZP glycoproteins (Pökkylä et al. 2011; Margalit et al. 2012; Sousa et al. 2015), in addition to contraception and control of mammalian/avian pest populations.

References

Adkins-Regan E, Banerjee SB, Correa SM, Schweitzer C. Maternal effects in quail and zebra finches: behavior and hormones. Gen Comp Endocrinol. 2013;190:34–41. doi:10.1016/j.ygcen.2013.03.002.

Altschul SF, Madden TL, Schaffer AA, Zhang J, Zhang Z, Miller W, Lipman DJ. Gapped BLAST and PSI-BLAST: a new generation of protein database search programs. Nucleic Acids Res. 1997;25(17):3389–402.

Avella MA, Baibakov B, Dean J. A single domain of the ZP2 zona pellucida protein mediates gamete recognition in mice and humans. J Cell Biol. 2014;205(6):801–9. doi:10.1083/jcb.201404025.

Baba T, Niida Y, Michikawa Y, Kashiwabara S, Kodaira K, Takenaka M, Kohno N, Gerton GL, Arai Y. An acrosomal protein, sp32, in mammalian sperm is a binding protein specific for two proacrosins and an acrosin intermediate. J Biol Chem. 1994;269(13):10133–40.

Bakst MR, Howarth B Jr. The fine structure of the hen's ovum at ovulation. Biol Reprod. 1977a;17(3):361–9.

Bakst MR, Howarth B Jr. Hydrolysis of the hen's perivitelline layer by cock sperm in vitro. Biol Reprod. 1977b;17(3):370–9.

Bansal P, Chakrabarti K, Gupta SK. Functional activity of human ZP3 primary sperm receptor resides toward its C-terminus. Biol Reprod. 2009;81(1):7–15. doi:10.1095/biolreprod.108.074716.

Bausek N, Waclawek M, Schneider WJ, Wohlrab F. The major chicken egg envelope protein ZP1 is different from ZPB and is synthesized in the liver. J Biol Chem. 2000;275(37):28866–72.

Benson AP, Christensen VL, Fairchild BD, Davis AJ. The mRNA for zona pellucida proteins B1, C and D in two genetic lines of turkey hens that differ in fertility. Anim Reprod Sci. 2009;111(2–4):149–59. doi:10.1016/j.anireprosci.2008.02.013.

Birkhead TR, Sheldon BC, Fletcher F. A comparative study of sperm–egg interactions in birds. J Reprod Fertil. 1994;101(2):353–61.

Blackmore DG, Baillie LR, Holt JE, Dierkx L, Aitken RJ, McLaughlin EA. Biosynthesis of the canine zona pellucida requires the integrated participation of both oocytes and granulosa cells. Biol Reprod. 2004;71(2):661–8. doi:10.1095/biolreprod.104.028779.

Bleil JD, Greve JM, Wassarman PM. Identification of a secondary sperm receptor in the mouse egg zona pellucida: role in maintenance of binding of acrosome-reacted sperm to eggs. Dev Biol. 1988;128(2):376–85.

Bleil JD, Wassarman PM. Mammalian sperm–egg interaction: identification of a glycoprotein in mouse egg zonae pellucidae possessing receptor activity for sperm. Cell. 1980;20(3):873–82.

Bokhove M, Nishimura K, Brunati M, Han L, de Sanctis D, Rampoldi L, Jovine L. A structured interdomain linker directs self-polymerization of human uromodulin. Proc Natl Acad Sci U S A. 2016;113(6):1552–7. doi:10.1073/pnas.1519803113.

Bork P, Sander C. A large domain common to sperm receptors (Zp2 and Zp3) and TGF-beta type III receptor. FEBS Lett. 1992;300(3):237–40.

Bramwell RK, Howarth B Jr. Preferential attachment of cock spermatozoa to the perivitelline layer directly over the germinal disc of the hen's ovum. Biol Reprod. 1992;47(6):1113–7.

Burkart AD, Xiong B, Baibakov B, Jimenez-Movilla M, Dean J. Ovastacin, a cortical granule protease, cleaves ZP2 in the zona pellucida to prevent polyspermy. J Cell Biol. 2012;197(1):37–44. doi:10.1083/jcb.201112094.

Callebaut I, Mornon JP, Monget P. Isolated ZP-N domains constitute the N-terminal extensions of Zona Pellucida proteins. Bioinformatics. 2007;23(15):1871–4. doi:10.1093/bioinformatics/btm265.

Clark GF. The role of carbohydrate recognition during human sperm–egg binding. Hum Reprod. 2013;28(3):566–77. doi:10.1093/humrep/des447.

Correa SM, Adkins-Regan E, Johnson PA. High progesterone during avian meiosis biases sex ratios toward females. Biol Lett. 2005;1(2):215–8. doi:10.1098/rsbl.2004.0283.

Darie CC, Biniossek ML, Jovine L, Litscher ES, Wassarman PM. Structural characterization of fish egg vitelline envelope proteins by mass spectrometry. Biochemistry. 2004;43(23):7459–78. doi:10.1021/bi0495937.

Dean J. Reassessing the molecular biology of sperm–egg recognition with mouse genetics. BioEssays. 2004;26(1):29–38. doi:10.1002/bies.10412.

Dias da Silva W, Tambourgi DV. IgY: a promising antibody for use in immunodiagnostic and in immunotherapy. Vet Immunol Immunopathol. 2010;135(3–4):173–80. doi:10.1016/j.vetimm.2009.12.011.

Diestel U, Resch M, Meinhardt K, Weiler S, Hellmann TV, Mueller TD, Nickel J, Eichler J, Muller YA. Identification of a novel TGF-beta-binding site in the Zona Pellucida C-terminal (ZP-C) domain of TGF-beta-Receptor-3 (TGFR-3). PLoS One. 2013;8(6):e67214. doi:10.1371/journal.pone.0067214.

Ensslin MA, Shur BD. Identification of mouse sperm SED1, a bimotif EGF repeat and discoidin-domain protein involved in sperm–egg binding. Cell. 2003;114(4):405–17.

Epifano O, Liang LF, Familari M, Moos MC Jr, Dean J. Coordinate expression of the three zona pellucida genes during mouse oogenesis. Development. 1995;121(7):1947–56.

Florman HM, Storey BT. Mouse gamete interactions: the zona pellucida is the site of the acrosome reaction leading to fertilization in vitro. Dev Biol. 1982;91(1):121–30.

Florman HM, Wassarman PM. O-linked oligosaccharides of mouse egg ZP3 account for its sperm receptor activity. Cell. 1985;41(1):313–24.

Gao Z, Garbers DL. Species diversity in the structure of zonadhesin, a sperm-specific membrane protein containing multiple cell adhesion molecule-like domains. J Biol Chem. 1998;273(6):3415–21.

Goudet G, Mugnier S, Callebaut I, Monget P. Phylogenetic analysis and identification of pseudogenes reveal a progressive loss of zona pellucida genes during evolution of vertebrates. Biol Reprod. 2008;78(5):796–806. doi:10.1095/biolreprod.107.064568.

Greve JM, Wassarman PM. Mouse egg extracellular coat is a matrix of interconnected filaments possessing a structural repeat. J Mol Biol. 1985;181(2):253–64.

Gwatkin RB, Williams DT. Receptor activity of the hamster and mouse solubilized zona pellucida before and after the zona reaction. J Reprod Fertil. 1977;49(1):55–9.

Gwatkin RB, Williams DT, Hartmann JF, Kniazuk M. The zona reaction of hamster and mouse eggs: production in vitro by a trypsin-like protease from cortical granules. J Reprod Fertil. 1973;32(2):259–65.

Han L, Monné M, Okumura H, Schwend T, Cherry AL, Flot D, Matsuda T, Jovine L. Insights into egg coat assembly and egg–sperm interaction from the X-ray structure of full-length ZP3. Cell. 2010;143(3):404–15. doi:10.1016/j.cell.2010.09.041.

Hanada A, Chang MC. Penetration of zone-free eggs by spermatozoa of different species. Biol Reprod. 1972;6(2):300–9.

Hanafy AM, Sasanami T, Mori M. Sensitivity of expression of perivitelline membrane glycoprotein ZP1 mRNA in the liver of Japanese quail (*Coturnix japonica*) to estrogenic compounds. Comp Biochem Physiol C Toxicol Pharmacol. 2007;144(4):356–62. doi:10.1016/j.cbpc.2006.11.007.

Hardy DM, Garbers DL. A sperm membrane protein that binds in a species-specific manner to the egg extracellular matrix is homologous to von Willebrand factor. J Biol Chem. 1995;270(44):26025–8.

Hartmann JF, Gwatkin RB. Alteration of sites on the mammalian sperm surface following capacitation. Nature. 1971;234(5330):479–81.

Hartmann JF, Gwatkin RB, Hutchison CF. Early contact interactions between mammalian gametes in vitro: evidence that the vitellus influences adherence between sperm and zona pellucida. Proc Natl Acad Sci U S A. 1972;69(10):2767–9.

Hoodbhoy T, Joshi S, Boja ES, Williams SA, Stanley P, Dean J. Human sperm do not bind to rat zonae pellucidae despite the presence of four homologous glycoproteins. J Biol Chem. 2005;280(13):12721–31. doi:10.1074/jbc.M413569200.

Hoodbhoy T, Talbot P. Mammalian cortical granules: contents, fate, and function. Mol Reprod Dev. 1994;39(4):439–48. doi:10.1002/mrd.1080390413.

Hughes GC. The population of germ cells in the developing female chick. J Embryol Exp Morphol. 1963;11:513–36.

Inoue N, Satouh Y, Ikawa M, Okabe M, Yanagimachi R. Acrosome-reacted mouse spermatozoa recovered from the perivitelline space can fertilize other eggs. Proc Natl Acad Sci U S A. 2011;108(50):20008–11. doi:10.1073/pnas.1116965108.

Izquierdo-Rico MJ, Jiménez-Movilla M, Llop E, Pérez-Oliva AB, Ballesta J, Gutiérrez-Gallego R, Jiménez-Cervantes C, Avilés M. Hamster zona pellucida is formed by four glycoproteins: ZP1, ZP2, ZP3, and ZP4. J Proteome Res. 2009;8(2):926–41. doi:10.1021/pr800568x.

Jin M, Fujiwara E, Kakiuchi Y, Okabe M, Satouh Y, Baba SA, Chiba K, Hirohashi N. Most fertilizing mouse spermatozoa begin their acrosome reaction before contact with the zona pellucida during in vitro fertilization. Proc Natl Acad Sci U S A. 2011;108(12):4892–6. doi:10.1073/pnas.1018202108.

Johnson AL. Ovarian follicle selection and granulosa cell differentiation. Poult Sci. 2015;94(4):781–5. doi:10.3382/ps/peu008.

Jones R. Evidence for boar sperm proacrosin as a carbohydrate binding protein. Cell Biol Int Rep. 1987;11(11):833.

Jovine L, Darie CC, Litscher ES, Wassarman PM. Zona pellucida domain proteins. Annu Rev Biochem. 2005;74:83–114. doi:10.1146/annurev.biochem.74.082803.133039.

Jovine L, Qi H, Williams Z, Litscher ES, Wassarman PM. A duplicated motif controls assembly of zona pellucida domain proteins. Proc Natl Acad Sci U S A. 2004;101(16):5922–7. doi:10.1073/pnas.0401600101.

Kanai S, Yonezawa N, Ishii Y, Tanokura M, Nakano M. Recombinant bovine zona pellucida glycoproteins ZP3 and ZP4 coexpressed in Sf9 cells form a sperm-binding active hetero-complex. FEBS J. 2007;274(20):5390–405. doi:10.1111/j.1742-4658.2007.06065.x.

Kim E, Baba D, Kimura M, Yamashita M, Kashiwabara S, Baba T. Identification of a hyaluronidase, Hyal5, involved in penetration of mouse sperm through cumulus mass. Proc Natl Acad Sci U S A. 2005;102(50):18028–33. doi:10.1073/pnas.0506825102.

Kinoshita M, Mizui K, Ishiguro T, Ohtsuki M, Kansaku N, Ogawa H, Tsukada A, Sato T, Sasanami T. Incorporation of ZP1 into perivitelline membrane after in vivo treatment with exogenous ZP1 in Japanese quail (*Coturnix japonica*). FEBS J. 2008;275(14):3580–9. doi:10.1111/j.1742-4658.2008.06503.x.

Kinoshita M, Rodler D, Sugiura K, Matsushima K, Kansaku N, Tahara K, Tsukada A, Ono H, Yoshimura T, Yoshizaki N, Tanaka R, Kohsaka T, Sasanami T. Zona pellucida protein ZP2 is expressed in the oocyte of Japanese quail (*Coturnix japonica*). Reproduction. 2010;139(2):359–71. doi:10.1530/REP-09-0222.

Koyanagi F, Masuda S, Nishiyama H. Acrosome reaction of cock spermatozoa incubated with the perivitelline layer of the hen's ovum. Poult Sci. 1988;67(12):1770–4.

Kudo K, Yonezawa N, Katsumata T, Aoki H, Nakano M. Localization of carbohydrate chains of pig sperm ligand in the glycoprotein ZPB of egg zona pellucida. Eur J Biochem/FEBS. 1998;252(3):492–9.

Lefiévre L, Conner SJ, Salpekar A, Olufowobi O, Ashton P, Pavlovic B, Lenton W, Afnan M, Brewis IA, Monk M, Hughes DC, Barratt CL. Four zona pellucida glycoproteins are expressed in the human. Hum Reprod. 2004;19(7):1580–6. doi:10.1093/humrep/deh301.

Leyton L, Saling P. Evidence that aggregation of mouse sperm receptors by ZP3 triggers the acrosome reaction. J Cell Biol. 1989;108(6):2163–8.

Li J, Leghari IH, He B, Zeng W, Mi Y, Zhang C. Estrogen stimulates expression of chicken hepatic vitellogenin II and very low-density apolipoprotein II through ER-alpha. Theriogenology. 2014;82(3):517–24. doi:10.1016/j.theriogenology.2014.05.003.

Liang LF, Chamow SM, Dean J. Oocyte-specific expression of mouse Zp-2: developmental regulation of the zona pellucida genes. Mol Cell Biol. 1990;10(4):1507–15.

Lin SJ, Hu Y, Zhu J, Woodruff TK, Jardetzky TS. Structure of betaglycan zona pellucida (ZP)-C domain provides insights into ZP-mediated protein polymerization and TGF-beta binding. Proc Natl Acad Sci U S A. 2011;108(13):5232–6. doi:10.1073/pnas.1010689108.

Lindsay LL, Wallace MA, Hedrick JL. A hatching enzyme substrate in the *Xenopus laevis* egg envelope is a high molecular weight ZPA homolog. Develop Growth Differ. 2001;43(3):305–13.

Lindsay LL, Yang JC, Hedrick JL. Identification and characterization of a unique *Xenopus laevis* egg envelope component, ZPD. Develop Growth Differ. 2002;44(3):205–12.

Louros NN, Chrysina ED, Baltatzis GE, Patsouris ES, Hamodrakas SJ, Iconomidou VA. A common "aggregation-prone" interface possibly participates in the self-assembly of human zona pellucida proteins. FEBS Lett. 2016; doi:10.1002/1873-3468.12099.

Maldera JA, Weigel Munoz M, Chirinos M, Busso D, G E Raffo F, Battistone MA, Blaquier JA, Larrea F, Cuasnicu PS. Human fertilization: epididymal hCRISP1 mediates sperm–zona pellucida binding through its interaction with ZP3. Mol Hum Reprod. 2014;20(4):341–9. doi:10.1093/molehr/gat092.

Mann K. Proteomic analysis of the chicken egg vitelline membrane. Proteomics. 2008;8(11):2322–32. doi:10.1002/pmic.200800032.

Mann K, Mann M. The chicken egg yolk plasma and granule proteomes. Proteomics. 2008;8(1):178–91. doi:10.1002/pmic.200700790.

Margalit M, Paz G, Yavetz H, Yogev L, Amit A, Hevlin-Schwartz T, Gupta SK, Kleiman SE. Genetic and physiological study of morphologically abnormal human zona pellucida. Eur J Obstet Gynecol Reprod Biol. 2012;165(1):70–6. doi:10.1016/j.ejogrb.2012.07.022.

Menkhorst E, Selwood L. Vertebrate extracellular preovulatory and postovulatory egg coats. Biol Reprod. 2008;79(5):790–7. doi:10.1095/biolreprod.108.068551.

Monné M, Han L, Jovine L. Tracking down the ZP domain: from the mammalian zona pellucida to the molluscan vitelline envelope. Semin Reprod Med. 2006;24(4):204–16. doi:10.1055/s-2006-948550.

Monné M, Han L, Schwend T, Burendahl S, Jovine L. Crystal structure of the ZP-N domain of ZP3 reveals the core fold of animal egg coats. Nature. 2008;456(7222):653–7. doi:10.1038/nature07599.

Nakamura Y, Kagami H, Tagami T. Development, differentiation and manipulation of chicken germ cells. Develop Growth Differ. 2013;55(1):20–40. doi:10.1111/dgd.12026.

Nishio S, Kohno Y, Iwata Y, Arai M, Okumura H, Oshima K, Nadano D, Matsuda T. Glycosylated chicken ZP2 accumulates in the egg coat of immature oocytes and remains localized to the germinal disc region of mature eggs. Biol Reprod. 2014;91(5):107. doi:10.1095/biolreprod.114.119826.

Ohtsuki M, Hanafy AM, Mori M, Sasanami T. Involvement of interaction of ZP1 and ZPC in the formation of quail perivitelline membrane. Cell Tissue Res. 2004;318(3):565–70. doi:10.1007/s00441-004-1000-9.

Okumura H, Aoki N, Sato C, Nadano D, Matsuda T. Heterocomplex formation and cell-surface accumulation of hen's serum zona pellucida B1 (ZPB1) with ZPC expressed by a mammalian cell line (COS-7): a possible initiating step of egg-envelope matrix construction. Biol Reprod. 2007a;76(1):9–18. doi:10.1095/biolreprod.106.056267.

Okumura H, Fukushima H, Momoda M, Ima Y, Matsuda T, Ujita M. Diverse lectin-binding specificity of four ZP3 glycoprotein isoforms with a discrete isoelectric point in chicken egg coat. Biochem Biophys Res Commun. 2012;424(3):586–92. doi:10.1016/j.bbrc.2012.06.157.

Okumura H, Kohno Y, Iwata Y, Mori H, Aoki N, Sato C, Kitajima K, Nadano D, Matsuda T. A newly identified zona pellucida glycoprotein, ZPD, and dimeric ZP1 of chicken egg envelope are involved in sperm activation on sperm–egg interaction. Biochem J. 2004;384(Pt 1):191–9. doi:10.1042/BJ20040299.

Okumura H, Okajima T, Nadano D, Matsuda T. Association of chicken zona pellucida glycoprotein (ZP) B1 with ZPC induces formation of ZPB1–ZPC fibrous aggregates containing disulfide-bridged ZPB1 dimer. Biochem Biophys Res Commun. 2007b;364(3):682–8. doi:10.1016/j.bbrc.2007.10.072.

Okumura H, Sato T, Sakuma R, Fukushima H, Matsuda T, Ujita M. Identification of distinctive interdomain interactions among ZP-N, ZP-C and other domains of zona pellucida glycoproteins underlying association of chicken egg-coat matrix. FEBS Open Bio. 2015;5:454–65. doi:10.1016/j.fob.2015.05.005.

Pan J, Sasanami T, Kono Y, Matsuda T, Mori M. Effects of testosterone on production of perivitelline membrane glycoprotein ZPC by granulosa cells of Japanese quail (*Coturnix japonica*). Biol Reprod. 2001;64(1):310–6.

Pan J, Sasanami T, Nakajima S, Kido S, Doi Y, Mori M. Characterization of progressive changes in ZPC of the vitelline membrane of quail oocyte following oviductal transport. Mol Reprod Dev. 2000;55(2):175–81. doi:10.1002/(SICI)1098-2795(200002)55:2<175::AID-MRD6>3.0.CO;2-6.

Perry MM. Nuclear events from fertilisation to the early cleavage stages in the domestic fowl (Gallus domesticus). J Anat. 1987;150:99–109.

Perry MM, Gilbert AB, Evans AJ. The structure of the germinal disc region of the hen's ovarian follicle during the rapid growth phase. J Anat. 1978;127(Pt 2):379–92.

Petrie M, Schwabl H, Brande-Lavridsen N, Burke T. Maternal investment. Sex differences in avian yolk hormone levels. Nature. 2001;412(6846):498–9. doi:10.1038/35087652.

Pike TW, Petrie M. Experimental evidence that corticosterone affects offspring sex ratios in quail. Proc Biol Sci. 2006;273(1590):1093–8. doi:10.1098/rspb.2005.3422.

Pökkylä RM, Lakkakorpi JT, Nuojua-Huttunen SH, Tapanainen JS. Sequence variations in human ZP genes as potential modifiers of zona pellucida architecture. Fertil Steril. 2011;95(8):2669–72. doi:10.1016/j.fertnstert.2011.01.168.

Rankin T, Dean J. The zona pellucida: using molecular genetics to study the mammalian egg coat. Rev Reprod. 2000;5(2):114–21.

Rankin T, Familari M, Lee E, Ginsberg A, Dwyer N, Blanchette-Mackie J, Drago J, Westphal H, Dean J. Mice homozygous for an insertional mutation in the Zp3 gene lack a zona pellucida and are infertile. Development. 1996;122(9):2903–10.

Rankin T, Talbot P, Lee E, Dean J. Abnormal zonae pellucidae in mice lacking ZP1 result in early embryonic loss. Development. 1999;126(17):3847–55.

Rankin TL, O'Brien M, Lee E, Wigglesworth K, Eppig J, Dean J. Defective zonae pellucidae in Zp2-null mice disrupt folliculogenesis, fertility and development. Development. 2001;128(7):1119–26.

Ringuette MJ, Sobieski DA, Chamow SM, Dean J. Oocyte-specific gene expression: molecular characterization of a cDNA coding for ZP-3, the sperm receptor of the mouse zona pellucida. Proc Natl Acad Sci U S A. 1986;83(12):4341–5.

Rodler D, Sasanami T, Sinowatz F. Assembly of the inner perivitelline layer, a homolog of the mammalian zona pellucida: an immunohistochemical and ultrastructural study. Cells Tissues Organs. 2012;195(4):330–9. doi:10.1159/000327013.

Säemann MD, Weichhart T, Horl WH, Zlabinger GJ. Tamm–Horsfall protein: a multilayered defence molecule against urinary tract infection. Eur J Clin Investig. 2005;35(4):227–35. doi:10.1111/j.1365-2362.2005.01483.x.

Saling PM, Sowinski J, Storey BT. An ultrastructural study of epididymal mouse spermatozoa binding to zonae pellucidae in vitro: sequential relationship to the acrosome reaction. J Exp Zool. 1979;209(2):229–38. doi:10.1002/jez.1402090205.

Saling PM, Storey BT. Mouse gamete interactions during fertilization in vitro. Chlortetracycline as a fluorescent probe for the mouse sperm acrosome reaction. J Cell Biol. 1979;83(3):544–55.

Sano K, Kawaguchi M, Watanabe S, Nagakura Y, Hiraki T, Yasumasu S. Inferring the evolution of teleostean zp genes based on their sites of expression. J Exp Zool B Mol Dev Evol. 2013;320(5):332–43. doi:10.1002/jez.b.22507.

Sano K, Kawaguchi M, Yoshikawa M, Iuchi I, Yasumasu S. Evolution of the teleostean zona pellucida gene inferred from the egg envelope protein genes of the Japanese eel, *Anguilla japonica*. FEBS J. 2010;277(22):4674–84. doi:10.1111/j.1742-4658.2010.07874.x.

Santambrogio S, Cattaneo A, Bernascone I, Schwend T, Jovine L, Bachi A, Rampoldi L. Urinary uromodulin carries an intact ZP domain generated by a conserved C-terminal proteolytic cleavage. Biochem Biophys Res Commun. 2008;370(3):410–3. doi:10.1016/j.bbrc.2008.03.099.

Sasanami T, Atsumi E, Toriyama M, Mori M. Asparagine-linked oligosaccharide-independent secretion of egg envelope glycoprotein ZPC of the Japanese quail (*Coturnix japonica*). Comp Biochem Physiol A Mol Integr Physiol. 2003a;134(3):631–8.

Sasanami T, Matsushima K, Ohtsuki M, Kansaku N, Hiyama G, Mori M. Vectorial secretion of perivitelline membrane glycoprotein ZPC of Japanese quail (*Coturnix japonica*) in polarized Madin–Darby canine kidney cells. Cells Tissues Organs. 2005;180(3):169–77. doi:10.1159/000088245.

Sasanami T, Murata T, Ohtsuki M, Matsushima K, Hiyama G, Kansaku N, Mori M. Induction of sperm acrosome reaction by perivitelline membrane glycoprotein ZP1 in Japanese quail (*Coturnix japonica*). Reproduction. 2007;133(1):41–9. doi:10.1530/REP-06-0104.

Sasanami T, Pan J, Doi Y, Hisada M, Kohsaka T, Toriyama M. Secretion of egg envelope protein ZPC after C-terminal proteolytic processing in quail granulosa cells. Eur J Biochem/FEBS. 2002;269(8):2223–31.

Sasanami T, Pan J, Mori M. Expression of perivitelline membrane glycoprotein ZP1 in the liver of Japanese quail (*Coturnix japonica*) after in vivo treatment with diethylstilbestrol. J Steroid Biochem Mol Biol. 2003b;84(1):109–16.

Sasanami T, Toriyama M, Mori M. Carboxy-terminal proteolytic processing at a consensus furin cleavage site is a prerequisite event for quail ZPC secretion. Biol Reprod. 2003c;68(5):1613–9. doi:10.1095/biolreprod.102.011841.

Sato T, Kinoshita M, Kansaku N, Tahara K, Tsukada A, Ono H, Yoshimura T, Dohra H, Sasanami T. Molecular characterization of egg envelope glycoprotein ZPD in the ovary of Japanese quail (*Coturnix japonica*). Reproduction. 2009;137(2):333–43. doi:10.1530/REP-08-0057.

Schneider WJ. Low density lipoprotein receptor relatives in chicken ovarian follicle and oocyte development. Cytogenet Genome Res. 2007;117(1–4):248–55. doi:10.1159/000103186.

Serizawa M, Kinoshita M, Rodler D, Tsukada A, Ono H, Yoshimura T, Kansaku N, Sasanami T. Oocytic expression of zona pellucida protein ZP4 in Japanese quail (*Coturnix japonica*). Anim Sci J. 2011;82(2):227–35. doi:10.1111/j.1740-0929.2010.00830.x.

Sousa M, Teixeira da Silva J, Silva J, Cunha M, Viana P, Oliveira E, Sa R, Soares C, Oliveira C, Barros A. Embryological, clinical and ultrastructural study of human oocytes presenting indented zona pellucida. Zygote. 2015;23(1):145–57. doi:10.1017/S0967199413000403.

Spargo SC, Hope RM. Evolution and nomenclature of the zona pellucida gene family. Biol Reprod. 2003;68(2):358–62.

Stetson I, Aviles M, Moros C, Garcia-Vazquez FA, Gimeno L, Torrecillas A, Aliaga C, Bernardo-Pisa MV, Ballesta J, Izquierdo-Rico MJ. Four glycoproteins are expressed in the cat zona pellucida. Theriogenology. 2015;83(7):1162–73. doi:10.1016/j.theriogenology.2014.12.019.

Stewart SG, Bausek N, Wohlrab F, Schneider WJ, Janet Horrocks A, Wishart GJ. Species specificity in avian sperm:perivitelline interaction. Comp Biochem Physiol A Mol Integr Physiol. 2004;137(4):657–63. doi:10.1016/j.cbpb.2004.01.027.

Stifani S, Barber DL, Nimpf J, Schneider WJ. A single chicken oocyte plasma membrane protein mediates uptake of very low density lipoprotein and vitellogenin. Proc Natl Acad Sci U S A. 1990;87(5):1955–9.

Takeuchi Y, Nishimura K, Aoki N, Adachi T, Sato C, Kitajima K, Matsuda T. A 42-kDa glycoprotein from chicken egg-envelope, an avian homolog of the ZPC family glycoproteins in mammalian Zona pellucida. Its first identification, cDNA cloning and granulosa cell-specific expression. Eur J Biochem/FEBS. 1999;260(3):736–42.

Tanghe S, Van Soom A, Nauwynck H, Coryn M, de Kruif A. Minireview: functions of the cumulus oophorus during oocyte maturation, ovulation, and fertilization. Mol Reprod Dev. 2002;61(3):414–24. doi:10.1002/mrd.10102.

Uto N, Yoshimatsu N, Lopata A, Yanagimachi R. Zona-induced acrosome reaction of hamster spermatozoa. J Exp Zool. 1988;248(1):113–20. doi:10.1002/jez.1402480115.

Visconti PE, Florman HM. Mechanisms of sperm–egg interactions: between sugars and broken bonds. Sci Signal. 2010;3(142):pe35. doi:10.1126/scisignal.3142pe35.

Waclawek M, Foisner R, Nimpf J, Schneider WJ. The chicken homologue of zona pellucida protein-3 is synthesized by granulosa cells. Biol Reprod. 1998;59(5):1230–9.

Wassarman PM. Zona pellucida glycoproteins. Annu Rev Biochem. 1988;57:415–42. doi:10.1146/annurev.bi.57.070188.002215.

Williams Z, Litscher ES, Jovine L, Wassarman PM. Polypeptide encoded by mouse ZP3 exon-7 is necessary and sufficient for binding of mouse sperm in vitro. J Cell Physiol. 2006;207(1):30–9. doi:10.1002/jcp.20532.

Wyburn GM, Aitken RN, Johnston HS. The ultrastructure of the zona radiata of the ovarian follicle of the domestic fowl. J Anat. 1965;99(Pt 3):469–84.

Yamaguchi R, Yamagata K, Ikawa M, Moss SB, Okabe M. Aberrant distribution of ADAM3 in sperm from both angiotensin-converting enzyme (Ace)- and calmegin (Clgn)-deficient mice. Biol Reprod. 2006;75(5):760–6. doi:10.1095/biolreprod.106.052977.

Yanagimachi R. Mammalian sperm acrosome reaction: where does it begin before fertilization? Biol Reprod. 2011;85(1):4–5. doi:10.1095/biolreprod.111.092601.

Fertilization 1: Sperm–Egg Interaction

6

Shunsuke Nishio and Tsukasa Matsuda

Abstract

In birds in the reproductive season, an egg is ovulated without cumulus cells from the largest follicle with the highest hierarchy in the ovary. The outermost part of the ovulated eggs is the perivitelline layer, a glycoprotein matrix consisting of a few ZP-glycoproteins. The fertilization starts from sperm penetration of the perivitelline layer predominantly in the germinal disc region, followed by uptake of the sperm into the egg, and goes through by the fusion of sperm male pronucleus with the female pronucleus in the egg. A series of these fertilization steps occurs in the infundibulum of the oviduct within a short period after ovulation. Some pioneering microstructural studies using electron microscopy and supporting biochemical data from later studies indicate that, in avian fertilization, sperm interacts with the perivitelline layer covering the germinal disc, locally degrade and dissolve the matrix of the perivitelline layer, and penetrate it through the hole made proteolytically at the sperm-binding site on the perivitelline layer. Several molecules and structures presumably involved in the sperm–perivitelline interaction have been characterized, especially sperm proteases and their targets in the egg perivitelline layer. On the other hand, no molecules involved in the sperm–egg membrane fusion for the male pronucleus uptake into the egg have yet been identified or characterized and, moreover, no orthologue but one have been annotated so far in the chicken genome for the mouse genes involved in the sperm–egg membrane fusion.

Keywords

Stigma • Perivitelline layer • Granulosa cell • Infundibulum • Acrosome reaction • Acrosin • Germinal disc • Membrane fusion

S. Nishio • T. Matsuda (✉)
Department of Applied Molecular Biosciences, Graduate School of Bioagricultural Sciences, Nagoya University, Nagoya 464-8601, Japan
e-mail: tmatsuda@agr.nagoya-u.ac.jp

T. Sasanami (ed.), *Avian Reproduction*, Advances in Experimental Medicine and Biology 1001,
DOI 10.1007/978-981-10-3975-1_6

6.1 Encounter of Ovulated Egg with Sperm and Subsequent Egg Movement in the Oviduct

The egg is ovulated from the follicle after the maturation of oocyte in the ovary. In the case of chicken, a typical domestic fowl, shortly after the laying of an egg, the next maturated egg is ovulated from the largest follicle with the highest hierarchy in the ovary (Gilbert 1979). A schematic drawing for the ovulation and the site of fertilization in chicken is shown in Fig. 6.1. The follicular capsule surrounding the egg ruptures along the specific non-vascularised region of the capsule, termed stigma, on the opposite site of the follicular pedicle. The egg, filled with yolk, is pushed out from the follicular capsule instantaneously. On this ovulation occasion, the perivitelline layer (*zona pellucida* of mammalian egg) surrounding the egg is detached from the granulosa cell layer adhering to the basal lamina of the follicular capsule tissue, resulting in the release of the egg with perivitelline layer and without granulosa cells (Romanoff 1960). In birds, therefore, ovulated eggs are surrounded only with the perivitelline layer, and the sperm directly encounter the perivitelline layer in the oviduct. This is a difference from mammalian eggs, which are released from the ovary together with cumulus cells surrounding the egg as cumulus oophorus, which is still present at the time of interaction with sperm (Tanghe et al. 2002). The absence of bird egg accessory cells equivalent to the mammalian cumulus oophorus may be relevant to the development of fertilized egg and embryo without implantation in uterus in birds.

The ovulated egg is released into a body cavity adjacent to the ovary and, subsequently, clasped and captured within a short time by the funnel-like open end, termed infundibulum, at the free end of the oviduct tube (Fig. 6.1) (Johnson 1986). This consecutive, rapid, and smooth process from ovulation to the capture by oviduct is considered to play a role in protection of the very large and mechanically weak egg from undesirable rupture in the intraperitoneal space of hens. The egg, captured into the infundibulum, encounters sperm migrating upstream from the distal oviduct and the fertilization takes place in the infundibulum lumen.

The sperm are transported to the upper end of the oviduct, the place of fertilization, through the very long oviduct tube of bird. In laying hens, for example, the length of oviduct is no less than 60–80 cm: the infundibulum (~7 cm), the magnum secreting albumen (~34 cm), the isthmus (8 cm), the uterus (~8 cm), and the vagina (~8 cm) (Fig. 6.1) (Romanoff and Romanoff 1949). Therefore, the sperm have to migrate a long distance from the vagina to the infundibulum for fertilization. In some old studies, such migration of the ejaculated sperm through the hens' oviduct was once reported to proceed with high speed and be completed in less than an hour (chicken) (Martin and Anderson 1918; Fronda 1926; Mimura 1939) or 5 h (pheasant) (Ivanoff 1913). However, mechanisms responsible for such high-speed migration of sperm in the bird oviduct remain to be investigated.The fertilization starts from sperm binding to and sperm penetration of the perivitelline layer, followed by uptake of the sperm into the egg, and goes through by the fusion of sperm male pronucleus with the female pronucleus in the egg. Such fertilization in the infundibulum completes within a short period, e.g., about 15 min in chicken (Olsen 1942, 1952) and turkey (Olsen and Fraps 1944) after the ovulation. After or concurrently with the fertilization, the egg is surrounded in sequence with two layers, termed the continuous layer and the extravitelline layer (outer layer of the vitelline membrane) (Bellairs et al. 1963), which consists of proteins secreted from the infundibulum epithelium. The covering of the perivitelline layer with the protein layers results in the physical blocking of further sperm-binding to the perivitelline layer. The fertilized egg is moved downstream in the oviduct by oviduct peristalsis, with further surrounding with egg white, shell membrane, and egg shell, and is laid through vagina and cloaca (Fig. 6.1) (Johnson 1986).

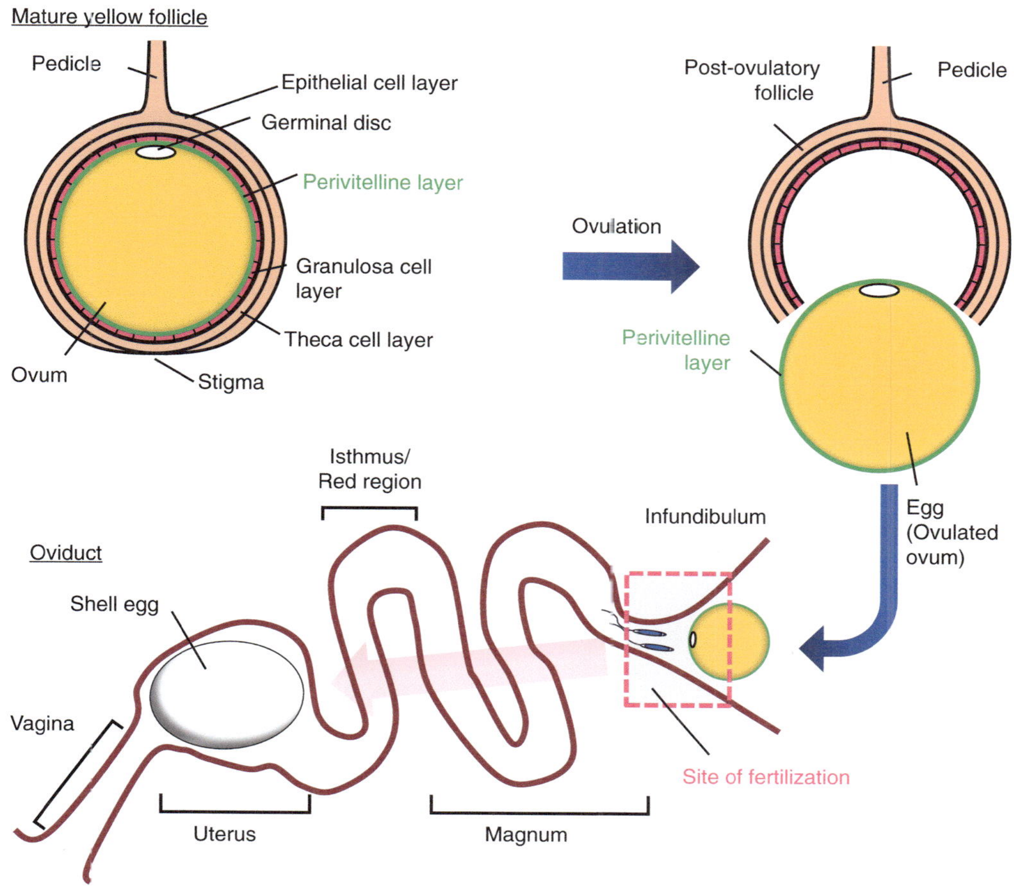

Fig. 6.1 The mature follicle, the ovulated ovum, and the ovum with sperm at the oviduct. The ovum in the preovulatory follicle is surrounded in order with the protein complex perivitelline layer, the granulosa cell layer, theca cell layer, and follicular epithelial cell layer. The follicular capsule consisting of the three cell layers ruptures along the stigma, and the ovum surrounded with the perivitelline layer is pushed out. Immediately after the ovulation, the ovum is clasped and captured by the infundibulum of oviduct. The ovum encounters sperm transported through the long oviduct. The ovulation occurs shortly after laying the previous egg and, therefore, this drawing is not the case of physiological conditions. An egg is drawn to show the shell egg in the uterus before the oviposition

6.2 Sperm Interaction with the Perivitelline Layer and the Sperm Acrosome Reaction

Direct observation and analysis of sperm interaction with the egg perivitelline layer is not easy, because such a series of sperm–perivitelline interactions occurs during a short period after ovulation and before coating with outer layer of the vitelline membrane, and because the sperm acrosome reaction and subsequent proteolytic degradation of the perivitelline layer are induced immediately after sperm–perivitelline interaction. The candidate molecules and structures for the interaction between sperm and perivitelline layer have been suggested mainly based on morphological and/or biochemical analysis of the inhibitory activity to the perivitelline layer degradation after sperm-perivitelline interaction. The sperm and egg of chicken are illustrated with the molecules presumably involved in the sperm acrosome reaction (Fig. 6.2).

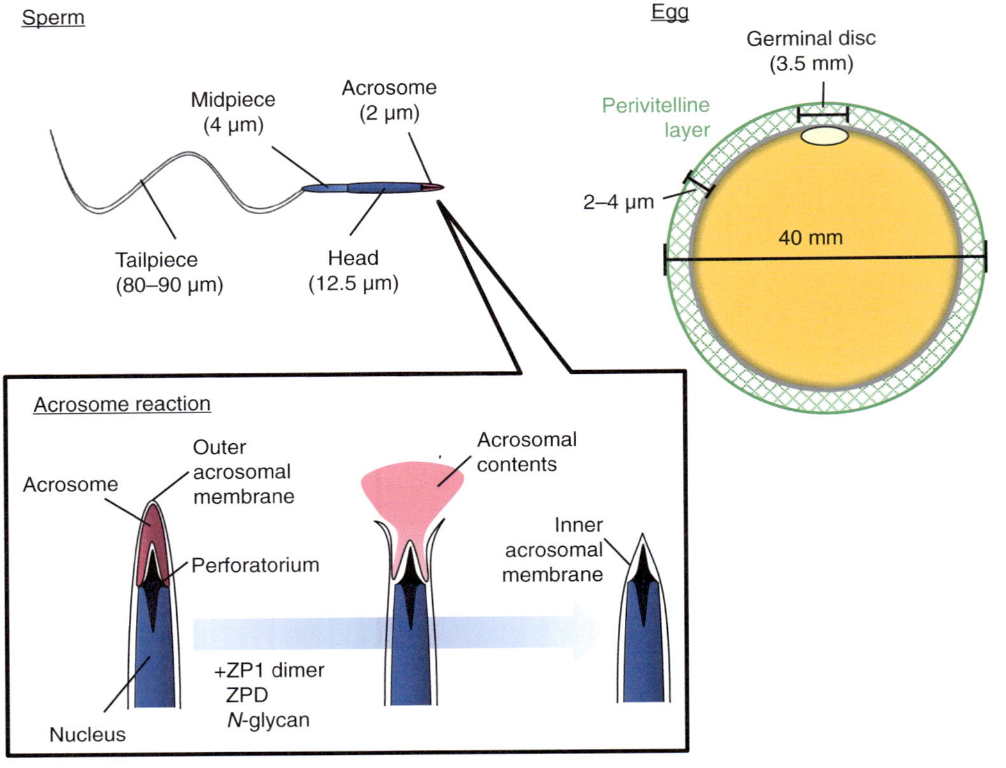

Fig. 6.2 Sperm, its acrosome reaction, and an ovulated egg of chicken. When the tip of sperm head interacts with *N*-glycan, ZP1, or ZPD of the egg perivitelline layer, the sperm acrosome reaction may presumably induced to release the acrosome contents including proteases for the local degradation of the perivitelline layer. The scheme of sperm acrosome reaction is drawn based on the electron microscopic images of chicken sperm reported by Okamura and Nishiyama (1978a)

The solubilized protein preparation, consisting of several ZP glycoproteins from the perivitelline layer, inhibits morphological changes, i.e., formation of holes for sperm penetration (Howarth 1990; Bramwell and Howarth 1992), suggesting that one or more of the ZP glycoproteins is the candidate of initial binding of sperm. When Japanese quail sperm was incubated with perivitelline pieces prepared from matured oocytes in the presence of trypsin inhibitors, proteolytic degradation of the perivitelline layer was inhibited and, thereby, the sperm retaining on the surface of intact perivitelline layer matrix can be observed under a microscope (Kuroki and Mori 1997). Pre-incubation of chicken perivitelline layer with either one of two monoclonal antibodies specific to ZP1 markedly inhibit the sperm-driven degradation of the perivitelline layer (Takeuchi et al. 2001), and the antiserum directed against chicken ZP1 also inhibits sperm-binding to the perivitelline layer (Bausek et al. 2004). On the other hand, a monoclonal antibody specific for the sperm protease, identified to be the bird homologue of mammalian acrosin, strongly inhibited sperm interaction with the perivitelline layer of Japanese quail (Sasanami et al. 2011).

Chemical removal of carbohydrate moiety from ZP glycoproteins diminishes the inhibitory activity remarkably (Howarth 1992). Such involvement of carbohydrate chains attached to the perivitelline components was also indicated by some more detailed analysis using lectins, sugars and deglycosylation enzymes (Robertson et al. 2000). Pretreatment of the perivitelline layer with wheat germ agglutinin (WGA, the *N*-acetyl-D-glucosamine specific lectin), but not several other lectins, inhibits sperm-binding to the perivitelline layer, and enzyme treatment with the Asn-linked carbohydrate chains (*N*-glycan)-specific glycosidase also inhibit the sperm-binding significantly. These biochemical data suggest that carbohydrate chain(s) of the ZP-glycoprotein(s) contribute, at least in part, to the sperm–egg interaction at the initial stages of fertilization in birds (Fig. 6.2).

Direct observation of the binding between protein components from the perivitelline layer and sperm have been shown in some previous studies. ZP3 prepared from chicken perivitelline layer binds to a ~180-kDa protein complex in chicken sperm extracts (Bausek et al. 2004). The recombinant chicken ZP3, but not the ZP3 mutant without the Ser/Thr-linked carbohydrate chains (*O*-glycans), binds to the tip of sperm head as observed by immuno-fluorescence microscopy (Han et al. 2010). A 45-kDa sperm protease, acrosin, in the membrane fraction of Japanese quail sperm binds to perivitelline layer components solubilized and immobilized onto PVDF membrane (Sasanami et al. 2011).

The sperm–perivitelline layer interaction has also been analyzed by evaluating the induction ability of acrosome reaction. Electron microscopic studies on the fertilized egg in the infundibulum indicate that the contact or approach of sperm acrosomal region to the egg perivitelline layer triggers acrosome reaction, i.e., fusion of the outer acrosomal membrane with the overlaying plasma membrane, opening of the acrosome apical region, and subsequent release of the acrosomal contents (Okamura and Nishiyama 1978a) (Fig. 6.2). The incubation of sperm with the suspension of isolated perivitelline layer also results in an increase in the number of sperm without the acrosome cap, and the release of proteolytic activity, suggesting that the perivitelline layer promotes the acrosome reaction (Koyanagi et al. 1988). The *N*-glycans with terminal *N*-acetyl-D-glucosamine prepared enzymatically from chicken perivitelline layer, but not *O*-glycans, has been reported to induce sperm acrosome reaction (Horrocks et al. 2000). Later, it was suggested that ZP1 and ZPD prepared from chicken perivitelline layer induced sperm acrosome reaction (Okumura et al. 2004) and, moreover, ZP1 purified from Japanese quail perivitelline layer, but not the enzymatically *N*-glycan-deglycosylated ZP1, has been reported to induce sperm acrosome reaction (Sasanami et al. 2007). Also, in mammalian fertilization, some zona pellucida components have previously reported to induce acrosome reaction in vitro, though the involved ZP-glycoprotein component(s) are not necessarily identical to those in bird fertilization. Mouse ZP3, both purified (Bleil and Wassarman 1983) and recombinant (Beebe et al. 1992) proteins, induced acrosome reaction, while human ZP1 and ZP4 in addition to ZP3 also did it, and glycosylation of the ZP proteins was required for the induction of acrosome reaction (Chakravarty et al. 2005, 2008; Bansal et al. 2009; Ganguly et al. 2010). In addition to these zona pellucida components, yet unidentified factors in cumulus oophorus surrounding the egg (Jin et al. 2011) and in the upper isthmus of the oviduct (La Spina et al. 2016) also induce acrosome reaction of mouse sperm. It is still unknown whether sperm of avian species undergo acrosome reaction in the oviduct and before encountering the perivitelline layer of the ovulated egg.

6.3 Perivitelline Layer Degradation and Sperm Penetration

Several earlier morphological and microstructural studies using intact egg demonstrate that bird sperm penetrate the egg perivitelline layer, with binding to and proteolytic degradation/dissolution of the perivitelline layer matrix (Fig. 6.3). In the first study, on in-vitro incubation of naturally ovulated egg with ejaculated sperm, the perivitelline layer, as well as egg plasma membrane, was ruptured by the incubation with sperm, and pre-treatment of sperm with a synthetic trypsin inhibitor (TLCK) completely inhibited egg rupture during incubation with sperm (Howarth and Digby 1973). Another study using artificially ovulated egg from excised follicles and partially purified sperm proteases, termed acrosin, showed that sperm protease, as

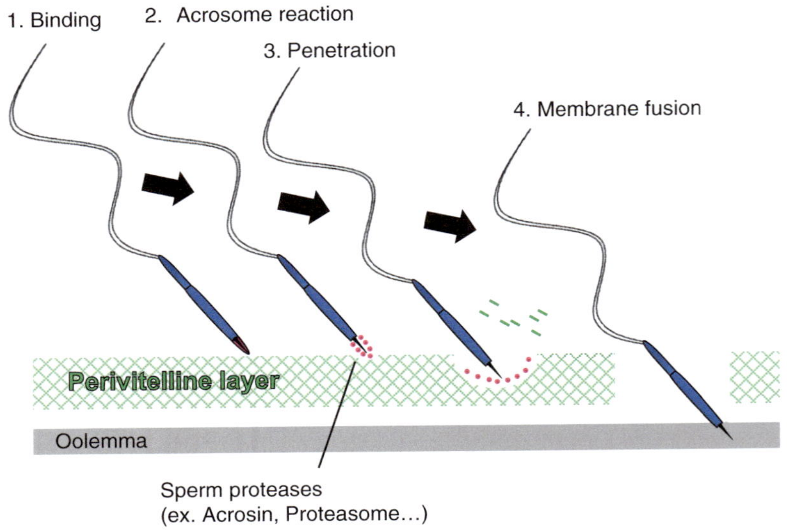

Fig. 6.3 Sperm binding to and penetration of the egg perivitelline layer with proteolytic degradation/dissolution of the perivitelline layer matrix. A series of sperm–perivitelline interactions presumably occur in the infundibulum of oviduct during a short period of time (~15 min in chicken) after ovulation. The scheme of sperm–perivitelline interaction is drawn based on the electron microscopic images of chicken sperm penetration of the perivitelline layer observed in vivo (Okamura et al. 1978a) and in vitro (Takeuchi et al. 2001)

well as some other endopeptidases including trypsin, hydrolyzed and dissolved the perivitelline layer surrounding the egg. The optimum pH for hydrolysis was 7~8, and soybean trypsin inhibitor (SBTI) inhibited perivitelline layer dissolution completely (Ho and Meizel 1975).

A microstructural study using electron microscopy clearly demonstrated the process of passage of sperm through the perivitelline layer (Okamura and Nishiyama 1978a). Within a few days of insemination, the eggs were removed at several different post-ovulation stages of movement from the anterior and posterior region of infundibulum and from the border region between the infundibulum and magnum of laying hens. The extravitelline layer (outer layer of the vitelline membrane) surrounding the perivitelline layer is observed more clearly in the eggs removed from the more posterior infundibulum, and sperm with intact acrosome in nearly parallel orientation with its perivitelline layer (inner layer) are detected in the growing extravitelline layer. In the electronmicroscopic images, the sperm in contact with the perivitelline layer induces acrosome reaction, and the network of the perivitelline layer around the tip of acrosome-reacted sperm appears to be dissolved. These microscopic images also show that sperm passing through the perivitelline layer diagonally through the hole created by dissolution in the region of the perivitelline layer (Fig. 6.3).

Perivitelline dissolution and resultant hole-formation in the perivitelline layer are reproducible in in-vitro experiments using isolated perivitelline layer and ejaculated sperm. The sperm-dependent hole formation on the perivitelline layer was inhibited completely by trypsin-specific inhibitors, suggesting that trypsin-like sperm protease is involved in perivitelline dissolution (Howarth and Digby 1973; Ho and Meizel 1975; Kuroki and Mori 1997; Takeuchi et al. 2001) Afterwards, addition of a proteasome inhibitor, MG132, or enzymatic depletion of extracellular ATP was shown to inhibit the sperm-dependent hole formation in the perivitelline layer of Japanese quail, indicating that sperm proteasome also plays a role in sperm penetration of the perivitelline layer (Sasanami et al. 2012). On the other hand, two monoclonal antibodies specific for ZP1, a component of the perivitelline layer, inhibit the degradation of not only ZP1 but also another component ZP3 during the in-vitro incubation of the perivitelline layer with sperm (Takeuchi et al. 2001). This result suggests that the ZP1 degradation by sperm protease(s) may trigger perivitelline layer dissolution, leading to the hole formation. Thus, in avian fertilization, it has clearly been demonstrated that sperm locally degrade and dissolve the matrix of the perivitelline layer and penetrate it through the hole made at the sperm-binding site on the perivitelline layer (Fig. 6.3).

A comparative study of the in-vitro sperm penetration of the egg perivitelline layer in various bird species showed that chicken sperm could interact and penetrate the perivitelline layer of not only chicken but also many other species, though the penetration efficiency to the perivitelline layer in some other species was lower that a half of that in chicken itself (Stewart et al. 2004). Chicken sperm almost equally penetrated the perivitelline layer of several species in Galliformes (chicken, turkey, quail, pheasant, peafowl and guinea fowl), whereas the efficiency of sperm penetration of perivitelline layer of some other species in Anseriformes (goose and duck) was less than half that of the perivitelline layer of chicken itself. Further lower penetration efficiency of less than 30% was observed for the egg perivitelline layer from some species in Passeriformes (Zebra Finch) and Columbiformes (collared-dove). Thus, such a wide spectrum of chicken sperm not only within but also outside Galliformes suggests that there is not necessarily a stringent species-specific recognition of egg perivitelline layer interaction in birds.

Degradation of zona pellucida by sperm proteases has been reported in several earlier studies on porcine fertilization. Boar sperm protease(s) degraded the zona pellucida in vitro by limited but specific proteolysis of ZP component glycoproteins (Urch et al. 1985a), and the limited proteolysis resulted only in the production of truncated glycoproteins, suggesting a small contribution of sperm protease(s) to sperm penetration of *zona pellucida* (Urch et al. 1985b). The other microstructural studies on porcine zona pellucida matrix indicated that the incubation of egg with sperm protease(s) did not induce dissolution and morphological changes of the zona pellucida matrix (Dunbar et al. 1985; Brown and Cheng 1985). On the other hand, a previous study on rabbit sperm demonstrated zona pellucida dissolution during the incubation of eggs with sperm trypsin-like protease(s) (Stambaugh and Buckley 1968). Thus, proteolytic degradation of zona pellucida matrix for sperm penetration appears different among species in mammals. By contrast, as described above, local dissolution (hole formation) of the perivitelline layer from sperm penetration seems common to avian species, Galliformes (chicken, turkey, quail, pheasant, peafowl and guinea fowl), Anseriformes (goose and duck), Passeriformes (Zebra Finch) and Columbiformes (collared-dove) (Stewart et al. 2004).

6.4 Sperm Proteases Involved in the Sperm Penetration of the Perivitelline Layer

The trypsin-like protease activity of bird sperm was first reported in 1956 (Buruiana 1956). Later, the biochemical properties of chicken sperm protease were characterized using native polyacrylamide gel (8%) electrophoresis of chicken sperm extract followed by activity staining using synthetic trypsin substrate, Arg and Lys derivatives, in combination with color-producing coupling-reaction (Ho and Meizel 1970). Three bands with trypsin-like activity were detected for chicken sperm extract, and the protease activity of the three bands was inhibited by treatment with soybean trypsin inhibitor. After these pioneer studies, sperm proteases of bird species have long waited to be investigated. Using zymography, with gelatin as a protein substrate, several forms of active proteases are detected clearly in the sperm extract of chicken, whereas there is a smaller number of activity bands in those of porcine and rat (Nishio and Matsuda, unpublished results) (Fig. 6.4).

Purification, characterization, and cDNA cloning of sperm protease were first reported in turkey sperm (Słowińska et al. 2010; Słowińska and Ciereszko 2012). The biochemical character-

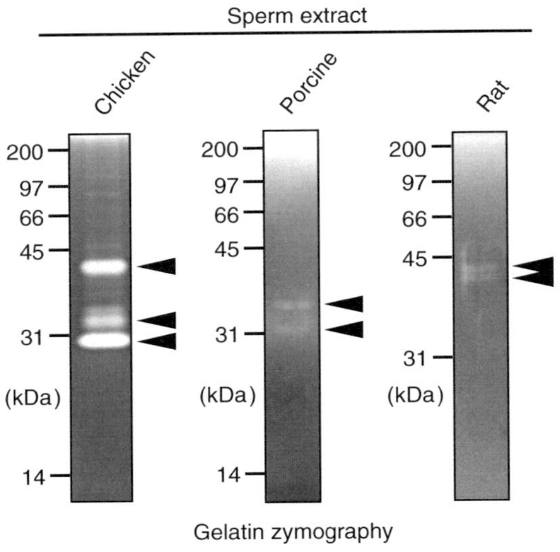

Fig. 6.4 Proteases detected in the extract from chicken, porcine, and rat sperm as analyzed by zymography. The zymography analysis of the sperm extracts was done in the SDS-PAGE gel, containing gelatin as a protein substrate (Nishio and Matsuda, unpublished results). The transparent bands (*white ones in the image*) indicate the proteins with proteolytic activity against the gelatin as a substrate

ization as well as the deduced amino acid sequence of the sperm protease suggested that the turkey sperm protease, named acrosin, is a glycoprotein synthesized as a pre-pro-protein, secreted and stored as pro-form within acrosome, and processed proteolytically into the mature and active form consisting of two polypeptide fragments, light and heavy chains. In addition to the turkey acrosin, cDNA sequences homologous to turkey acrosin are registered on Japanese quail (ID: XM_015878977) and chicken (ID: AB769484, AB769485) in databases. Moreover, genomic DNA sequences predicted to be the acrosin gene are also found in the databases for a variety of bird species.

6.5 Predominant Sperm Penetration Around a Germinal Disc

A small whitish-yellow area (about 3 mm in diameter in chicken eggs) visibly distinguishable on the surface of an ovulated egg is the germinal disc, where female pronucleus is situated underneath with cytoplasmic inclusions (Fig. 6.5). Scanning and transmission electron microscopy showed that during the incubation with sperm, the fibrous meshwork of the chicken egg perivitelline layer was broken down more considerably around the region on the germinal disc (Bakst and Howarth 1977). Later, the number of holes formed presumably by sperm penetration was counted under a microscope after staining the carbohydrate moieties of glycoproteins with a chemical reagent (Bramwell and Howarth 1992) or fluorescent-labeled lectin (Steele et al. 1994). More holes were observed on the perivitelline layer over the germinal disc area of in-vivo fertilized eggs (Steele 1994). Moreover, a comparative study using the eggs of 27 bird species showed that in most species, the holes were observed around the germinal disc at higher density than in other regions of the egg (Birkhead et al. 1994). On the other hand, when perivitelline layer fragments prepared from a freshly ovulated egg were incubated with sperm in vitro, the holes were observed with only a slightly more (Bramwell and Howarth 1992) or similar (Steele et al. 1994) frequency in the germinal disc region as compared with other areas of the egg.

Another study using fertilized and oviposited chicken eggs from naturally mated hens showed more quantitative data on the preferential sperm penetration of the perivitelline layer overlying the germinal disc region (Bramwell et al. 1995). Sperm penetration was analyzed for 41 eggs laid over a 3-day period from three hens, and the mean numbers of sperm penetrating the germinal disc region and non-germinal disc region were 162.8 and 8.4 respectively. In this experiment using naturally mated hens, the sperm penetration of the perivitelline layer in the germinal disc region positively correlated with fertility of the eggs ($r = 0.89$, $P < 0.001$). Another quantitative analysis of sperm penetration of the perivitelline layer around the germinal disc was reported in chicken and turkey (Wishart 1997). The sperm penetration as estimated from hole-formation is evenly distributed in the perivitelline layer of most regions of the egg. Exceptionally in the perivitelline layer over the germinal disc region,

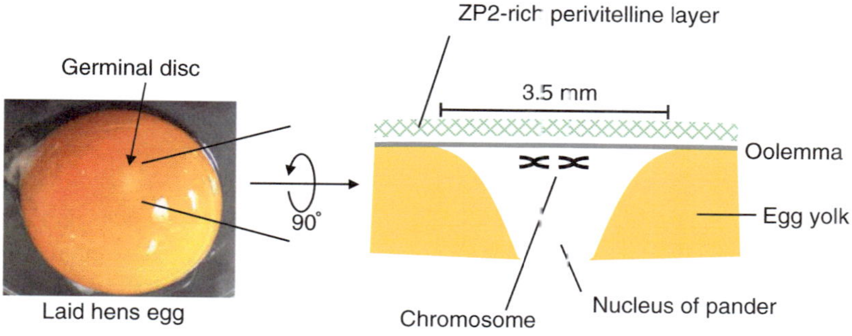

Fig. 6.5 The germinal disc of bird eggs: a typical picture of an egg with egg white and a schematic drawing of the cross-section of a germinal disc. The germinal disc of an oocyte or an egg is located at the region of animal pole, which is formed during the oocyte development at the site of follicular pedicle (see Fig. 6.1). It is the small whitish-yellow area consisting of cytoplasm without yolk derived from the developing primary oocyte. The female pronucleus is situated underneath the egg plasma membrane of the germinal disc. A heavily *N*-glycosylated ZP-glycoprotein, ZP2, is rich in the perivitelline layer covering the germinal disc region of an egg

however, sperm penetration was detected about 25 times more than that in other regions of the perivitelline layer. Eggs with about three sperm that had penetrated the perivitelline layer over the germinal disc region showed a 50% probability of subsequent fertilization, while those with penetration by more than six sperm in the germinal disc region showed maximum fertility.

In a previous study, probably due to the not very high detection sensitivity of protein analysis in those days, no difference in the protein components between the perivitelline layer regions over the germinal disc area and the other part of the egg was reported as analyzed by gel electrophoresis (Steele et al. 1994), suggesting that preferential binding and penetration of sperm around the germinal disc region may not associate with the perivitelline layer itself. Moreover, although three major glycoprotein components, ZP1, ZP2, and ZP3 (Bleil and Wassarman 1980), have been known to form the mammalian egg coat, *zona pellucida*, that is equivalent to the bird perivitelline layer, the major glycoprotein components of bird egg perivitelline layer have been reported on ZP1 [Bausek et al. 2000; Takeuchi et al. 2001 (Chicken); Sasanami et al. 2003 (Quail)] and ZP3 [Waclawek et al. 1998; Takeuchi et al. 1999 (Chicken); Pan et al. 2001 (Quail)], but not on ZP2, known as the second sperm receptor in mammals (Bleil et al. 1988). Afterwards, however, a glycoprotein component of the perivitelline layer matrix, ZP2, was shown to localize predominantly in the perivitelline layer over the germinal disc region, using biochemical and immuno-histological methods (Nishio et al. 2014) (Fig. 6.5). The germinal disc of a mature egg is derived from the cytoplasmic part of the developing primary oocyte and forms only a small part of cytosol at the animal pole of the further developing oocyte. Therefore, the perivitelline layer matrix over the germinal disc of a matured egg is likely to originate from the perivitelline layer of the immature primary oocyte. Thus, in chicken follicles, ZP2 is suggested to accumulate in the oocyte perivitelline layer of immature follicles at the early developmental stages, and remains localized at the germinal disc region as the oocyte is growing. Biological roles of ZP2 localized in the germinal disc region in the preferential penetration of sperm at the germinal disc region in bird fertilization remain to be investigated.

6.6 Absence of Bird Homologues of Mammalian Molecules Involved in Sperm–Egg Membrane Fusion

After penetrating the perivitelline layer and arriving in the perivitelline space of the germinal disc region, the acrosome-reacted sperm interacts and fuses with the egg to deliver its nuclear contents to the female pronucleus (Okamura and Nishiyama 1978b). The inner acrosomal membrane of acrosome-reacted sperm fuses with the vitelline membrane, and the male pronucleus is released, with its sub-acrosomal rod into the egg. Several sperm (~20 depending on bird species) among those, which penetrated into the perivitelline space of the germinal disc region, succeed in fusing with the egg plasma membrane. However, only one sperm among them can fertilize the egg. Thus, morphological and histological observation by electron microscopy clearly indicates that the plasma membrane fusion occurs between the acrosome-reacted sperm and the egg, in a way similar to that of mammals. However, as far as we know, no studies have been reported so far on the molecules involved in the sperm–egg fusion in bird fertilization. Moreover, searching in gene databases retrieved no sequences for bird homologues of mammalian genes, such as Izumo1 (Inoue et al. 2005) and Equatorin (Toshimori et al. 1998) of sperm and Juno (Bianchi et al. 2014) of egg, responsible for the mouse sperm–egg membrane fusion. Comparison among human-, mouse- and chicken-genomic regions containing Juno loci shows conserved synteny not only between mouse and human but also between birds and mammals, except for the Juno loci in the chicken genome (Fig. 6.6). In the chicken genome, no gene is annotated to the region between PANX1 and GPR83, where Juno is located in mouse and human genomes. The physical distance between PANX1 and GPR83 in the human or mouse genome is much longer than that of chicken (*Gallus gallus*) and, moreover, Juno is not present between PANX1 and GPR83 in the genome of Chinese soft-shell turtle (*Pelodiscus sinensis*). In mammals, DNA segments including Juno might be inserted between PANX1 and GRP83, possibly by gene duplication (Grayson 2015), after the separation of mammals from the other Amniotes including birds and turtles. Only

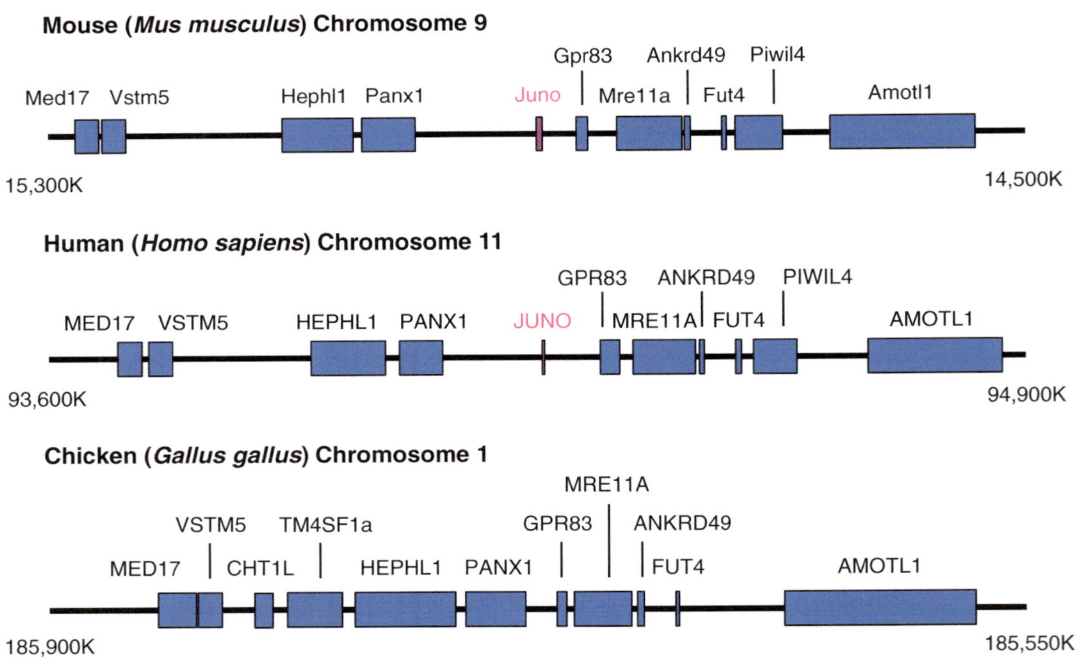

Fig. 6.6 Schematic drawing of several annotated genes around the locus for Juno in the mouse, human and chicken genomes. The genomic region around Juno appears to be syntenic among the three, though the segment size of for the syntenic genes rather different among the three, i.e., ~800 kb, ~1300 kb, and ~350 kb for mouse, human, and chicken respectively

one chicken gene (ID: AB032767) registered in the database is a homologue of mouse CD9, the egg plasma–membrane protein required for mammalian fertilization (Miyado et al. 2000).

References

Bakst MR, Howarth B Jr. Hydrolysis of the hen's perivitelline layer by cock sperm in vitro. Biol Reprod. 1977;17:370–9.

Bansal P, Chakrabarti K, Gupta SK. Functional activity of human ZP3 primary sperm receptor resides toward its C-terminus. Biol Reprod. 2009;81:7–15.

Bausek N, Waclawek M, Schneider WJ, Wohlrab F. The major chicken egg envelope protein ZP1 is different from ZPB and is synthesized in the liver. J Biol Chem. 2000;275:28866–72.

Bausek N, Ruckenbauer HH, Pfeifer S, Schneider WJ, Wohlrab F. Interaction of sperm with purified native chicken ZP1 and ZPC proteins. Biol Reprod. 2004;71:684–90.

Beebe SJ, Leyton L, Burks D, Ishikawa M, Fuerst T, Dean J, Saling P. Recombinant mouse ZP3 inhibits sperm binding and induces the acrosome reaction. Dev Biol. 1992;151:48–54.

Bellairs R, Harkness ML, Harkness RD. The vitelline membrane of the hen's egg: a chemical and ultrastructural study. J Ultrastruct Res. 1963;8:339–59.

Bianchi E, Doe B, Goulding D, Wright GJ. Juno is the egg Izumo receptor and is essential for mammalian fertilization. Nature. 2014;508:483–7.

Birkhead TR, Sheldon BC, Fletcher F. A comparative study of sperm–egg interactions in birds. J Reprod Fertil. 1994;101:353–61.

Bleil JD, Wassarman PM. Structure and function of the zona pellucida: identification and characterization of the proteins of the mouse oocyte's zona pellucida. Dev Biol. 1980;76:185–202.

Bleil JD, Wassarman PM. Sperm–egg interactions in the mouse: sequence of events and induction of the acrosome reaction by a zona pellucida glycoprotein. Dev Biol. 1983;95:317–24.

Bleil JD, Greve JM, Wassarman PM. Identification of a secondary sperm receptor in the mouse egg zona pellucida: role in maintenance of binding of acrosome-reacted sperm to eggs. Dev Biol. 1988;128:376–85.

Bramwell RK, Howarth B Jr. Preferential attachment of cock spermatozoa to the perivitelline layer directly over the germinal disc of the hen's egg. Biol Reprod. 1992;47:1113–7.

Bramwell RK, Marks HL, Howarth B. Quantitative determination of spermatozoa penetration of the perivitelline layer of the hen's egg as assessed on oviposited eggs. Poult Sci. 1995;74:1875–83.

Brown CR, Cheng WT. Limited proteolysis of the porcine zona pellucida by homologous sperm acrosin. J Reprod Fertil. 1985;74:257–60.

Buruiana LM. Sur l'activité hyaluronidasique et trypsinique du sperme. Naturwissenschaften. 1956;43:523.

Chakravarty S, Suraj K, Gupta SK. Baculovirus-expressed recombinant human zona pellucida glycoprotein-B induces acrosomal exocytosis in capacitated spermatozoa in addition to zona pellucida glycoprotein-C. Mol Hum Reprod. 2005;11:365–72.

Chakravarty S, Kadunganattil S, Bansal P, Sharma RK, Gupta SK. Relevance of glycosylation of human zona pellucida glycoproteins for their binding to capacitated human spermatozoa and subsequent induction of acrosomal exocytosis. Mol Reprod Dev. 2008;75:75–88.

Dunbar BS, Dudkiewicz AB, Bundman DS. Proteolysis of specific porcine zona pellucida glycoproteins by boar acrosin. Biol Reprod. 1985;32:619–30.

Fronda FM. Studies on the fertility of the hen's egg. Philippine Agr. 1926;15:349–60.

Ganguly A, Bukovsky A, Sharma RK, Bansal P, Bhandari B, Gupta SK. In humans, zona pellucida glycoprotein-1 binds to spermatozoa and induces acrosomal exocytosis. Hum Reprod. 2010;25:1643–56.

Gilbert AB. Female genital organs. In: King AS, McLelland J, editors. Form and function in birds, vol. 1. London: Academic Press; 1979.

Grayson P. Izumo1 and Juno: the evolutionary origins and coevolution of essential sperm–egg binding partners. R Soc Open Sci. 2015;2:150296.

Han L, Monné M, Okumura H, Schwend T, Cherry AL, Flot D, Matsuda T, Jovine L. Insights into egg coat assembly and egg–sperm interaction from the X-ray structure of full-length ZP3. Cell. 2010;143:404–15.

Ho JJ, Meizel S. Electrophoretic detection of multiple forms of trypsin-like activity in spermatozoa of the domestic fowl. J Reprod Fertil. 1970;23:177–9.

Ho JJ, Meizel S. Hydrolysis of the hen egg vitelline membrane by cock sperm acrosin and other enzymes. J Exp Zool. 1975;194:429–37.

Horrocks AJ, Stewart S, Jackson L, Wishart GJ. Induction of acrosomal exocytosis in chicken spermatozoa by inner perivitelline-derived *N*-linked glycans. Biochem Biophys Res Commun. 2000;278:84–9.

Howarth B. Avian sperm–egg interaction: perivitelline layer possesses receptor activity for spermatozoa. Poult Sci. 1990;69:1012–5.

Howarth B. Carbohydrate involvement in sperm–egg interaction in the chicken. J Recept Res. 1992;12:255–65.

Howarth B Jr, Digby ST. Evidence for the penetration of the vitelline membrane of the hen's egg by a trypsin-like acrosomal enzyme. J Reprod Fertil. 1973;33:123–5.

Inoue N, Ikawa M, Isotani A, Okabe M. The immunoglobulin superfamily protein Izumo is required for sperm to fuse with eggs. Nature. 2005;434:234–8.

Ivanoff EJ. Compt Rend Soc Biol. 1913;75:371–4.

Jin M, Fujiwara E, Kakiuchi Y, Okabe M, Satouh Y, Baba SA, Chiba K, Hirohashi N. Most fertilizing mouse spermatozoa begin their acrosome reaction before contact with the zona pellucida during in vitro fertilization. Proc Natl Acad Sci U S A. 2011;108:4892–6.

Johnson AL. Reproduction in the female. In: Sturkie PD, editor. Avian physiology. 4th ed. New York: Springer-Verlag; 1986. p. 403–31.

Koyanagi F, Masuda S, Nishiyama H. Acrosome reaction of cock spermatozoa incubated with the perivitelline layer of the hen's egg. Poult Sci. 1988;67:1770–4.

Kuroki M, Mori M. Binding of spermatozoa to the perivitelline layer in the presence of a protease inhibitor. Poult Sci. 1997;76:748–52.

La Spina FA, Puga Molina LC, Romarowski A, Vitale AM, Falzone TL, Krapf D, Hirohashi N, Buffone MG. Mouse sperm begin to undergo acrosomal exocytosis in the upper isthmus of the oviduct. Dev Biol. 2016;411:172–82.

Martin JH, Anderson WB. J Am Assoc Instr Invest Poultry Husbandry. 1918;5:22–3.

Mimura H. On the mechanism of travel of spermatozoa through the oviduct in the domestic fowl, with special reference to the artificial insemination. Okajimas Folia Anat Jpn. 1939;17:459–76.

Miyado K, Yamada G, Yamada S, Hasuwa H, Nakamura Y, Ryu F, Suzuki K, Kosai K, Inoue K, Ogura A, Okabe M, Mekada E. Requirement of CD9 on the egg plasma membrane for fertilization. Science. 2000;287:321–4.

Nishio S, Kohno Y, Iwata Y, Arai M, Okumura H, Oshima K, Nadano D, Matsuda T. Glycosylated chicken ZP2 accumulates in the egg coat of immature oocytes and remains localized to the germinal disc region of mature eggs. Biol Reprod. 2014;91:107.

Okamura F, Nishiyama H. The passage of spermatozoa through the vitelline membrane in the domestic fowl, *Gallus gallus*. Cell Tissue Res. 1978a;188:497–508.

Okamura F, Nishiyama H. Penetration of spermatozoon into the egg and transformation of the sperm nucleus into the male pronucleus in the domestic fowl, *Gallus gallus*. Cell Tissue Res. 1978b;190:89–98.

Okumura H, Kohno Y, Iwata Y, Mori H, Aoki N, Sato C, Kitajima K, Nadano D, Matsuda T. A newly identified zona pellucida glycoprotein, ZPD, and dimeric ZP1 of chicken egg envelope are involved in sperm activation on sperm–egg interaction. Biochem J. 2004;384:191–9.

Olsen MW. Maturation, fertilization, and early cleavage in the hen's egg. J Morphol. 1942;70:513–33.

Olsen MW. Intra-ovarian insemination in the domestic fowl. J Exp Zool. 1952;119:461–81.

Olsen MW, Fraps RM. Maturation, fertilization, and early cleavage of the egg of the domestic turkey. J Morphol. 1944;74:297–309.

Pan J, Sasanami T, Kono Y, Matsuda T, Mori M. Effects of testosterone on production of perivitelline membrane glycoprotein ZPC by granulosa cells of Japanese quail (*Coturnix japonica*). Biol Reprod. 2001;64:310–6.

Robertson L, Wishart GJ, Horrocks AJ. Identification of perivitelline N-linked glycans as mediators of sperm–egg interaction in chickens. J Reprod Fertil. 2000;120:397–403.

Romanoff AL. Fertilization and fertility. In: The avian embryo: structural and functional development. New York: Macmillan; 1960.

Romanoff AL, Romanoff AJ. The avian egg. New York: John Wiley & Sons; 1949.

Sasanami T, Pan J, Mori M. Expression of perivitelline membrane glycoprotein ZP1 in the liver of Japanese quail (*Coturnix japonica*) after in vivo treatment with diethylstilbestrol. J Steroid Biochem Mol Biol. 2003;84:109–16.

Sasanami T, Murata T, Ohtsuki M, Matsushima K, Hiyama G, Kansaku N, Mori M. Induction of sperm acrosome reaction by perivitelline membrane glycoprotein ZP1 in Japanese quail (Coturnix japonica). Reproduction. 2007;133:41–9.

Sasanami T, Yoshizaki N, Dohra H, Kubo H. Sperm acrosin is responsible for the sperm binding to the egg envelope during fertilization in Japanese quail (*Coturnix japonica*). Reproduction. 2011;142:267–76.

Sasanami T, Sugiura K, Tokumoto T, Yoshizaki N, Dohra H, Nishio S, Mizushima S, Hiyama G, Matsuda T. Sperm proteasome degrades egg envelope glycoprotein ZP1 during fertilization of Japanese quail (*Coturnix japonica*). Reproduction. 2012;144:423–31.

Słowińska M, Ciereszko A. Identification of the second form of acrosin in Turkey spermatozoa. Reprod Domest Anim. 2012;47:849–55.

Słowińska M, Olczak M, Liszewska E, Watorek W, Ciereszko A. Isolation, characterization and cDNA sequencing of acrosin from turkey spermatozoa. Comp Biochem Physiol B Biochem Mol Biol. 2010;157:127–36.

Stambaugh R, Buckley J. Zona pellucida dissolution enzymes of the rabbit sperm head. Science. 1968;161:585–6.

Steele MG, Meldrum W, Brillard JP, Wishart GJ. The interaction of avian spermatozoa with the perivitelline layer in vitro and in vivo. J Reprod Fertil. 1994;101:599–603.

Stewart SG, Bausek N, Wohlrab F, Schneider WJ, Janet Horrocks A, Wishart GJ. Species specificity in avian sperm: perivitelline interaction. Comp Biochem Physiol A Mol Integr Physiol. 2004;137:657–63.

Takeuchi Y, Nishimura K, Aoki N, Adachi T, Sato C, Kitajima K, Matsuda T. A 42-kDa glycoprotein from chicken egg-envelope, an avian homolog of the ZPC family glycoproteins in mammalian Zona pellucida. Its first identification, cDNA cloning and granulosa cell-specific expression. Eur J Biochem. 1999;260:736–42.

Takeuchi Y, Cho R, Iwata Y, Nishimura K, Kato T, Aoki N, Kitajima K, Matsuda T. Morphological and biochemical changes of isolated chicken egg-envelope during sperm penetration: degradation of the 97-kilodalton glycoprotein is involved in sperm-driven hole formation on the egg-envelope. Biol Reprod. 2001;64:822–30.

Tanghe S, Van Soom A, Nauwynck H, Coryn M, de Kruif A. Minireview: Functions of the cumulus oophorus during oocyte maturation, ovulation, and fertilization. Mol Reprod Dev. 2002;61:414–24.

Toshimori K, Saxena DK, Tanii I, Yoshinaga K. An MN9 antigenic molecule, equatorin, is required for successful sperm–oocyte fusion in mice. Biol Reprod. 1998;59:22–9.

Urch UA, Wardrip NJ, Hedrick JL. Limited and specific proteolysis of the zona pellucida by acrosin. J Exp Zool. 1985a;233:479–83.

Urch UA, Wardrip NJ, Hedrick JL. Proteolysis of the zona pellucida by acrosin: the nature of the hydrolysis products. J Exp Zool. 1985b;236:239–43.

Waclawek M, Foisner R, Nimpf J, Schneider WJ. The chicken homologue of zona pellucida protein-3 is synthesized by granulosa cells. Biol Reprod. 1998;59:1230–9.

Wishart GJ. Quantitative aspects of sperm: egg interaction in chickens and turkeys. Anim Reprod Sci. 1997;48:81–92.

Fertilization 2: Polyspermic Fertilization

7

Shusei Mizushima

Abstract

During fertilization in animals, a haploid egg nucleus fuses with a haploid sperm nucleus to restore the diploid genome. In most animals including mammals, echinoderms, and teleostei, the penetration of only one sperm into an egg is ensured at fertilization because the entry of two or more sperm is prevented by polyspermy block systems in these eggs. On the other hand, several animals such as birds, reptiles, and most urodele amphibians exhibit physiological polyspermy, in which the entry of several sperm into one egg is permitted. However, in these polyspermic eggs, only one sperm nucleus is involved in zygotic formation with a female nucleus, thereby avoiding syngamy with multiple sperm nuclei. In the chicken, 20–60 sperm are generally found within the egg cytoplasm at fertilization and this number is markedly higher than that of other polyspermic species; however, avian-specific events such as the degeneration and mitosis of supernumerary sperm nuclei during early embryo development allow a polyspermic egg to develop normally. This chapter describes current knowledge on polyspermy-related events in avian eggs during fertilization, and is characterized by a comparison to the fertilization modes of other vertebrates. The close relationship between sperm numbers and egg sizes, and the movement of supernumerary sperm nuclei towards the periphery of the egg cytoplasm and their degeneration are summarized. The molecular mechanisms by which polyspermy initiates egg activation to start embryo development are also discussed.

Keywords

Diploid genome • Polyspermy block • Physiological polyspermy • Supernumerary sperm • Fertilization modes • Sperm numbers • Egg sizes • Egg activation

S. Mizushima
Department of Biological Sciences,
Faculty of Science, Hokkaido University,
Kita 10, Nishi 8, Kita-ku, Sapporo,
Hokkaido 060-0810, Japan
e-mail: smizus@sci.hokudai.ac.jp

T. Sasanami (ed.), *Avian Reproduction*, Advances in Experimental Medicine and Biology 1001,
DOI 10.1007/978-981-10-3975-1_7

7.1 Polyspermy and Embryo Development

Fertilization is an indispensable event in the restoration of diploid genomes, and the initiation of several reactions that lead to embryonic development in all animals. Most animals exhibit monospermy, in which a single sperm has the ability to activate the reaction important for early development (Swann 1996; Stricker 1999; Runft et al. 2002; Yanagimachi 2005). Conversely, the simultaneous fertilization of an egg by two or more sperm is a lethal condition leading to aneuploidy and developmental arrest. Polyspermy in humans mostly results in spontaneous abortion. Although the birth of triploid or tetraploid children has been reported, severe malformations and multiple abnormalities are associated with polyploidy births (Uchida and Freeman 1985; Sherard et al. 1986; Shiono et al. 1988; Roberts et al. 1996; Dean et al. 1997; Sun 2003). On the other hand, several sperm have been observed in porcine eggs at a high incidence under physiological conditions or in-vitro fertilization (Sun and Nagai 2003); however, porcine egg cytoplasm has the ability to remove successive sperm (Xia et al. 2001), such that some polyspermic pig eggs may develop to term if successive sperm do not interrupt the diploid zygotic genome (Han et al. 1999). Polyspermic fertilization in this species is an extraordinary case, and, in most cases of monospermic fertilization in mammals, defenses against polyspermy are rapidly established after penetration of the first sperm (Galeati et al. 1991; Yanagimachi 1994).

The number of sperm reaching the egg surface is reduced during passage through the female reproductive tract and the egg extracellular matrix, zona pellucida, jelly layer, and vitelline envelope. In mice, only 100 or 200 out of the 50 million sperm ejaculated were previously shown to reach the ovulated egg (Wassarman 1994). Although the possibility of the penetration of multiple sperm into an egg decreases at fertilization, eggs still remain at risk of polyspermy. Therefore, most eggs have established a membrane block and zona pellucida block in order to prevent polyspermy (Yanagimachi 1994; Abbott and Ducibella 2001). Following sperm penetration, the rapid and transient depolarization of membrane potential at the level of the plasma membrane (membrane block) is elicited in order to prevent the fusion of additional sperm in the frog, *Xenopus laevis*, and several marine invertebrates (Jaffe and Gould 1985; Gould and Stephano 2003). This change in membrane potential is not observed in fertilized mouse, hamster, rabbit, or bony fish eggs, in which the membrane block employs a different mechanism to changes in membrane potential (Nuccitelli 1980; Miyazaki and Igusa 1981; Igusa et al. 1983; Jaffe et al. 1983; McCulloh et al. 1983). After a temporal fast block, the contents (e.g., ovastacin in human and mouse; Quesada et al. 2004; Liu 2011; Burkart et al. 2012) of cortical granules, a special organelle in eggs, are released into the perivitelline space via their exocytosis (a cortical reaction). Although the structure and molecules involved in the zona reaction differ among species, this reaction makes the zona pellucida refractory to the binding and fusion of a second sperm (the zona reaction), which is common to all vertebrate species (Wong and Wessel 2006). An egg must recognize the binding and fusion of the first fertilizing sperm and rapidly establish a polyspermy block. Therefore, the fertilizing sperm must immediately activate and propagate a signal cascade throughout the whole egg. Although the molecular mechanisms underlying the activation of a polyspermy block have not yet been elucidated in detail, fertilization-mediated Ca^{2+} release from intracellular stores and the activation of protein kinase C (PKC) are known to be involved in the establishment of two different polyspermy blocks (Sun 2003; Gardner et al. 2007).

The eggs of some species, including ctenophores, elasmobranchs, urodele amphibians, reptiles, and birds, physiologically permit the penetration of more than one sperm into the ooplasm at fertilization (Elinson 1986; Iwao 2000; Wong and Wessel 2006; Snook et al. 2011; Mizushima 2012). Unlike monospermic eggs, neither a membrane block nor intracellular organelles similar to cortical granules have been observed in these polyspermic eggs; therefore,

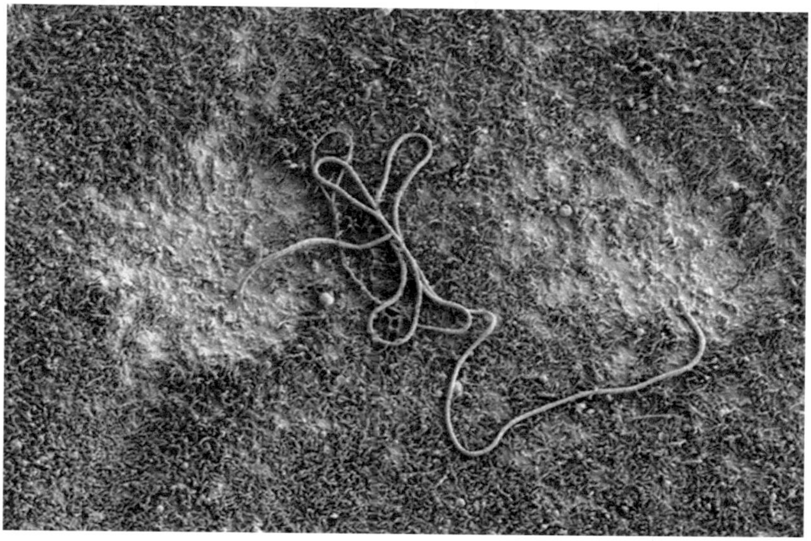

Fig. 7.1 Scanning electron microscopic observation of two sperm and egg plasma membrane fusion in Japanese quail (photograph provided by Dr. Norio Yoshizaki, Gifu University). Usually, 2–60+ sperm penetrate egg cytoplasm in birds

there is no block against the entry of a second or more sperm after the first sperm is incorporated into the egg cytoplasm. Since only one sperm is eventually involved in zygotic formation with a female pronucleus, embryo development with a diploid configuration is ensured. During natural fertilization, the avian egg receives a markedly higher number of sperm than other polyspermic species (Fig. 7.1). Although it is technically difficult to calculate the total number of sperm in the germinal disc, it appears likely that, at the lowest estimate, 20 sperm are typically incorporated into an egg (Harper 1904; Patterson 1910, Fofanova 1965; Nakanishi et al. 1990; Waddington et al. 1998). The maximum number of sperm found within the germinal disc has previously been reported to be 25 for the pigeon and 62 for the chicken (Harper 1904; Nakanishi et al. 1990). Furthermore, more than 70 sperm have been detected in the quail egg within 1 h of fertilization (our unpublished data).

The reason why the avian ovum may accept the entry of numerous sperm remained unclear until recently. The findings of recent experiments on birds suggest that the number of sperm incorporated into an egg cytoplasm affects the fate of the egg. The rate of chicken and turkey embryos developing to the blastoderm stage was found to be approximately 50% when approximately three sperm penetrated the inner perivitelline layer over the germinal disc region (in this case, the number of sperm that penetrated the egg cytoplasm was estimated to be less than three), while the probability of egg development was almost 100% when more than six sperm penetrated the region (Bramwell et al. 1995; Wishart 1997). This finding suggests that the fertilization rate increases when the number of sperm incorporated into an egg is higher. In addition, the direct injection of sperm into the cytoplasm of a mature quail egg, namely, an intracytoplasmic sperm injection (ICSI), directly revealed that a single sperm was insufficient for a high rate of fertilization and subsequent blastoderm development (approximately 20%; Hrabia et al. 2003; Takagi et al. 2007a; Mizushima et al. 2008, 2009; Mizushima 2012; Shimada et al. 2014; Kang et al. 2015). It has been proposed that the purpose of polyspermy is to increase the opportunity of a sperm nucleus migrating to the center of the germinal disc and making contact with the female nucleusm because the surface area of the avian germinal disc in relation to that of sperm is very large. However, the ICSI technique revealed that the microinjection of a single sperm together with soluble protein equivalent to multiple sperm induced fertilization and subsequent development up to the blastoderm stage in 70% of eggs tested (Mizushima et al. 2014). Therefore, instead of the frequency of migrating sperm nuclei, an alternative hypothesis in which many spermatozoa are necessary to provide sufficient amounts of sperm-derived egg-activating proteins to ensure the successful initiation of egg development in birds is currently considered the most likely theory (see Sect. 7.2 for more details). On the other hand, there may be a physiological limitation to excessive polyspermy in the avian egg. For example, an increase in the number of sperm that make contact with the ovum by intramagnal insemination, which is greater than the normal numbers of sperm, induces an increase in the frequency of early blastoderm developmental arrest (Van Krey et al. 1966). More critically, a recent study showed that cytoplasmic segmentation occurs at a high rate in the absence of nuclear divisions in the germinal discs of quail and chicken ova inseminated in vitro using 1–2 × 10^4 semen, and eggs underwent development to the early pseudo-blastoderm stage (Olszanska et al. 2002; Batellier et al. 2003), indicating that an excessive number of supernumerary sperm interferes with normal fertilization and subsequent embryo development (Fechheimer 1981; Olszanska et al. 2002; Mizushima et al. 2009). A novel structure coated with calcium carbonate in the infundibulum part of the female reproductive tract in which fertilization occurs and sperm are stored until fertilization, termed the sperm-associated body (SB), was identified in domestic birds, and SB-accompanied sperm only were found to pass through the vitelline membrane (Sultana et al. 2004; Rabbani et al. 2006, 2007). This SB may contribute physiologically to reducing the number of excessive fertilizing sperm.

7.2 Egg Activation

7.2.1 Ca^{2+} Increase During Fertilization

Physiological polyspermy occurs in oviparous species that exhibit internal fertilization, such as cartilaginous fishes, urodele amphibians, reptiles, and birds (Elinson 1986; Iwao 2000; Wong and Wessel 2006; Snook et al. 2011). Fertilizing sperm provide a signal to trigger the initiation of egg development (egg activation) as well as the nucleus, because the development of animal eggs is arrested at the species-specific phase of meiosis until fertilization, e.g., at metaphase of second meiosis (metaphase II) in most vertebrate eggs (Stricker 1999). In all monospermic and polyspermic animals studied to date, an increase in intracellular Ca^{2+} concentrations ($[Ca^{2+}]i$) has been observed immediately after sperm–egg binding or fusion during fertilization (Stricker 1999; Runft et al. 2002), and this increase in $[Ca^{2+}]i$ plays a pivotal role in restarting cell cycle events in the egg, which comprise the resumption of meiosis and extrusion of the second polar body (Ducibella et al. 2002; Miyazaki 2006). Furthermore, in mammalian species, the establishment of polyspermy prevention is also evoked by an increase in $[Ca^{2+}]i$ (McAvey et al. 2002; Gardner et al. 2007). In spite of its universality, the spatiotemporal patterns of the Ca^{2+} signal associated with egg activation vary widely among species (Stricker 1999). In monospermic fishes and the frog, *Xenopus laevis*, a transient increase in $[Ca^{2+}]i$ induced by a single sperm has been shown to propagate throughout the whole egg from the sperm-entering position as a Ca^{2+} wave (Ridgway et al. 1977; Busa and Nuccitelli 1985; Fluck et al. 1991; Abraham et al. 1993; Keating et al. 1994; Creton et al. 1998). An increase in $[Ca^{2+}]i$ during the activation of mammalian eggs is known to occur periodically in the form of long-lasting oscillations, which are known as Ca^{2+} oscillations (Miyazaki et al. 1993; Jones et al. 1995), and continues to approximately the first interphase stage, similar to that in mouse eggs (Jones et al. 1995; Marangos et al. 2003), or to the first mitotic cell cycle, as observed in bovine eggs (Nakada et al. 1995).

In mice, artificial stimuli such as electrical pulses or exposure to ethanol have revealed that a single Ca^{2+} pulse induces second polar body extrusion, but causes only partial egg activation due to the incomplete inactivation of cytostatic factor, which is the cytosolic protein responsible for meiotic arrest at metaphase II (Tatone et al. 1999; Ducibella et al. 2002; Jones 2005). In contrast, continuous stimuli for 24 h fully activate eggs, which includes an increase in cortical granule exocytosis and the formation of both polar bodies (Deguchi and Osanai 1995; Lawrence et al. 1998; Ozil 1998; Ducibella et al. 2002; Jones 2005, 2007). Accordingly, the mammalian egg needs to be exposed to a series of repetitive Ca^{2+} pulses in order to ensure it escapes meiosis, because a number of hours are needed for a lapse between meiotic resumption and the interphase stage. Alternatively, unlike mammals, fertilization in fishes and *Xenopus laevis* triggers a rapid transition to pronuclear formation (Rugh 1951; Iwamatsu and Ohta 1978). The short post-fertilization phase in these zygotes may obviate the second or more Ca^{2+} oscillations (Jones 1998).

In physiological polyspermic eggs, a few sperm successively enter at different points on the egg surface, and increases in $[Ca^{2+}]i$ spread concentrically into the egg cytoplasm at each sperm entry site as small waves (Harada et al. 2011; Iwao 2012). However, unlike monospermic eggs such as those of fishes, *Xenopus laevis*, and mammals, each Ca^{2+} wave does not reach the opposite site of the egg. The $[Ca^{2+}]i$ intensity in the entire egg region of polyspermic eggs, as revealed by a Ca^{2+}-sensitive fluorescence dye, shows a slow increase that continues for approximately 40 min, with the peak level reached being markedly lower that than in *Xenopus laevis* eggs (Fontanilla and Nuccitelli 1998). In addition, a slow wave-like increase in $[Ca^{2+}]i$ is also caused by a single injection of sperm extract (SE) containing soluble proteins into the newt egg cytoplasm, which propagates concentrically from the injection site through the entire egg without a local small wave, but is also able to evoke meiotic resumption (Harada et al. 2007). In polyspermy of the frog, *Discoglossus*, an increase in $[Ca^{2+}]i$ lasts for 50 min after fertilization (Nuccitelli et al. 1988). However, several rapid spike-like depolarizations caused by sperm entry precede the major depolarization mediated by Ca^{2+}-activated Cl^- efflux (Talevi 1989), suggesting that a non-propagative small Ca^{2+} increase occurs at each sperm entry point in advance of the major Ca^{2+} wave (Iwao 2012).

In birds, a microinjection of SE into an ovulated egg retrieved from the infundibulum or upper part of the magnum revealed a very unique pattern for the increase in $[Ca^{2+}]i$ in the quail egg, which has not been observed in other species (Fig. 7.2). Its Ca^{2+} signal pattern is classified as two different kinds of Ca^{2+} waves, namely, a transient, slow wave and multiple, spiral-like oscillations (Mizushima et al. 2014). The slow Ca^{2+} wave was observed immediately after the microinjection; an increase in $[Ca^{2+}]i$ was initiated at the injection site of the germinal disc, and its Ca^{2+} wave spread concentrically into the egg cytoplasm, but was

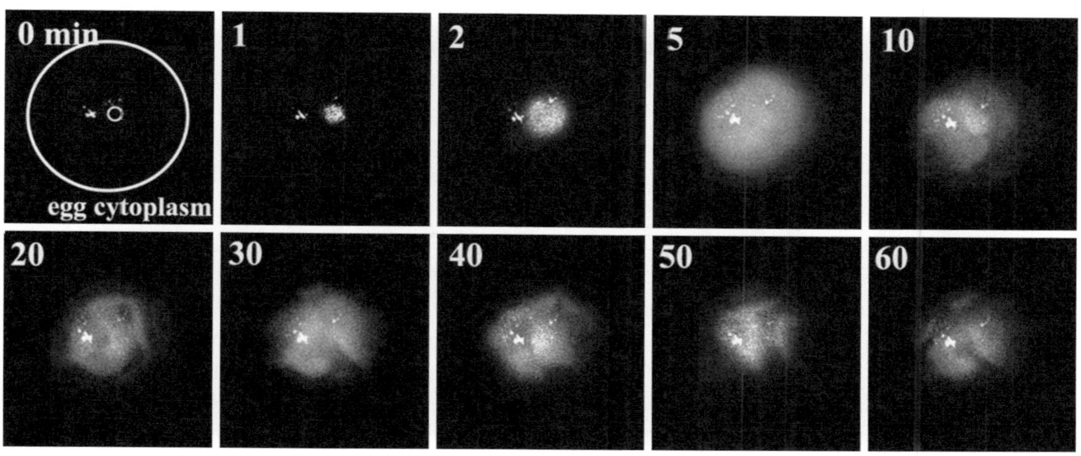

Fig. 7.2 Ca^{2+} rise in quail egg cytoplasm after microinjection of sperm-derived egg activating factors. An intracellular Ca^{2+} increased immediately from injecting site and the Ca^{2+} signal propagated over the egg cytoplasm. The Ca^{2+} rise continues for about 5 min. Thereafter, long-lasting spiral-like Ca^{2+} oscillations are generated and continue every 1 min for over 1 h. *Small circle* indicates injecting site

restricted to the germinal disc. The increase in $[Ca^{2+}]i$ at the injection site continued for approximately 5 min, and then decreased gradually before returning to the basal level within 30 min of the microinjection. On the other hand, an initial spiral-like Ca^{2+} signal occurred at the injection site 10–15 min after the microinjection, which was prior to the restoration of increases in $[Ca^{2+}]i$ by the Ca^{2+} wave to the basal level. Second and further spiral-like Ca^{2+} waves originated successively before the disappearance of the previous Ca^{2+} spiral, with a mean interspike interval of less than 1 min, and the oscillation lasted for at least 3 h. Furthermore, a few Ca^{2+} spirals overlapping partially in one egg were found to have complicated waveforms, which markedly differed from the Ca^{2+} oscillations observed during mammalian egg activation. It is important to note that these two different $[Ca^{2+}]i$ patterns play different roles in the activation of the quail egg. A slow Ca^{2+} wave has the ability to evoke the resumption of meiosis and subsequent zygote formation (Mizushima et al. 2007, 2008, 2014), whereas spiral-like Ca^{2+} oscillations are not involved in these events. However, the induction of spiral-like Ca^{2+} oscillations in a slow Ca^{2+} wave-generated ICSI quail egg showing developmental arrest at the early embryo stage (Mizushima et al. 2008, 2010, 2014) activates full-term development to hatching, suggesting that spiral-like Ca^{2+} oscillations are necessary for cell cycle progression in the advanced zygote stage. More than 20 sperm generally enter chicken and quail eggs and then progressively form male pronuclei between 1 and 3 h (Perry 1987; Nakanishi et al. 1990). Although the spatiotemporal changes that occur in $[Ca^{2+}]i$ in the avian egg following natural fertilization or in-vitro insemination still remain unknown, multiple Ca^{2+} spirals appear to be necessary, in part, for a continuous stimulation to complete pronuclei formation, besides the cell cycle stimulation (Fig. 7.4).

7.2.2 Egg Size and Sperm Number

The microinjection of a whole sperm into a mammalian egg (ICSI), which avoids any membrane contact between sperm and the egg, generates a Ca^{2+} oscillation similar to that observed during fertilization, and this Ca^{2+} oscillation is also triggered by a microinjection of SE corresponding to the content of a single sperm (Homa and Swann 1994; Wu et al. 1997, 1998; Dong et al. 2000; Parrington et al. 2000; Tang et al. 2000). Thus, the quantity of egg-activating proteins contained in a single sperm is necessary and sufficient for fully activating the eggs of monospermic species. It is important to note that avian SE equivalent to approximately one sperm or a whole avian sperm possess the ability to initiate Ca^{2+} oscillations and subsequent zygotic formation in a mouse egg (Dong et al. 2000; Takagi et al. 2007b), but have a negligible effect on the activation of an avian egg (Hrabia et al. 2003; Takagi et al. 2007a; Mizushima et al. 2009, 2014). These findings led to the hypothesis that a large number of sperm may be necessary in order to provide egg-activating factors not fulfilled by a single sperm, due to the large size of an avian egg in comparison to that of a mammalian egg. Indeed, the volume of quail egg cytoplasm is more than 1–2 μl, whereas that of the mouse is approximately 200 pl. Egg-activating proteins equivalent to 100–200 sperm are required for fertilization and full-term development to hatching in quails, which supports the strong correlation between egg size and polyspermy (Birkhead et al. 1994; Mizushima 2012; Mizushima et al. 2014; Shimada et al. 2014; Kang et al. 2015).

Unlike mammals, in which a sperm-triggered Ca^{2+} wave propagates through the entire mammalian egg, the Ca^{2+} wave generated by a single sperm only propagates in one-eighth to one-quarter of the egg surface in physiologically polyspermic newts (Harada et al. 2011; Iwao 2012). Consistent with the egg size theory, several sperm must enter from different points in order to increase $[Ca^{2+}]i$ throughout the large eggs of newts. Similarly, in polyspermy in birds, the propagation of a Ca^{2+} signal by a single sperm may not reach the whole egg, and, as such, a larger number of sperm than that needed by newts is required for the activation of the whole egg because the size of an avian egg is markedly larger than those of newts (Fig. 7.4). Although little is known about the discrepancy that a larger amount of SE (equivalent to 100–200 sperm) than that from the 2 to 60+ observed during natural fertilization is required for the full-term development of quail eggs following ICSI with a single sperm (Mizushima et al. 2014), these differences lie in the fact that 2–20 sperm typically penetrate newt eggs in situ (Iwao et al. 1985, 1993), whereas complete egg activation by a microinjection of newt SE requires a protein content equivalent to 330 sperm (Harada et al. 2011). With regard to egg activation, although limited information is available on the minimum number of sperm needed for the full activation of an avian egg, at least 20 sperm may be required to increase $[Ca^{2+}]i$ throughout the entire avian egg. Otherwise, egg size may influence where and when sperm enters, and, thus, potentially affect the timing of sperm aster formation and migration of potential microtubule organization centers (MTOCs), which are responsible for the movement of the pronucleus to karyogamy. The extra space in the egg cytoplasm of large eggs is one of the most important factors for eliminating supernumerary sperm nuclei in the same egg cytoplasm (Elinson 1986; Iwao 2012). Based on various aspects of internal fertilization, centrosome dynamics, evolutionary history, and egg cytological polyspermy blocks (see Sect. 7.3 for details), the close relationship between increased egg sizes and the number of sperm in the acquisition of polyspermy needs to be discussed in more detail (Snook et al. 2011).

7.2.3 Molecular Mechanisms Underlying Egg Activation in Physiological Polyspermy

An increase in $[Ca^{2+}]i$ in the eggs of vertebrates is initiated by Ca^{2+} release via the inositol 1,4,5-trisphosphate (IP_3) receptor (IP_3R). A large number of studies on vertebrate eggs have demonstrated that an injection of IP_3 induces the release of Ca^{2+} from Ca^{2+} stores, mainly the endoplasmic reticulum, in the egg (Miyazaki 1988; Miyazaki et al. 1992; Fissore and Robl 1993; Swann and Ozil 1994; Wang et al. 1999; Amano et al. 2004; Lee et al. 2010; Mizushima et al. 2014). The repetitive increases observed in $[Ca^{2+}]i$ (Ca^{2+} oscillations) during mammalian fertilization are no exception. IP_3-induced Ca^{2+} release as well as Ca^{2+}-induced Ca^{2+} release, which operates by the sensitizing effects of Ca^{2+} on the IP_3R, contribute to the regenerative process of Ca^{2+} release (Miyazaki et al. 1993). Since IP_3 is generated by enzymes of the phospholipase C (PLC) family, which catalyze the hydrolysis of phosphatidylinositol 4,5-bisphosphate (PIP_2) into IP_3 and diacylglycerol (Rhee 2001), two main molecular signaling models have been proposed for the production of IP_3 in the egg. The first suggestion proposes that sperm–egg fusion induces the activation of a receptor on the egg plasma membrane and couples to either a G-protein or tyrosine kinase (Src kinase), and this is followed by the activation of PLCβ or PLCγ respectively (Runft et al. 2002). In *Xenopus laevis*, Src kinase is phosphorylated and then stimulates PLCγ at the time of sperm–egg contact (Sato et al. 1999, 2000, 2001, 2003). Sperm induces the transient phosphorylation of uroplakin III (UPIII) on the egg membrane, which is a substrate of the egg cytoplasmic Src kinase, and Src (Sakakibara et al. 2005; Mahbub Hasan et al. 2014). In addition, a sperm-derived protease associated with a sperm surface glycoprotein (Nagai et al. 2009) serves as a target of the extracellular domain of UPIII (Mahbub Hasan et al. 2005; Sakakibara et al. 2005). Although the molecular signaling pathway for egg activation in the monospermy of the primitive jawless fish lamprey remains unknown, the involvement of a receptor on the egg membrane, and not sperm–egg fusion, has been postulated because most eggs clamped at a positive potential (+20 to +40 mV) were found to undergo egg activation (Kobayashi et al. 1994).

An alternative hypothesis has been proposed for increases in $[Ca^{2+}]i$, in which the sperm itself contains soluble egg-activating factors that diffuse directly into the cytoplasm of the egg after fusion. Since fertilization-induced increases in $[Ca^{2+}]i$ are not prevented in the mouse egg in spite of the introduction of a G-protein-specific inhibitor or PLCγ (Moore et al. 1993; Mehlmann et al. 1998; Williams et al. 1998), this appears to be the case in most mammalian species (Swann 1996; Stricker 1999; Runft et al. 2002). The most likely candidate for the initiator of an increase in $[Ca^{2+}]i$ in mammalian eggs is a sperm-specific member of the PLC isozyme, PLCzeta (PLCζ; Saunders et al. 2002). Although other sperm-inducing factors have been reported (Sette et al. 2002; Perry et al. 1999, 2000), they have not yet been substantiated, and important evidence now exists to show that PLCζ alone has the ability to generate Ca^{2+} oscillations in mouse eggs at an amount equivalent to the content in a single sperm (Saunders et al. 2002; Kouchi et al. 2004). Furthermore, a PLCζ orthologue has also been identified in humans, monkeys, rats, pigs, bovines, and horses as a common egg-activating factor in mammals (Cox et al. 2002; Yoneda et al. 2006; Ito et al. 2008; Ross et al. 2008; Sato et al. 2013). The mouse PLCζ contains a nuclear localization signal that promotes its accumulation in the pronuclei (Larman et al. 2004; Yoda et al. 2004; Kuroda et al. 2006). The nuclear accumulation of PLCζ appears to terminate long-lasting Ca^{2+} oscillations. In addition, the sperm of the teleost fish, tilapia, contains an egg-activating factor for increasing Ca^{2+} in mouse eggs or sea urchin egg homogenates (Coward et al. 2003), and the medaka testes also contain PLCζ that initiates Ca^{2+} oscillations in mouse eggs (Coward et al. 2011). However, PLCζ in the pufferfish, Fugu, is expressed in the ovary and brain, but not in the testis, and its ovarian form does not have the ability to trigger Ca^{2+} oscillations in mouse eggs (Coward et al. 2011). Therefore, PLCζ does not appear to be involved in the sperm-induced activation of all fish eggs.

On the other hand, citrate synthase (CS) is the most likely candidate for an initiator of increases in $[Ca^{2+}]i$ in newt eggs (Harada et al. 2007). Although the molecular mechanisms underlying the initiation of an increase in $[Ca^{2+}]i$ by sperm CS have not yet been elucidated in detail, an experimental hypothesis was recently proposed in which CS released from sperm forms a complex with maternal microtubules and the ER with the IP_3R as well as PLCγ in the midpiece region of the sperm, and this complex acts on small Ca^{2+} wave propagation by the sequential activation of PLCγ in order to stimulate the IP_3R (Ueno et al. 2014). Or possibly paternal CS cleaves citrate into acetyl-CoA and oxaloacetate in the egg cytoplasm, and the former and latter then trigger Ca^{2+} release from the ER and mitochondria respectively (Harada et al. 2011). In somatic cells, CS produces citrate from acetyl-CoA and oxaloacetate in the mitochondrial tricarboxylic acid (TCA) cycle, but may inversely cleave citrate, which is abundant in the egg cytoplasm, to produce acetyl-CoA and oxaloacetate (Srere 1992; Iwao, 2012). Increases induced in acetyl-CoA in the egg cytoplasm by sperm CS may sensitize IP_3Rs on the ER (Missiaen et al. 1997). Oxaloacetate has been suggested to have the potential to induce the release of Ca^{2+} from mitochondria (Leikin et al. 1993). Therefore, paternal CS functions as an enzyme in the egg in order to produce acetyl-CoA and oxaloacetate and/or as a PLCγ and IP_3R stimulator to generate the release of Ca^{2+} from Ca^{2+} stores. Furthermore, since SE derived from newts cannot generate Ca^{2+} oscillations in the mouse egg, the mechanism responsible for egg activation in newts differs from PLCζ signaling (Harada et al. 2007). A PLCζ orthologue has not yet been identified in newts.

In contrast to PLCζ and CS alone being sufficient to evoke the release of Ca^{2+} in mammalian and newt eggs, at least three egg-activating factors, namely PLCζ, CS, and aconitate hydratase (AH), are essential for the complete release of Ca^{2+} in order to activate the full-term development of avian eggs to hatching (Fig. 7.3; Mizushima et al. 2014). Even though avian PLCζ has the ability to generate Ca^{2+} oscillations in the mouse egg (Coward et al. 2005), it is only involved in an initial slow Ca^{2+} wave in quail eggs (Mizushima et al. 2014). This difference in responses to PLCζ in these eggs may be associated with the desensitization of the IP_3R channel to IP_3 and IP_3-induced Ca^{2+} resulting from PLCζ activity, and not to any reductions in enzyme activity. The IP_3Rs in mammalian eggs are progressively desensitized by ubiquitination and subsequent proteasome activity, which corresponds to the termination of Ca^{2+} oscillations at the interphase stage (Zhu et al. 1999;

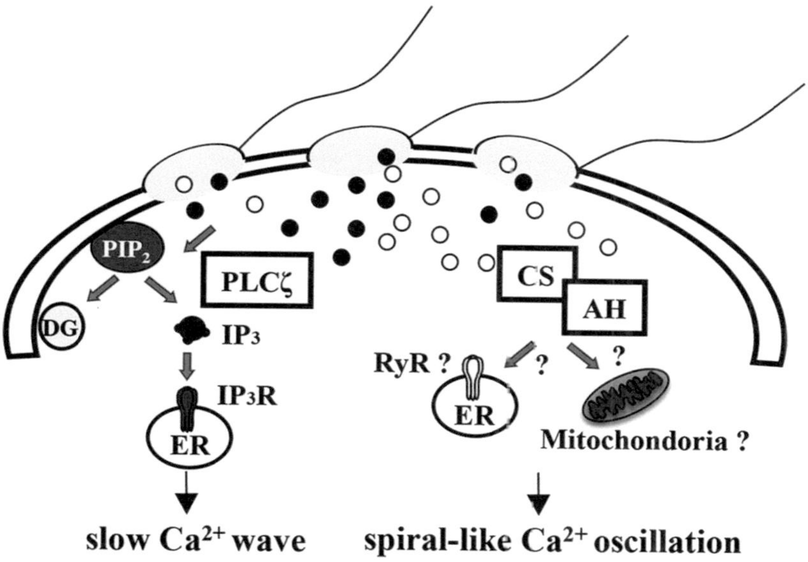

Fig. 7.3 Schematic signaling pathway in avian egg activation. Sperm-specific PLCζ introduced from each sperm induces a slow Ca^{2+} wave, while sperm-specific CS and AH induces a periodic spiral-like Ca^{2+} spike. Inositol 1,4,5-trisphosphate (IP_3) hydrolyzed from phosphatidylinositol 4,5-bisphosphate by PLCζ activity evokes Ca^{2+} release from endoplasmic reticulum (ER) via IP_3 receptor. CS and AH may induce Ca^{2+} release from ER or mitochondria. In a part, another molecule and ryanodin receptor may be involved in Ca^{2+} release

Brind et al. 2000; Zhu and Wojcikiewicz 2000; Malcuit et al. 2005; Lee et al. 2010), suggesting that IP_3 binding to IP_3Rs initiates the down-regulation of IP_3Rs to terminate PLCζ-induced egg activation immediately after fertilization in birds. Although limited information is available for the down-regulation of maternal IP3Rs during avian fertilization, the long-lasting spiral-like Ca^{2+} oscillation initiated 10–15 min after the PLCζ-induced Ca^{2+} wave slowly occurs irrespective of the absence of IP_3Rs, because the introduction of an antagonist of IP_3R did not prevent Ca^{2+} oscillations in the quail (Mizushima et al. 2014). On the other hand, CS and AH are both responsible for the generation of spiral-like Ca^{2+} oscillations. This spiral-like Ca^{2+} oscillation is not induced by CS or AH alone. Egg-activating activity in SE was found to be abolished by a treatment with each antibody, which supports these findings. Spiral-like Ca^{2+} oscillations induced by CS and AH may be partly ascribed to the release of Ca^{2+} from ryanodine receptors on the ER, because the microinjection of an agonist of ryanodine receptors into quail eggs was found to initiate repetitive Ca^{2+} spikes similar to CS- and AH-induced spike-like Ca^{2+} oscillations (Fig. 7.3; Mizushima et al. 2014). Previous studies reported that ryanodine receptors participated in the Ca^{2+} waves observed in sea urchin eggs during egg activation (Galione et al. 1993; Lee et al. 1993; Miyazaki 2006). The elucidation of variations in egg activation systems and particularly increases in Ca^{2+} at fertilization in vertebrates may contribute to a more detailed understanding of the evolutionary history of egg activation concomitant with the acquisition of polyspermy (Table 7.1).

7.2.4 Variations in and Evolution of the Sperm Factor

The egg-activating factors responsible for increases in $[Ca^{2+}]i$ in eggs are characterized as sperm-specific molecular triggers in all vertebrates studied. Although AH and CS are originally mitochondrial genes involved in the TCA cycle, the molecular weights of proteins detected in sperm are slightly higher for CS and lower for AH than

Table 7.1 Comparison of fertilization mode, pattern of Ca^{2+} rise in egg activation, and sperm-derived egg-activating factor among vertebrates

Species	Fertilization mode	Ca^{2+} pattern	Ca^{2+} store	Sperm factor
Fish				
Fugi	Monospermy	Single	Endoplasmic reticulum	?
Medaka	Monospermy	Single	Endoplasmic reticulum	PLCζ?
Discoglossus	Polyspermy	Multiple	Endoplasmic reticulum	?
Amhibian				
Anuran (*Xenopus*)	Monospermy	Single	Endoplasmic reticulum	Receptor
Anuran (*Discoglossus*)	Polyspermy	?	?	?
Urodele (*Cynopus*)	Polyspermy	Multiple	Endoplasmic reticulum, mitochondria?	CS
Bird				
Japanese quail	Polyspermy	Multiple	Endoplasmic reticulum, mitochondria?	PLCζ, CS, AH
Chicken	Polyspermy	?	?	PLCζ?
Mammal				
Platypus	Polyspermy	?	?	?
Higher eutherians	Monospermy	Multiple	Endoplasmic reticulum	PLCζ

those in body cells (Mizushima et al. 2014). Since their predicted amino acid sequences are partially different from those in somatic cells, their sperm-specific isoforms appear to be diversified and specified by gene duplication as an egg-activating factor. The molecular weight of sperm-specific newt CS has also previously been shown to be slightly higher than that of other tissues (Harada et al. 2007), and this difference has been attributed to the hyper-phosphorylation of sperm CS (Ueno et al. 2014). Unlike PLCζ in monospermic species, the acquisition of novel egg-activating factors in the polyspermic species, quails and newts, may have played a pivotal role in the evolution of slower activation in polyspermic eggs, and may also have promoted the reproductive isolation necessary for speciation in vertebrates. However, a review of variations in sperm-specific CS and AH between quails and newts indicates divergent evolution in the molecular mechanisms underlying increases in $[Ca^{2+}]i$ concomitant with a species-specific transition in the mode of polyspermic fertilization. A previous study showed that a microinjection of quail CS alone into a homogeneous egg did not have the ability to generate Ca^{2+} waves (Mizushima et al. 2014).

On the other hand, PLCζ is solely expressed in the sperm of most animals such as mammals (Cox et al. 2002; Saunders et al. 2002; Yoneda et al. 2006, Young et al. 2009), birds (Coward et al. 2005; Mizushima et al. 2008, 2009), and medaka (Ito et al. 2008). This ensures the sperm-specific enrichment of egg-activating factors, consistent with a gamete-specific role. Genomic DNA analyses have revealed that the PLCζ gene is located back to back with another testis-specific gene, the actin-capping protein gene *CAPZA3* (Hurst et al. 1998; Yoshimura et al. 1999; Miyagawa et al. 2002), which is inserted in the 5′-region of PLCζ and shares a bidirectional promoter with PLCζ (Coward et al. 2005, 2011), suggesting that the expression of CAPZA3 and PLCζ is male germ cell- or testis-specifically transcripted in the same process. However, the PLCζ of puffer fish was expressed in the brain and ovaries, but not in the testes. More interestingly, the CAPZA3 genes of puffer fish and medaka were not adjacent to the PLCζ genes (Coward et al. 2011). Although difficulties are associated with accounting for variations in the tissue-specific expression patterns of PLCζ in these fishes, during the rapid evolution of egg activation mechanisms, PLCζ seemed to be the first important differentiation factor for sperm function in the vertebrate species. Pufferfish PLCζ does not have the ability to activate the release of Ca^{2+} in mouse eggs; however, this does not exclude the possibility that the activation of PLCζ by a fertilizing sperm functions as an egg-derived Ca^{2+}-releasing factor and/or plays some other role in the egg. Further investigations will provide an insight into the functional role and genesis of PLCζ.

PLCζ has not yet been detected in intermediate species between fishes and higher eutherian mammals, such as primitive mammals including the *monotrematous* platypus and small marsupial mammals (Table 7.1). The platypus, *Ornithorhynchus anatinus*, lays large yolky eggs that exhibit polyspermy, and several sperm have been suggested to enter the egg cytoplasm (Gatenby and Hill 1924; Hughes and Hall 1998). Furthermore, the eggs of the small marsupial mammal, *Sminthopsis crassicaudata*, are relatively small, but occasionally exhibit polyspermy (Breed and Leigh 1990). Although changes in $[Ca^{2+}]i$ at fertilization have not yet been examined in primitive mammals, egg activation in the ancestors of mammals may be achieved by polyspermy, with eggs showing primitive Ca^{2+} oscillations. Since the platypus, the most primitive mammal, shares common molecular, genetic, and morphological features with birds and reptiles (Rens et al. 2007) and they are in a comparatively close cluster taxonomically (Warren et al. 2008), the spiral-like Ca^{2+} oscillations observed in avian egg activation may be closely related to an ancestor of mammalian Ca^{2+} oscillations. In this respect, these primitive mammalian species may possess a PLCζ orthologue or intermediate molecules similar to three egg-activating factors.

7.3 Syngamy and Elimination of Supernumerary Sperm in the Egg Cytoplasm

In monospermic species, frog, and *Hynobium* salamander eggs, increases in Ca^{2+} open Ca^{2+}-dependent Cl^- channels on the egg plasma membrane in order to produce rapid depolarization, which prevents the penetration of additional fertilizing sperm (Cross and Elinson 1980; Iwao 1989; Iwao and Jaffe 1989). Previous studies demonstrated that rapid depolarization or a cortical reaction does not occur in polyspermic urodeles (Charbonneau et al. 1983; Iwao 2012). In polyspermic frog, *Discoglossus* eggs exhibit a fertilization-activated membrane potential due to the opening of Cl^- channels; however, it does not block additional sperm entry (Talevi 1989). On the other hand, hyperpolarization mediated by Na^+ channels in response to each sperm penetration has been indicated in polyspermic newt eggs (Iwao et al. 1985), but does not prevent second sperm penetration (Iwao and Jaffe 1989). Another polyspermy block system, the cortical reaction described above, is well-developed in mammalian eggs (Yanagimachi 1994; Quesada et al. 2004; Wong and Wessel 2006; Liu 2011; Burkart et al. 2012), whereas the dynamic movement and exocytosis of cortical granules are not likely in newt eggs (Iwao 2000). Although electrical responses in avian eggs have not yet been examined, a review of findings obtained using several amphibian eggs suggests that there is neither electrical regulation on the surface of egg plasma membranes nor alterations in the extracellular matrix that prevent polyspermy in all polyspermic species.

Even though numerous sperm are incorporated into polyspermic eggs, the egg nucleus proceeds to karyogamy with a single sperm nucleus because of the presence of an ooplasmic block to escape polyploidy. In birds, a large number of sperm penetrate the egg cytoplasm, most of which undergo transformation into male pronuclei after swelling, a change from the elongated form to the spherical head, chromatin decondensation, and reconstitution of the nuclear envelope (Fofanova 1965; Okamura and Nishiyama 1978; Perry 1987; Nakanishi et al. 1990; Waddington et al. 1998). However, supernumerary sperm (accessory sperm) nuclei, except for the one (principal sperm) that unites with the female pronucleus to form a zygote, move towards the periphery of the germinal disc; it is not in the yolk, but in the vicinity of cytoplasm, and supernumerary sperm nuclei undergo one or two mitoses (Perry 1987). This dynamic movement of accessory sperm in avian eggs has not been observed in other polyspermic species. The mitosis of accessory sperm occurs in synchrony with the zygotic nucleus, and the resultant cell division at the peripheral region of the egg cytoplasm disappears during the early cleavage stage (Patterson 1910; Eyal-Giladi and Kochav 1976). Such nuclear division without DNA synthesis may represent a deviation from periodic nuclear activation (Emanuelsson 1965; Gurdon and Woodland 1968). It is important to note that, in addition to this, since ICSI with a single sperm, without accessory sperm-activating temporal cleavage, has been shown to develop quail eggs to the late blastoderm stage (Hrabia et al. 2003; Takagi et al. 2007a; Mizushima et al. 2007, 2008, 2014; Mizushima 2012; Kang et al. 2015), accessory sperm nuclei are not essential for any cellular events related to zygotic development. Although the molecular mechanisms suppressing supernumerary sperm nuclei have not been elucidated fully, the involvement of maternal-derived deoxyribonucleases (DNase I and II) in birds, which are not expressed in monospermic mammalian eggs (Stepinska and Olszanska 2001, 2003; Olszanska and Stepinska 2008), is under debate.

On the other hand, the mechanisms by which the movement of the male and female pronuclei forms the zygote nucleus remain obscure in birds. This assures that only one sperm is selected as the principal sperm by some unknown mechanisms, and is subsequently paired with the female pronucleus in the center of the germinal disc 3–4 h after fertilization (Fig. 7.4; Perry 1987). Female metaphase II chromatin is localized in the superficial cytoplasm near the center of the germinal disc when the avian egg is ovulated (Perry 1987); however, even in the presence of numerous sperm, the female pronucleus still occupies a central position (Perry 1987), indicating that the selected principal sperm pronucleus appears to move toward the female pronucleus. In fertilized newt eggs, only one sperm nucleus, possibly that nearest to the female nucleus, forms a larger sperm aster than that of accessory sperm, and then makes contact with the female pronucleus in the center of the animal hemisphere (Iwao et al. 2002). The maternal γ-tubulin predominantly distributed in the animal hemisphere strongly accumulates in the centrosomes of the one principal sperm nucleus and subsequent zygote nucleus in order to promote microtubule polymerization, whereas only a small amount of γ-tubulin is associated with those of other sperm nuclei (Iwao et al. 2002; Morito et al. 2005). γ-tubulin is a well-known

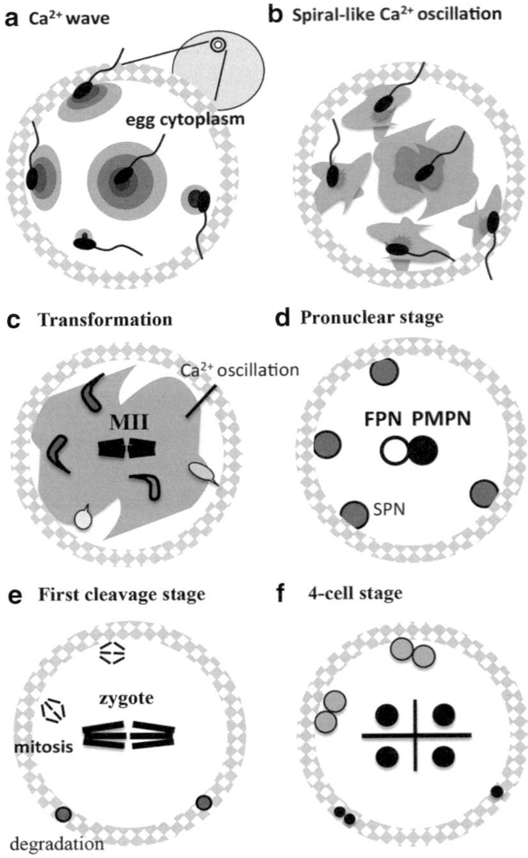

Fig. 7.4 Ca^{2+} rise and zygote formation in avian egg cytoplasm. **a** Immediately after sperm penetration, Ca^{2+} waves spreads in the whole egg cytoplasm from each sperm entry site. **b** At 15 min, a spiral-like Ca^{2+} spikes from each sperm entry site are generated and it continues to pronuclear stage. **c** At 1.5 h, the female chromatin is in metaphase II (MII) of the second meiosis and the incorporated sperm are starting to transform to pronuclei in order of entry into egg cytoplasm. **d** At 3 h, female (FPN) and principal male pronucleus (PMPN) places in the center of egg cytoplasm, whilst supernumerary sperm pronuclei (SPN) move towards the the vicinity of cytoplasm. **e** At 4.0 h, the zygotic nucleus and several supernumerary sperm nuclei undergo mitosis. **f** At 5 h, a zygotic nucleus progresses towards 4-cell stages and supernumerary sperm are degraded by DNases. Modified from J Anat 150:99–109 (Perry 1987)

major component of the MTOC, and is involved in the movement of the male pronucleus to fuse with the female pronucleus (Haren et al. 2006; Eot-Houllier et al. 2010), suggesting that γ-tubulin is one of the key factors in the event related to the selection of the principal sperm nucleus in newts (Reinsch and Karsenti 1997; Iwao et al. 2002). Furthermore, since the principal sperm and female nuclei enter the DNA synthesis phase of the first cleavage earlier than the accessory sperm nuclei, zygotic nuclei can enter the mid-term of the first cleavage. This is because cyclin B that forms the MPF complex accumulates in and disappears from the zygotes earlier than accessory sperm cells (Iwao and Elinson 1990; Iwao et al. 2002). The failed progression of cyclic nuclear activities in accessory sperm may be due to insufficient exposure of the MPF complex (Iwao et al. 1993; Sakamoto et al. 1998). In addition to the possible participation of the motor proteins, dynein, dynactin, and kinesin in opposite movements between the principal and accessory sperm nuclei (Payne et al. 2003; Waitzman and Rice 2014), further investigations on the sequential distribution of avian cdc2 (Mori et al. 1991) and cyclin B as well as γ-tubulin in the egg will provide insights into not only the selection of principal sperm nuclei, but also the cellular process of zygotic fusion.

Conclusion

Development of avian ICSI technique has brought us a new schematic diagram in avian fertilization (Fig. 7.3). However, new findings such as novel sperm-derived egg-activating factors are only a part of the mysterious events of polyspermic fertilization. Therefore, more information will be needed in order to understand the comprehensive molecular mechanism of avian fertilization. Experiments using gene-disrupted animals are very powerful tools for validating which factors are essential, which has also contributed to finding novel genes in many species. In particular, in-vitro fertilization study is one of the suitable research fields to use gene-manipulated animals. Fortunately, the TALEN and Crisper/Cas 9 systems have opened a new door for avian gene-disruption (see Chap. 12 for more details). The combination of avian ICSI and gene-manipulation systems will make a significant progress in our understanding of the avian fertilization system.

References

Abbott AL, Ducibella T. Calcium and the control of mammalian cortical granule exocytosis. Front Biosci. 2001;6:D792–806. doi:10.2741/A643.

Abraham VC, Gupta S, Fluck RA. Ooplasmic segregation in the medaka (*Oryzias latipes*) egg. Biol Bull. 1993;184(2):115–24.

Amano T, Mori T, Watanabe T. Activation and development of porcine oocytes matured in vitro following injection of inositol 1,4,5-trisphosphate. Anim Reprod Sci. 2004;80(1–2):101–12. doi:10.1016/S0378-4320(03)00115-5.

Batellier F, Couty I, Olszanska B, et al. In vitro fertilization of chicken oocytes after in vitro ovulation. Br Poult Sci. 2003;44(5):819–20. doi:10.1080/00071660410001667014.

Birkhead TR, Sheldon BC, Fletcher F. A comparative study of sperm–egg interactions in birds. J Reprod Fertil. 1994;101(2):353–61. doi:10.1530/jrf.0.1010353.

Bramwell RK, Marks HL, Howarth B. Quantitative determination of spermatozoa penetration of the perivitelline layer of the hen's ovum as assessed on oviposited eggs. Poult Sci. 1995;74(11):1875–83. doi:10.3382/ps.0741875.

Breed WG, Leigh CM. Morphological changes in the oocyte and its surrounding vestments during in vivo fertilization in the dasyurid marsupial Sminthopsis crassicaudata. J Morphol. 1990;204(2):177–96. doi:10.1002/jmor.1052040207.

Brind S, Swann K, Carroll J. Inositol 1,4,5-trisphosphate receptors are downregulated in mouse oocytes in response to sperm or adenophostin A but not to increases in intracellular Ca^{2+} or egg activation. Dev Biol. 2000;223(2):251–65. doi:10.1006/dbio.2000.9728.

Burkart AD, Xiong B, Baibakov B, et al. Ovastacin, a cortical granule protease, cleaves ZP2 in the zona pellucida to prevent polyspermy. J Cell Biol. 2012;197(1):37–44. doi:10.1083/jcb.201112094.

Busa WB, Nuccitelli R. An elevated free cytosolic Ca^{2+} wave follows fertilization in eggs of the frog, *Xenopus laevis*. J Cell Biol. 1985;100(4):1325–9. doi:10.1083/jcb.100.4.1325.

Charbonneau M, Moreau M, Picheral B, et al. Fertilization of amphibian eggs: a comparison of electrical responses between anurans and urodeles. Dev Biol. 1983;98(2):304–18. doi:10.1016/0012-1606(83)90361-5.

Coward K, Campos-Mendoza A, Larman M, et al. Teleost fish spermatozoa contain a cytosolic protein factor that induces calcium release in sea urchin egg homogenates and triggers calcium oscillations when injected into mouse oocytes. Biochem Biophys Res Commun. 2003;305(2):299–304. doi:10.1016/S0006-291X(03)00753-8.

Coward K, Ponting CP, Chang HY, et al. Phospholipase Czeta, the trigger of egg activation in mammals, is present in a non-mammalian species. Reproduction. 2005;130(2):157–63. doi:10.1530/rep.1.00707.

Coward K, Ponting CP, Zhang N, et al. Identification and functional analysis of an ovarian form of the egg activation factor phospholipase Czeta (PLCζ) in pufferfish. Mol Reprod Dev. 2011;78(1):48–56. doi:10.1002/mrd.21262.

Cox LJ, Larman MG, Saunders CM, et al. Sperm phospholipase Cζ from humans and cynomolgus monkeys triggers Ca^{2+} oscillations, activation and development of mouse oocytes. Reproduction. 2002;124(5):611–23. doi:10.1530/rep.0.1240611.

Creton R, Speksnijder JE, Jaffe LF. Patterns of free calcium in zebrafish embryos. J Cell Sci. 1998;111(Pt 12):1613–22.

Cross NL, Elinson RP. A fast block to polyspermy in frogs mediated by changes in the membrane potential. Dev Biol. 1980;75(1):187–98. doi:10.1016/0012-1606(80)90154-2.

Dean J, Cohen G, Kemp J, et al. Karyotype 69, XXX/47, XX, +15 in a 2 1/2 year old child. J Med Genet. 1997;34(3):246–9. doi:10.1136/jmg.34.3.246.

Deguchi R, Osanai K. Serotonin-induced meiosis reinitiation from the first prophase and from the first metaphase in oocytes of the marine bivalve Hiatella flaccida: respective changes in intracellular Ca^{2+} and pH. Dev Biol. 1995;171(2):483–96. doi:10.1006/dbio.1995.1298.

Dong JB, Tang TS, Sun FZ. *Xenopus* and chicken sperm contain a cytosolic soluble protein factor which can trigger calcium oscillations in mouse eggs. Biochem Biophys Res Commun. 2000;268(3):947–51. doi:10.1006/bbrc.2000.2218.

Ducibella T, Huneau D, Angelichio E, et al. Egg-to-embryo transition is driven by differential responses to Ca^{2+} oscillation number. Dev Biol. 2002;250(2):280–91. doi:10.1006/dbio.2002.0788.

Elinson RP. Fertilization in amphibians: the ancestry of the block to polyspermy. Int Rev Cytol. 1986;101:59–100. doi:10.1016/S0074-7696(08)60246-6.

Emanuelsson H. Cell manipulation in the chick blastoderm up to the time of laying. Exp Cell Res. 1965;39(2):386–99. doi:10.1016/0014-4827(65)90042-X.

Eot-Houllier G, Venoux M, Vidal-Eychenie S, et al. Plk1 regulates both ASAP localization and its role in spindle pole integrity. J Biol Chem. 2010;285(38):29556–68. doi:10.1074/jbc.M110.144220.

Eyal-Giladi H, Kochav S. From cleavage to primitive streak formation: a complementary normal table and a new look at the first stages of the development of the chick. Dev Biol. 1976;49(2):321–37. doi:10.1016/0012-1606(76)90178-0.

Fechheimer NS. Origins of heteroploidy in chicken embryos. Poult Sci. 1981;60(7):1365–71. doi:10.3382/ps.0601365.

Fissore RA, Robl JM. Sperm, inositol trisphosphate, and thimerosal-induced intracellular Ca^{2+} elevations in rabbit eggs. Dev Biol. 1993;159(1):122–30. doi:10.1006/dbio.1993.1226.

Fluck RA, Miller AL, Jaffe LF. Slow calcium waves accompany cytokinesis in medaka fish eggs. J Cell Biol. 1991;115(5):1259–65. doi:10.1083/jcb.115.5.1259.

Fofanova KA. Morphologic data on polyspermy in chickens. Fed Proc Transl Suppl. 1965;24:239–47.

Fontanilla RA, Nuccitelli R. Characterization of the sperm-induced calcium wave in *Xenopus* eggs using confocal microscopy. Biophys J. 1998;75(4):2079–87. doi:10.1016/S0006-3495(98)77650-7.

Galeati G, Modina S, Lauria A, et al. Follicle somatic cells influence pig oocyte penetrability and cortical granule distribution. Mol Reprod Dev. 1991;29(1):40–6. doi:10.1002/mrd.1080290107.

Galione A, McDougall A, Busa W, et al. Redundant mechanisms of calcium-induced calcium release underlying calcium waves during fertilization of sea urchin eggs. Science. 1993;261(5119):348–52. doi:10.1126/science.8392748.

Gardner AJ, Williams CJ, Evans JP. Establishment of the mammalian membrane block to polyspermy: evidence for calcium-dependent and -independent regulation. Reproduction. 2007;133(2):383–93. doi:10.1530/REP-06-0304.

Gatenby JB, Hill JP. On an ovum of Ornithorhynchus exhibiting polar bodies and polyspermy. J Cell Sci. 1924;s2–68(270):229–38.

Gould MC, Stephano JL. Polyspermy prevention in marine invertebrates. Microsc Res Tech. 2003;61(4):379–88. doi:10.1002/jemt.10351.

Gurdon JB, Woodland HR. The cytoplasmic control of nuclear activity in animal development. Biol Rev Camb Philos Soc. 1968;43(2):233–67. doi:10.1111/j.1469-185X.1968.tb00960.x.

Han YM, Wang WH, Abeydeera LR, et al. Pronuclear location before the first cell division determines ploidy of polyspermic pig embryos. Biol Reprod. 1999;61(5):1340–6. doi:10.1095/biolreprod61.5.1340.

Harada Y, Matsumoto T, Hirahara S, et al. Characterization of a sperm factor for egg activation at fertilization of the newt *Cynops pyrrhogaster*. Dev Biol. 2007;306(2):797–808. doi:10.1016/j.ydbio.2007.04.019.

Harada Y, Kawazoe M, Eto Y, et al. The Ca^{2+} increase by the sperm factor in physiologically polyspermic newt fertilization: its signaling mechanism in egg cytoplasm and the species-specificity. Dev Biol. 2011;351(2):266–76. doi:10.1016/j.ydbio.2011.01.003.

Haren L, Remy MH, Bazin I, et al. NEDD1-dependent recruitment of the γ-tubulin ring complex to the centrosome is necessary for centriole duplication and

spindle assembly. J Cell Biol. 2006;172(4):505–15. doi:10.1083/jcb.200510028.
Harper EH. The fertilization and early development of the pigeon's egg. Am J Anat. 1904;3(4):349–86.
Homa ST, Swann K. A cytosolic sperm factor triggers calcium oscillations and membrane hyperpolarizations in human oocytes. Hum Reprod. 1994;9(12):2356–61. doi:10.1042/bj3130369.
Hrabia A, Takagi S, Ono T, et al. Fertilization and development of quail oocytes after intracytoplasmic sperm injection. Biol Reprod. 2003;69(5):1651–7. doi:10.1095/biolreprod.103.019315.
Hughes RL, Hall LS. Early development and embryology of the platypus. Philos Trans R Soc Lond B Biol Sci. 1998;353(1372):1101–14. doi:10.1098/rstb.1998. 0269.
Hurst S, Howes EA, Coadwell J, et al. Expression of a testis-specific putative actin-capping protein associated with the developing acrosome during rat spermiogenesis. Mol Reprod Dev. 1998;49(1):81–91. doi:10.1002/(SICI)1098-2795(199801)49:1<81::AID-MRD9>3.0.CO;2-K.
Igusa Y, Miyazaki S, Yamashita N. Periodic hyperpolarizing responses in hamster and mouse eggs fertilized with mouse sperm. J Physiol. 1983;340:633–47. doi:10.1113/jphysiol.1983.sp014784.
Ito M, Shikano T, Oda S, et al. Difference in Ca^{2+} oscillation-inducing activity and nuclear translocation ability of PLCZ1, an egg-activating sperm factor candidate, between mouse, rat, human, and medaka fish. Biol Reprod. 2008;78(6):1081–90. doi:10.1095/biolreprod.108.067801.
Iwamatsu T, Ohta T. Electron microscopic observation on sperm penetration and pronuclear formation in the fish egg. J Exp Zool. 1978;205(2):157–80. doi:10.1002/jez.1402050202.
Iwao Y. An electrically mediated block to polyspermy in the primitive urodele Hynobius nebulosus and phylogenetic comparison with other amphibians. Dev Biol. 1989;134(2):438–45. doi:10.1016/0012-1606(89)90116-4.
Iwao Y. Fertilization in amphibians. In: Tarin JJ, Cano A, editors. Fertilization in Protozoa and Metazoan Animal. Berlin: Springer; 2000. p. 147–91.
Iwao Y. Egg activation in physiological polyspermy. Reproduction. 2012;144(1):11–22. doi:10.1530/REP-12-0104.
Iwao Y, Elinson RP. Control of sperm nuclear behavior in physiologically polyspermic newt eggs: possible involvement of MPF. Dev Biol. 1990;142(2):301–12. doi:10.1016/0012-1606(90)90351-I.
Iwao Y, Jaffe LA. Evidence that the voltage-dependent component in the fertilization process is contributed by the sperm. Dev Biol. 1989;134(2):446–51. doi:10.1016/0012-1606(89)90117-6.
Iwao Y, Yamasaki H, Katagiri C. Experiments pertaining to the suppression of accessory sperm in fertilized newt eggs. Dev Growth Differ. 1985;27(3):323–31. doi:10.1111/j.1440-169X.1985. 00323.x.
Iwao Y, Sakamoto N, Takahara K, et al. The egg nucleus regulates the behavior of sperm nuclei as well as cycling of MPF in physiologically polyspermic newt eggs. Dev Biol. 1993;160(1):15–27. doi:10.1006/dbio.1993.1282.
Iwao Y, Murakawa T, Yamaguchi J, et al. Localization of γ-tubulin and cyclin B during early cleavage in physiologically polyspermic newt eggs. Dev Growth Differ. 2002;44(6):489–99. doi:10.1046/j.1440-169X.2002.00661.x.
Jaffe LA, Gould M. Polyspermy-preventing mechanisms. In: Metz CB, Monroy A, editors. Biology of fertilization: the fertilization response of the egg. Orlando: Academic Press; 1985. p. 223–50.
Jaffe LA, Sharp AP, Wolf DP. Absence of an electrical polyspermy block in the mouse. Dev Biol. 1983;96(2):317–23. doi:10.1016/0012-1606(83)90168-9.
Jones KT. Ca^{2+} oscillations in the activation of the egg and development of the embryo in mammals. Int J Dev Biol. 1998;42(1):1–10.
Jones KT. Mammalian egg activation: from Ca^{2+}spiking to cell cycle progression. Reproduction. 2005;130(6):813–23. doi:10.1530/rep.1.00710.
Jones KT. Intracellular calcium in the fertilization and development of mammalian eggs. Clin Exp Pharmacol Physiol. 2007;34(10):1084–9. doi:10.1111/j.1440-1681.2007.04726.x.
Jones KT, Carroll J, Merriman JA, et al. Repetitive sperm-induced Ca^{2+} transients in mouse oocytes are cell cycle dependent. Development. 1995;121(10):3259–66.
Kang KS, Park TS, Rengaraj D, et al. Fertilization of cryopreserved sperm and unfertilized quail ovum by intracytoplasmic sperm injection. Reprod Fertil Dev. 2016;28(12):1974–1981. doi:10.1071/RD15126.
Keating TJ, Cork RJ, Robinson KR. Intracellular free calcium oscillations in normal and cleavage-blocked embryos and artificially activated eggs of *Xenopus laevis*. J Cell Sci. 1994;107(Pt 8):2229–37.
Kobayashi W, Baba Y, Shimozawa T, et al. The fertilization potential provides a fast block to polyspermy in lamprey eggs. Dev Biol. 1994;161(2):552–62. doi:10.1006/dbio.1994.1053.
Kouchi Z, Fukami K, Shikano T, et al. Recombinant phospholipase Cζ has high Ca^{2+} sensitivity and induces Ca^{2+} oscillations in mouse eggs. J Biol Chem. 2004;279(11):10408–12. doi:10.1074/jbc.M313 301200.
Kuroda K, Ito M, Shikano T, et al. The role of X/Y linker region and N-terminal EF-hand domain in nuclear translocation and Ca^{2+} oscillation-inducing activities of phospholipase Czeta, a mammalian egg-activating factor. J Biol Chem. 2006;281(38):27794–805. doi:10.1074/jbc.M603473200.
Larman MG, Saunders CM, Carroll J, et al. Cell cycle-dependent Ca^{2+} oscillations in mouse embryos are regulated by nuclear targeting of PLCζ. J Cell Sci. 2004;117(Pt 12):2513–21. doi:10.1242/jcs.01109.

Lawrence Y, Ozil JP, Swann K. The effects of a Ca^{2+} chelator and heavy-metal-ion chelators upon Ca^{2+} oscillations and activation at fertilization in mouse eggs suggest a role for repetitive Ca^{2+} increases. Biochem J. 1998;335(Pt 2):335–42. doi:10.1042/bj3350335.

Lee HC, Aarhus R, Walseth TF. Calcium mobilization by dual receptors during fertilization of sea urchin eggs. Science. 1993;261(5119):352–5. doi:10.1126/science.8392749.

Lee B, Yoon SY, Malcuit C, et al. Inositol 1,4,5-trisphosphate receptor 1 degradation in mouse eggs and impact on $[Ca^{2+}]i$ oscillations. J Cell Physiol. 2010;222(1):238–47. doi:10.1002/jcp.21945.

Leikin YN, Zharova TV, Tjulina OV. Novel oxaloacetate effect on mitochondrial Ca^{2+} movement. FEBS Lett. 1993;331 (1–2):35–7. doi:10.1016/0014-5793(93)80292-3.

Liu M. The biology and dynamics of mammalian cortical granules. Reprod Biol Endocrinol. 2011;9:149. doi:10.1186/1477-7827-9-149.

Mahbub Hasan AK, Hashimoto A, Maekasa Y, et al. The egg membrane microdomain-associated uroplakin III-Src system becomes functional during oocyte maturation and is required for bidirectional gamete signaling at fertilization in *Xenopus Laevis*. Development. 2014a;141(8):1705–14. doi:10.1242/dev.105510.

Mahbub Hasan AK, Sato K, Sakakibara K, et al. Uroplakin III, a novel Src substrate in *Xenopus* egg rafts, is a target for sperm protease essential for fertilization. Dev Biol. 2005;286(2):483–92. doi:10.1016/j.ydbio.2005.08.020.

Mahbub Hasan AK, Hashimoto A, Maekasa Y, et al. The egg membrane microdomain-associated uroplakin III-Src system becomes functional during oocyte maturation and is required for bidirectional gamete signaling at fertilization in *Xenopus Laevis*. Development. 2014b;141(8):1705–14. doi:10.1242/dev.105510.

Malcuit C, Knott JG, He C, et al. Fertilization and inositol 1,4,5-trisphosphate (IP3)-induced calcium release in type-1 inositol 1,4,5-trisphosphate receptor down-regulated bovine eggs. Biol Reprod. 2005;73(1):2–13. doi:10.1095/biolreprod.104.037333.

Marangos P, FitzHarris G, Carroll J. Ca^{2+} oscillations at fertilization in mammals are regulated by the formation of pronuclei. Development. 2003;130(7):1461–72. doi:10.1242/dev.00340.

McAvey BA, Wortzman GB, Williams CJ, et al. Involvement of calcium signaling and the actin cytoskeleton in the membrane block to polyspermy in mouse eggs. Biol Reprod. 2002;67(4):1342–52. doi:10.1095/biolreprod.102.004630.

McCulloh DH, Rexroad CE Jr, Levitan H. Insemination of rabbit eggs is associated with slow depolarization and repetitive diphasic membrane potentials. Dev Biol. 1983;95(2):372–7. doi:10.1016/0012-1606(83)90038-6.

Mehlmann LM, Carpenter G, Rhee SG, et al. SH2 domain-mediated activation of phospholipase Cγ is not required to initiate Ca^{2+} release at fertilization of mouse eggs. Dev Biol. 1998;203(1):221–32. doi:10.1006/dbio.1998.9051.

Missiaen L, Parys JB, Smedt HD, et al. Effect of adenine nucleotides on myo-inositol-1, 4, 5-trisphosphate-induced calcium release. Biochem J. 1997;325(Pt 3):661–6. doi:10.1042/bj3250661.

Miyagawa Y, Tanaka H, Iguchi N, et al. Molecular cloning and characterization of the human orthologue of male germ cell-specific actin capping protein alpha3 (CPalpha3). Mol Hum Reprod. 2002;8(6):531–93. doi:10.1093/molehr/8.6.531.

Miyazaki S. Inositol 1,4,5-trisphosphate-induced calcium release and guanine nucleotide-binding protein-mediated periodic calcium rises in golden hamster eggs. J Cell Biol. 1988;106(2):345–53.

Miyazaki S. Thirty years of calcium signals at fertilization. Semin Cell Dev Biol. 2006;17(2):233–43. doi:10.1016/j.semcdb.2006.02.007.

Miyazaki S, Igusa Y. Fertilization potential in golden hamster eggs consists of recurring hyperpolarizations. Nature. 1981;290(5808):702–4. doi:10.1038/290702a0.

Miyazaki S, Yuzaki M, Nakada K, et al. Block of Ca^{2+} wave and Ca^{2+} oscillation by antibody to the inositol 1,4,5-trisphosphate receptor in fertilized hamster eggs. Science. 1992;257(5067):251–5. doi:10.1126/science.1321497.

Miyazaki S, Shirakawa H, Nakada K, et al. Essential role of the inositol 1,4,5,-trisphosphate receptor/Ca^{2+} release channel in Ca^{2+} waves and Ca^{2+} oscillations at fertilization of mammalian eggs. Dev Biol. 1993;158(1):62–78. doi:10.1006/dbio.1993.1168.

Mizushima S. Establishment of intracytoplasmic sperm injection technique in Japanese quail and its possible application for poultry resources and transgenic birds. J Poult Sci. 2012;49(4):225–30. doi:10.2141/jpsa.0120042.

Mizushima S, Takagi S, Ono T, et al. Possible role of calcium on oocyte development after intracytoplasmic sperm injection in quail (Coturnix japonica). J Exp Zool A Ecol Genet Physiol. 2007;307(11):647–53. doi:10.1002/jez.a.418.

Mizushima S, Takagi S, Ono T, et al. Developmental enhancement of intracytoplasmic sperm injection (ICSI)-generated quail embryos by phospholipase Cζ cRNA. J Poult Sci. 2008;45(2):152–8. doi:10.2141/jpsa.45.152.

Mizushima S, Takagi S, Ono T, et al. Phospholipase Cζ mRNA expression and its potency during spermatogenesis for activation of quail oocytes a sperm factor. Mol Reprod Dev. 2009;76(12):1200–7. doi:10.1002/mrd.21097.

Mizushima S, Takagi S, Ono T, et al. Novel method of gene transfer in birds: intracytoplasmic sperm injection for green fluorescent protein expression in quail blastoderms. Biol Reprod. 2010;83(6):965–9. doi:10.1095/biolreprod.110.085860.

Mizushima S, Hiyama G, Shiba K, et al. The birth of quail chicks after intracytoplasmic sperm injection. Development. 2014;141(19):3799–806. doi:10.1242/dev.111765.

Moore GD, Kopf GS, Schultz RM. Complete mouse egg activation in the absence of sperm by stimulation of an exogenous G protein-coupled receptor. Dev Biol. 1993;159(2):669–78. doi:10.1006/dbio.1993.1273.

Mori M, Yamashita M, Yoshikuni M, et al. Maturation-promoting factor and p34cdc2 kinase during oocyte maturation of the Japanese quail. Dev Biol. 1991;146(1):246–9. doi:10.1016/0012-1606(91)90465-F.

Morito Y, Terada Y, Nakamura S, et al. Dynamics of microtubules and positioning of female pronucleus during bovine parthenogenesis. Biol Reprod. 2005;73(5):935–41. doi:10.1095/biolreprod.105.042366.

Nagai K, Ishida T, Hashimoto T, et al. The sperm-surface glycoprotein, SGP, is necessary for fertilization in the frog, *Xenopus laevis*. Dev Growth Differ. 2009;51(5):499–510. doi:10.1111/j.1440-169X.2009.01112.x.

Nakada K, Mizuno J, Shiraishi K, et al. Initiation, persistence, and cessation of the series of intracellular Ca^{2+} responses during fertilization of bovine eggs. J Reprod Dev. 1995;41(1):77–84. doi:10.1262/jrd.41.77.

Nakanishi A, Utsumi K, Iritani A. Early nuclear events of in vitro fertilization in the domestic fowl (Gallus domesticus). Mol Reprod Dev. 1990;26(3):217–21. doi:10.1002/mrd.1080260304.

Nuccitelli R. The fertilization potential is not necessary for the block to polyspermy or the activation of development in the medaka egg. Dev Biol. 1980;76(2):499–504. doi:10.1016/0012-1606(80) 90397-8.

Nuccitelli R, Kline D, Busa WB, et al. A highly localized activation current yet widespread intracellular calcium increase in the egg of the frog, Discoglossus pictus. Dev Biol. 1988;130(1):120–32. doi:10.1016/0012-1606(88)90419-8.

Okamura F, Nishiyama H. Penetration of spermatozoa into the ovum and transformation of the sperm nucleus into the male pronucleus in the domestic fowl, Gallus gallus. Cell Tissue Res. 1978;190(1):89–98. doi:10.1007/BF00210039.

Olszanska B, Stepinska U, Perry MM. Development of embryos from in vitro ovulated and fertilized oocytes of the quail (*Coturnix coturnix japonica*). J Exp Zool. 2002;292(6):580–6. doi:10.1002/jez.10096.

Olszanska B, Stepinska U. Molecular aspects of avian oogenesis and fertilisation. The Int J Dev Bio. 2008;52(2-3):187–194.

Ozil JP. Role of calcium oscillations in mammalian egg activation: experimental approach. Biophys Chem. 1998;72(1-2):141–52. doi:10.1016/S0301-4622(98)00130-6.

Parrington J, Lai FA, Swann K. The soluble mammalian sperm factor protein that triggers Ca^{2+} oscillations in eggs: Evidence for expression of mRNA(s) coding for sperm factor protein(s) in spermatogenetic cells. Biol Cell. 2000;92(3-4):267–75. doi:10.1016/S0248-4900(00)01064-9.

Patterson JT. Studies on the early development of the hen's egg. I History of the early cleavage and of the accessory cleavage. J Morphol. 1910;21(1):101–34.

Payne C, Rawe V, Ramalho-Santos J, et al. Preferentially localized dynein and perinuclear dynactin associate with nuclear pore complex proteins to mediate genomic union during mammalian fertilization. J Cell Sci. 2003;116(Pt 23):4727–38. doi:10.1242/jcs.00784.

Perry MM Nuclear events from fertilization to the early cleavage stages in the domestic fowl (*Gallus domesticus*). J Anat. 1987;150:99–109.

Perry AC, Wakayama T, Yanagimachi R. A novel trans-complementation assay suggests full mammalian oocyte activation is coordinately initiated by multiple, submembrane sperm components. Biol Reprod. 1999;60(3):747–55. doi:10.1095/biolreprod60.3.747.

Perry AC, Wakayama T, Cooke IM. Mammalian oocyte activation by the synergistic action of discrete sperm head components: induction of calcium transients and involvement of proteolysis. Dev Biol. 2000;217(2):386–93. doi:10.1006/dbio.1999.9552.

Quesada V, Sanchez LM, Alvarez J, et al. Identification and characterization of human and mouse ovastacin: a novel metalloproteinase similar to hatching enzymes from arthropods, birds, amphibians, and fish. J Biol Chem. 2004;279(25):26627–34. doi:10.1074/jbc.M401588200.

Rabbani MG, Sasanami T, Mori M, et al. Sperm–egg interaction is mediated by a sperm-associated body in quail. Dev Growth Differ. 2006;48(1):33–40. doi:10.1111/j.1440-169X.2006.00842.x.

Rabbani MG, Sasanami T, Mori M, et al. Characterization of the sperm-associated body and its role in the fertilization of the chicken Gallus domestics. Dev Growth Differ. 2007;49(1):39–48. doi:10.1111/j.1440-169X.2007.00903.x.

Reinsch S, Karsenti E. Movement of nuclei along microtubules in *Xenopus* egg extracts. Curr Biol. 1997;7(3):211–4. doi:10.1016/S0960-9822(97)70092-7.

Rens W, O'Brien PC, Grutzner F, et al. The multiple sex chromosomes of platypus and echidna are not completely identical and several share homology with the avian Z. Genome Biol. 2007;8(11):R243. doi:10.1186/gb-2007-8-11-r243.

Rhee SG. Regulation of phosphoinositide-specific phospholipase C. Annu Rev Biochem. 2001;70:281–312. doi:10.1146/annurev.biochem.70.1.281.

Ridgway EB, Gilkey JC, Jaffe LF. Free calcium increases explosively in activating medaka eggs. Proc Natl Acad Sci U S A. 1977;74(2):623–7.

Roberts FE, Saxe DF, Muralidharan K, et al. Unique mosaicism of tetraploidy and trisomy 8: clinical, cytogenetic, and molecular findings in a live-born infant. Am J Med Genet. 1996;62(3):243–6. doi:10.1002/(SICI)1096-8628(19960329)62:3<243::AID-AJMG8>3.0.CO;2-U.

Ross PJ, Beyhan Z, Iager AE, et al. Parthenogenetic activation of bovine oocytes using bovine and murine phospholipase C zeta. BMC Dev Biol. 2008;8:16. doi:10 1186/1471-213X-8-16.

Rugh R. The frog. Its reproduction and Development. Yale J Biol Med. 1951;23(5):441–2.

Runft LL, Jaffe LA, Mehlmann LM. Egg activation at fertilization: where it all begins. Dev Biol. 2002;245(2):237–54. doi:10.1006/dbio.2002.0600.

Sakakibara K, Sato K, Yoshino K, et al. Molecular identification and characterization of *Xenopus* egg uroplakin III, an egg raft-associated transmembrane protein that is tyrosine-phosphorylated upon fertilization. J Biol

Chem. 2005;280(15):15029–37. doi:10.1074/jbc.M410538200.

Sakamoto I, Takahara K, Yamashita M, et al. Changes in cyclin B during oocyte maturation and early embryonic cell cycle in the newt, Cynops pyrrhogaster: requirement of germinal vesicle for MPF activation. Dev Biol. 1998;195(1):60–9. doi:10.1006/dbio.1997.8835.

Sato K, Iwao Y, Fujimura T, et al. Evidence for the involvement of a Src-related tyrosine kinase in *Xenopus* egg activation. Dev Biol. 1999;209(2):308–20. doi:10.1006/dbio.1999.9255.

Sato K, Tokmakov AA, Iwasaki T, et al. Tyrosine kinase-dependent activation of phospholipase Cγ is required for calcium transient in *Xenopus* egg fertilization. Dev Biol. 2000;224(2):453–69. doi:10.1006/dbio.2000.9782.

Sato K, Ogawa K, Tokmakov AA, et al. Hydrogen peroxide induces Src family tyrosine kinase-dependent activation of *Xenopus* eggs. Dev Growth Differ. 2001;43(1):55–72. doi:10.1046/j.1440-169x.2001.00554.x.

Sato K, Tokmakov AA, He CL, et al. Reconstitution of Src- dependent phospholipase Cγ phosphorylation and transient calcium release by using membrane rafts and cell-free extracts from *Xenopus* eggs. J Biol Chem. 2003;278(40):38413–20. doi:10.1074/jbc.M302617200.

Sato K, Wakai T, Seita Y, et al. Molecular characteristics of horse phospholipase Czeta (PLCζ). Anim Sci J. 2013;84(4):359–68. doi:10.1111/asj.12044.

Saunders CM, Larman MG, Parrington J, et al. PLCζ: a sperm-specific trigger of Ca^{2+} oscillations in eggs and embryo development. Development. 2002;129(15):3533–44.

Sette C, Paronetto MP, Barchi M, et al. Tr-kit-induced resumption of the cell cycle in mouse eggs requires activation of a Src-like kinase. EMBO J. 2002;21(20):5386–95. doi:10.1093/emboj/cdf553.

Sherard J, Bean C, Bove B, et al. Long survival in a 69XXY triploid male. Am J Med Genet. 1986;25(2):307–12. doi:10.1002/ajmg.1320250216.

Shimada K, Ono T, Mizushima S. Application of intracytoplasmic sperm injection (ICSI) for fertilization and development in birds. Gen Comp Endocrinol. 2014;196:100–5. doi:10.1016/j.ygcen.2013.11.001.

Shiono H, Azumi J, Fujiwara M, et al. Tetraploidy in a 15-month old girl. Am J Med Genet. 1988;29(3):543–7. doi:10.1002/ajmg.1320290311.

Snook RR, Hosken DJ, Karr TL. The biology and evolution of polyspermy: insights from cellular and functional studies of sperm and centrosomal behavior in the fertilized egg. Reproduction. 2011;142(6):779–92. doi:10.1530/REP-11-0255.

Srere PA. The molecular physiology of citrate. Curr Top Cell Regul. 1992;33:261–75.

Stepinska U, Olszanska B. Detection of deoxyribonuclease I and II activities in Japanese quail oocytes. Zygote. 2001;9(1):1–7. doi:10.1017/S0967199401001010.

Stepinska U, Olszanska B. DNase I and II present in avian oocytes: a possible involvement in sperm degradation at polyspermic fertilisation. Zygote. 2003;11(1):35–42. doi:10.1017/S0967199403001059.

Stricker SA. Comparative biology of calcium signaling during fertilization and egg activation in animals. Dev Biol. 1999;211(2):157–76. doi:10.1006/dbio.1999.9340.

Sultana F, Mao KM, Yoshizaki N. Possible involvement of a sperm-associated body in the process of fertilization in quail. Zoolog Sci. 2004;21(8):851–8. doi:10.2108/zsj.21.851.

Sun QY. Cellular and molecular mechanisms leading to cortical reaction and polyspermy block in mammalian eggs. Microsc Res Tech. 2003;61(4):342–8. doi:10.1002/jemt.10347.

Sun QY, Nagai T. Molecular mechanism underlying pig oocyte maturation and fertilization. J Reprod Dev. 2003;49(5):347–59. doi:10.1262/jrd.49.347.

Swann K. Soluble sperm factors and Ca^{2+} release in eggs at fertilization. Rev Reprod. 1996;1(1):33–9. doi:10.1530/ror.0.0010033.

Swann K, Ozil JP. Dynamics of the calcium signal that triggers mammalian egg activation. Int Rev Cytol. 1994;152:183–222.

Takagi S, Ono T, Tsukada A, et al. Fertilization and blastoderm development of quail oocytes after intracytoplasmic injection of chicken sperm bearing the W chromosome. Poult Sci. 2007a;86(5):937–43. doi:10.1093/ps/86.5.937.

Takagi S, Tsukada A, Saito N, et al. Fertilizing ability of chicken sperm bearing the W chromosome. Poult Sci. 2007b;86(4):731–8. doi:10.1093/ps/86.4.731.

Talevi R. Polyspermic eggs in the anuran Discoglossus pictus develop normally. Development. 1989;105:343–9.

Tang TS, Dong JB, Huang XY, et al. Ca(2+) Oscillations induced by a cytosolic sperm protein factor are mediated by a maternal machinery that functions only once in mammalian eggs. Development. 2000;127(5):1141–50.

Tatone C, Iorio R, Francione A, et al. Biochemical and biological effects of KN-93, an inhibitor of calmodulin-dependent protein kinase II, on the initial events of mouse egg activation induced by ethanol. J Reprod Fertil. 1999;115(1):151–7. doi:10.1530/jrf.0.1150151.

Uchida IA, Freeman VC. Triploidy and chromosomes. Am J Obstet Gynecol. 1985;151(1):65–9. doi:10.1016/0002-9378(85)90426-0.

Ueno T, Ohgami T, Harada Y, et al. Egg activation in physiologically polyspermic newt eggs: involvement of IP_3 receptor, PLCγ, and microtubules in calcium wave induction. Int J Dev Biol. 2014;58(5):315–23. doi:10.1387/ijdb.130333yi.

Van Krey HP, Ogasawara FX, Lorenz FW. Distribution of spermatozoa in the oviduct and fertility in domestic birds IV. Fertility of spermatozoa from infundibular and uteroviginal glands. J Reprod Fertil. 1966;11(2):257–62. doi:10.1530/jrf.0.0110257.

Waddington D, Gribbin C, Sterling RJ, et al. Chronology of events in the first cell cycle of the polyspermic egg of the domestic fowl (*Gallus domesticus*). Int J Dev Biol. 1998;42(4):625–8.

Waitzman JS, Rice SE. Mechanism and regulation of kinesin-5, an essential motor for the mitotic spindle. Biol Cell. 2014;106(1):1–12. doi:10.1111/boc.201300054.

Wang WH, Machaty Z, Ruddock N, et al. Activation of porcine oocytes with calcium ionophore: effects of extracellular calcium. Mol Reprod Dev. 1999;53(1):99–107. doi:10.1002/(SICI)1098-2795(199905)53:1<99::AID-MRD12>3.0.CO;2-G.

Warren WC, Hillier LW, Marshall Graves JA, et al. Genome analysis of the platypus reveals unique signatures of evolution. Nature. 2008;453(7192):175–83. doi:10.1038/nature06936.

Wassarman PM. Gamete interactions during mammalian fertilization. Theriogenology. 1994;41(1):31–44. doi:10.1016/S0093-691X(05)80046-9.

Williams CJ, Mehlmann LM, Jaffe LA, et al. Evidence that Gq family G proteins do not function in mouse egg activation at fertilization. Dev Biol. 1998;198(1):116–27. doi:10.1006/dbio.1998.8892.

Wishart GJ. Quantitative aspects of sperm: egg interaction in chickens and turkeys. Anim Reprod Sci. 1997;48(1):81–92. doi:10.1016/S0378-4320(97)00042-0.

Wong JL, Wessel GM. Defending the zygote: search for the ancestral animal block to polyspermy. Curr Top Dev Biol. 2006;72:1–151. doi:10.1016/S0070-2153(05)72001-9.

Wu H, He CL, Fissore RA. Injection of a porcine sperm factor triggers calcium oscillations in mouse oocyte and bovine eggs. Mol Reprod Dev. 1997;46(2):176–89. doi:10.1002/(SICI)1098-2795(199702)46:2<176::AID-MRD8>3.0.CO;2-N.

Wu H, He CL, Jehn B, et al. Partial characterization of the calcium-releasing activity of porcine sperm cytosolic extracts. Dev Biol. 1998;203(2):369–81. doi:10.1006/dbio.1998.9070.

Xia P, Wang Z, Yang Z, et al. Ultrastructural study of polyspermy during early embryo development in pigs, observed by scanning electron microscope and transmission electron microscope. Cell Tissue Res. 2001;303(2):271–5. doi:10.1007/s004410000315.

Yanagimachi R. Mammalian fertilization. In: Knobil E, Neill JD, editors. The physiology of reproduction. 2nd ed. New York: Raven Press; 1994. p. 189–319.

Yanagimachi R. Intracytoplasmic injection of spermatozoa and spermatogenic cells: its biology and applications in humans and animals. Reprod Biomed Online. 2005;10(2):247–88. doi:10.1016/S1472-6483(10)60947-9.

Yoda A, Oda S, Shikano T, Kouchi Z, et al. Ca^{2+} oscillation-inducing phospholipase C zeta expressed in mouse eggs is accumulated to the pronucleus during egg activation. Dev Biol. 2004;268(2):245–57. doi:10.1016/j.ydbio.2003.12.028.

Yoneda A, Kashima M, Yoshida S, et al. Molecular cloning, testicular postnatal expression, and oocyte-activating potential of porcine phospholipase Cζ. Reproduction. 2006;132(3):393–401. doi:10.1530/rep.1.01018.

Yoshimura Y, Tanaka H, Nozaki M, et al. Genomic analysis of male germ cell-specific actin capping protein alpha. Gene. 1999;237(1):193–9. doi:10.1016/S0378-1119(99)00287-5.

Young C, Grasa P, Coward K, et al. Phospholipase C zeta undergoes dynamic changes in its pattern of localization in sperm during capacitation and the acrosome reaction. Fertil Steril. 2009;91(5 Suppl):2230–42. doi:10.1016/j.fertnstert.2008.05.021.

Zhu CC, Wojcikiewicz RJ. Ligand binding directly stimulates ubiquitination of the inositol 1,4,5-trisphosphate receptor. Biochem J. 2000;248(Pt 3):551–6. doi:10.1042/bj3480551.

Zhu CC, Furuichi T, Mikoshiba K, et al. Inositol 1,4,5-trisphosphate receptor down-regulation is activated directly by inositol 1,4,5-trisphosphate binding. Studies with binding-defective mutant receptors. J Biol Chem. 1999;274(6):3476–84. doi:10.1074/jbc.274.6.3476.

Molecular and Neuroendocrine Mechanisms of Avian Seasonal Reproduction

8

T. Katherine Tamai and Takashi Yoshimura

Abstract

Animals living outside tropical zones experience seasonal changes in the environment and accordingly, adapt their physiology and behavior in reproduction, molting, and migration. Subtropical birds are excellent models for the study of seasonal reproduction because of their rapid and dramatic response to changes in photoperiod. For example, testicular weight typically changes by more than a 100-fold. In birds, the eyes are not necessary for seasonal reproduction, and light is instead perceived by deep brain photoreceptors. Functional genomic analysis has revealed that long day (LD)-induced thyrotropin from the pars tuberalis of the pituitary gland causes local thyroid hormone (TH) activation within the mediobasal hypothalamus. This local bioactive TH, triiodothyronine (T_3), appears to regulate seasonal gonadotropin-releasing hormone (GnRH) secretion through morphological changes in neuro-glial interactions. GnRH, in turn, stimulates gonadotropin secretion and hence, gonadal development under

T.K. Tamai
Institute of Transformative Bio-Molecules (WPI-ITbM), Nagoya University, Furo-cho, Chikusa-ku, Nagoya, 464-8601 Japan

T. Yoshimura (✉)
Institute of Transformative Bio-Molecules (WPI-ITbM), Nagoya University, Furo-cho, Chikusa-ku, Nagoya, 464-8601 Japan

Laboratory of Animal Physiology, Graduate School of Bioagricultural Sciences, Nagoya University, Furo-cho, Chikusa-ku, Nagoya 464-8601, Japan

Avian Bioscience Research Center, Graduate School of Bioagricultural Sciences, Nagoya University, Furo-cho, Chikusa-ku, Nagoya 464-8601, Japan

National Institute for Basic Biology, 38 Nishigonaka, Myodaiji, Okazaki 444-8585, Japan
e-mail: takashiy@agr.nagoya-u.ac.jp

T. Sasanami (ed.), *Avian Reproduction*, Advances in Experimental Medicine and Biology 1001,
DOI 10.1007/978-981-10-3975-1_8

LD conditions. In marked contrast, low temperatures accelerate short day (SD)-induced testicular regression in winter. Interestingly, low temperatures increase circulating levels of T_3 to support adaptive thermogenesis, but this induction of T_3 also triggers the apoptosis of germ cells by activating genes involved in metamorphosis. This apparent contradiction in the role of TH has recently been clarified. Central activation of TH during spring results in testicular growth, while peripheral activation of TH during winter regulates adaptive thermogenesis and testicular regression.

Keywords
Photoperiodism • Circadian rhythm • Mediobasal hypothalamus • Pars tuberalis • Thyrotropin • Thyroid hormone • Deep brain photoreceptor • Opsin

8.1 Introduction

Organisms living outside the tropics experience seasonal changes in the environment and accordingly, adapt their physiology and behavior in growth, migration, hibernation, and molting. In temperate zones, temperature and precipitation show dynamic annual variation, but these changes are unreliable because there are occasional cold summers or dry seasons. In contrast, changes in photoperiod are the most reliable seasonal cue, because solstices and equinoxes occur at nearly identical times each year. The primary role of photoperiod (day length) in seasonal time measurement was conclusively demonstrated in various organisms in the 1920s (e.g., plants: Garner and Allard 1920; insects: Marcovitch 1923; birds: Rowan 1925). Therefore, these day length-dependent responses are collectively referred to as photoperiodism. Seasonal changes in reproductive activity occur to maximize the survival of offspring. For example, most birds breed during the spring and summer and are therefore called long-day (LD) breeders. The incubation period of most birds last only a few weeks, and their offspring are born during the spring and summer. In contrast, relatively large mammals, such as goats and sheep, breed during fall and are called short-day (SD) breeders. Since these animals have a gestation period of approximately 6 months, their offspring are also born and raised during spring and summer. Thus, the offspring of both LD and SD breeders can grow when the climate is moderate and food is abundant.

Seasonal breeders experience annual cycles of reproductive quiescence and recrudescence. In vertebrates, reproductive activity is regulated by the hypothalamic–pituitary–gonadal (HPG) axis. Gonadotropin-releasing hormone (GnRH) secretion from the hypothalamus induces the secretion of gonadotropins (luteinizing hormone [LH] and follicle-stimulating hormone [FSH]) from the anterior pituitary gland, and gonadotropin stimulates growth and activity of the testis and ovary. Thus, the HPG axis is activated during the breeding season, resulting in a dramatic increase in gonadal size, particularly in birds (Dawson et al. 2001). In birds, gonadal weight changes by more than a 100-fold. In addition, when birds are transferred from SD to LD conditions, LD-induced gonadotropin secretion can be observed as early as the end of the first LD (Dawson et al. 2001; Follett et al. 1998). Therefore, birds have a highly sensitive and sophisticated photoperiodic mechanism in comparison to other vertebrate species (Follett et al. 1998). In addition to the dramatic gonadal response, birds have very short breeding seasons. After breeding, the HPG axis is automatically switched off, and their gonads start to regress, even though photoperiod is still increasing. This phenomenon is known as photorefractoriness, and its underlying mechanism is unknown. Gonadal weight is significantly reduced during the non-breeding season (Hahn and MacDougall-Shackleton 2008; Nicholls et al. 1988). Such dramatic anatomical responses and the restricted breeding season appear to be adaptations of birds to flight, that is, they are lighter and more able to fly during the non-breeding season.

Accordingly, birds have often been used to study seasonal reproduction.

8.2 Components Involved in Photoperiodic Time Measurement

Organisms show seasonal responses when the photoperiod reaches or exceeds a so-called "critical day length." Most seasonally breeding organisms have a highly accurate mechanism for photoperiodic time measurement and show dramatic changes in response to small changes in photoperiod. For example, in Japanese quail, days that are longer than 11.5 h induce testicular growth, while days that are shorter than 11.5 h do not induce testicular growth (Follett and Maung 1978). Thus, animals can recognize differences in photoperiod as small as 30 min, and the minimum day length required for gonadal growth is called the "critical day length". The photoperiodic time measurement system consists of three components, as is the case for the circadian timekeeping system. These components include: (1) a light input pathway that perceives external light–dark information, (2) a biological clock that measures day length, and (3) an output pathway that regulates various aspects of physiology and behavior. Although the underlying mechanism that determines the critical day length is at the "heart of the photoperiodic time measurement", it remains a mystery. Here, we present an overview of the signal transduction pathways that regulate avian seasonal reproduction.

8.3 Birds Have Multiple Circadian Oscillators

Organisms have an endogenous biological oscillator with a period of about 24 h, known as the circadian clock. The existence of a circadian clock has been demonstrated in almost all tissues and cell types. In mammals, a master circadian pacemaker that synchronizes peripheral tissue clocks is localized to the suprachiasmatic nucleus (SCN) of the hypothalamus. Unlike mammals, the master circadian pacemaker of birds is located not only in the SCN, but also in the eyes and pineal organ (Ebihara and Kawamura 1981; Menaker 1968; Steele et al. 2003; Takahashi and Menaker 1982; Yoshimura et al. 2001). The relative importance of these three pacemakers differs significantly between various avian species. The pineal organ of birds is photoreceptive and plays a critical role in circadian timekeeping in the house sparrow (Takahashi and Menaker 1982), while the eyes play a significant role in quail (Steele et al. 2003). In contrast, the SCN, pineal organ, and eyes all play a significant role in the pigeon circadian system (Yoshimura et al. 2001).

In marked contrast to mammals, the photoperiodic response of birds is not affected by removal of specific structures, such as the eyes in duck (Benoit 1935) and lesions around the SCN (Davies and Follett 1975) or pinealectomy (Siopes and Wilson 1974) in quail. Circadian rhythmicity is regulated by a transcriptional–translational feedback loop of circadian clock genes (Young and Kay 2001; Reppert and Weaver 2002), and expression of clock genes is observed not only in master circadian pacemakers, but also in most cells throughout the body (Balsalobre 2002; Yamazaki et al. 2000). It is well established that the circadian clock is involved in photoperiodic time measurement (Follett and Sharp 1969) (Fig. 8.1). However, it is still unclear which tissue clocks are responsible for the photoperiodic response. In quail, the mediobasal hypothalamus (MBH) is considered to be at the center of seasonal reproduction (Fig. 8.2). This is because MBH (1) lesions block the photoperi-

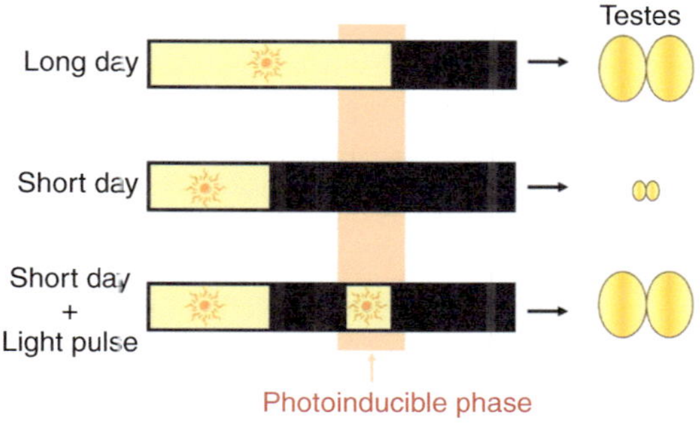

Fig. 8.1 Involvement of the circadian clock in photoperiodic time measurement. A light pulse given at a specific time of day, called the "photoinducible phase" (highlighted in *orange*), induces testicular development. The circadian clock is known to be involved in determining this photoinducible phase, but by an unknown mechanism

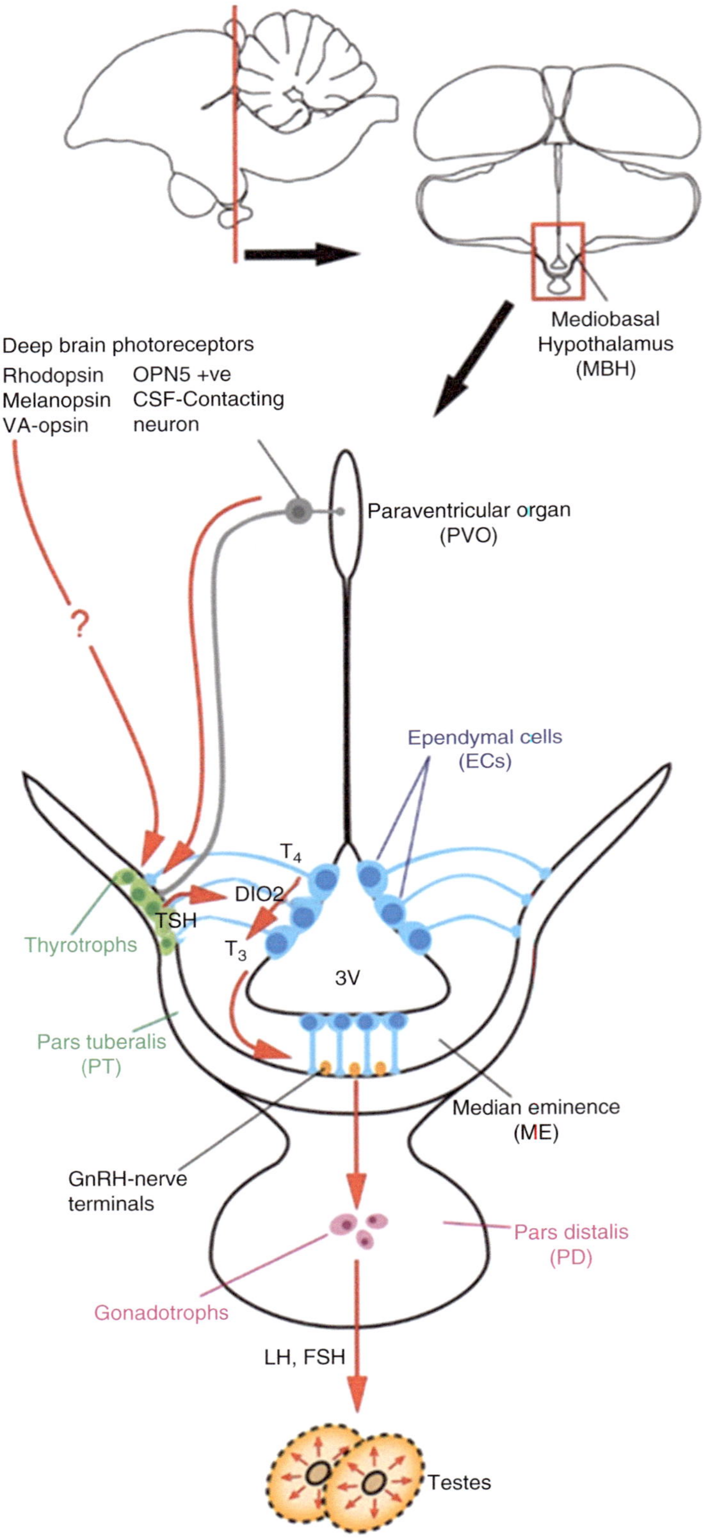

Fig. 8.2 Photoperiodic signal transduction cascade in birds. Light information detected by OPN5-positive cerebrospinal fluid (CSF)-contacting neurons is transmitted to the pars tuberalis of the pituitary gland, inducing thyroid-stimulating hormone (TSH). TSH in the pars tuberalis acts on the ependymal cells to induce thyroid hormone-activating enzyme, type 2 deiodinase (DIO2). Triiodothyronine (T_3) converted from thyroxine (T_4) by DIO2 regulates neuro-glial interactions between GnRH nerve terminals and the endfeet of glial processes. This morphological change regulates GnRH secretion and hence, gonadotropin secretion from the anterior pituitary gland. Modified from Nakane and Yoshimura (2014)

odic response of LH secretion and gonadal development (Sharp and Follett 1969; Juss et al. 1993), (2) electrical stimulation induces LH secretion (Konishi et al. 1987), and (3) expression of c-Fos, a neuronal activation marker, is induced by LD stimulus (Meddle and Follett 1997). Rhythmic expression of clock genes and proteins has been observed in both master circadian pacemakers, the SCN and the pineal gland, and the MBH (Ikegami et al. 2009; Yasuo et al. 2003). In the SCN and the pineal organ, changes in the expression profiles of clock genes have been observed after exposure to various photoperiodic schedules (e.g., SD, LD, and night interruption schedules). However, the expression profiles of clock genes are not consistent with photoperiodic inducibility. In contrast, temporal clock genes expression profiles in the MBH are stable under various photoperiodic schedules and seem to enable animals to maintain a steady-state photosensitive phase (Yasuo et al. 2003).

8.4 Deep Brain Photoreceptors in Birds

In contrast to mammals, non-mammalian vertebrates have extra-ocular photoreceptors in the pineal organ and the deep brain. Apart from the eye, the pineal organ is the best known photoreceptive organ, and the photo-pigment pinopsin (also named P-opsin) has been isolated from chicken pineal (Max et al. 1995; Okano et al. 1994). Karl von Frisch is thought to be the first to report the existence of a deep brain photoreceptor that mediates skin color changes in minnows (*Phoxinus laevis*; von Frisch 1911). Benoit (1935) first demonstrated that blind ducks show gonadal development after photostimulation, and this response disappears when black caps are placed on their heads (Benoit 1935; Oliver and Bayle 1982). Pinealectomized and blind quail have also been shown to induce gonadal development (Siopes and Wilson 1974). Injection of India ink under the scalp of blinded and pinealectomized sparrows blocked photoperiodic responses (Menaker et al. 1970). These results suggest that photoreceptors in the eyes and the pineal organ are not necessary for the regulation of seasonal reproduction in birds. Furthermore, local illumination of the septal region of the telencephalon or the MBH using radioluminous-painted beads caused testicular growth in quail (Homma et al. 1979). Therefore, these two regions are candidate sites for deep brain photoreceptors in birds. Since these studies were published, the expression of multiple photoreceptors has been reported in the avian brain. Expression of rhodopsin was first reported in the cerebrospinal fluid (CSF)-contacting neurons in the lateral septum region in ring doves and pigeons, and this rhodopsin appears to act as a deep brain photoreceptor (Silver et al. 1988; Wada et al. 1998). CSF-contacting neurons are bipolar neurons with knob-like terminals that bear a cilia and penetrate into the lumen of the third ventricle (Lamb 2009). These CSF-contacting neurons structurally resemble pineal photoreceptors and photoreceptor cells in the developing retina. Therefore, these neurons have been considered for several decades as candidate deep brain photoreceptors (Vigh and Vigh-Teichmann 1998). However, because doves and pigeons are not seasonal breeders in urban environments (Lofts et al. 1967), the physiological significance of this extra-ocular rhodopsin remains unclear. Interestingly, little or no expression of rhodopsin is observed in the quail brain (Foster et al. 1987). However, expression of melanopsin in the brain, pineal organ, and retina of chickens and turkeys has been described, although the reported distribution of its expression is inconsistent in several papers (Bailey and Cassone 2005; Chaurasia et al. 2005; Kang et al. 2010; Tomonari et al. 2005, 2007). The vertebrate ancient (VA) opsin has also been shown to be expressed in the hypothalamus of chickens and quail (Davies et al. 2012; Halford et al. 2009). In addition, localization of OPN5 (also known as Opsin5 or neuropsin; Tarttelin et al. 2003) in CSF-contacting neurons in the paraventricular organ (PVO) has been reported in quail and chicken (Fig. 8.2; Nakane et al. 2010; Yamashita et al. 2010). Functional analysis demonstrated that OPN5 is a short wavelength-sensitive photopigment involved in the seasonal regulation of reproduction (Nakane et al. 2010; Stevenson and Ball 2012). In general, longer wavelengths of light are thought to penetrate the brain more effectively than shorter wavelengths. However, it has been demonstrated that the latter can penetrate the quail hypothalamus (Foster and Follett 1985). Foster et al. (1985) reported an action spectrum for photoperiodic responses and suggested the involvement of photoreceptors that have high sensitivity at around 480 nm. Furthermore, LD stimulation with short wavelengths ranging from blue light to UV-B causes testicular growth in eye-patched, pinealectomized quail. These findings suggest that OPN5-positive CSF-contacting neurons of the PVO are the long-sought deep brain photoreceptors that regulate seasonal reproduction in birds (Nakane et al. 2010, 2014). It is also noteworthy that lesions around the PVO block the photoperiodic response of gonads in quail (Sharp and Follett 1969). However, it is also possible that the reported sensitivity to 480 nm by Foster et al. can be accounted for by the presence of melanopsin and/or VA opsin. Because sunlight is composed of a wide range of wavelengths, it is possible that birds utilize multiple wavelengths to measure photoperiod.

8.5 Identification of Key Genes Regulating the Photoperiodic Response

As shown in Fig. 8.1, light pulses delivered during the photoinducible phase stimulate testicular development (Follett and Sharp 1969). Therefore, we predicted that molecular events must be taking place within the MBH in response to light pulses delivered in this phase. By using differential subtractive hybridization analysis, we discovered light induction of the type 2 deiodinase gene (*DIO2*) (Yoshimura et al. 2003). Importantly, this induction of *DIO2* is not observed when light pulses are given outside the photoinducible phase. *DIO2* encodes the thyroid hormone (TH)-activating enzyme that converts the prohormone thyroxine (T_4) to bioactive triiodothyronine (T_3) (Bernal 2002; Fig. 8.3). *DIO2* expression is upregulated under LD conditions and downregulated under SD conditions in the ependymal cells (ECs; also known as tanycytes) that line the ven-

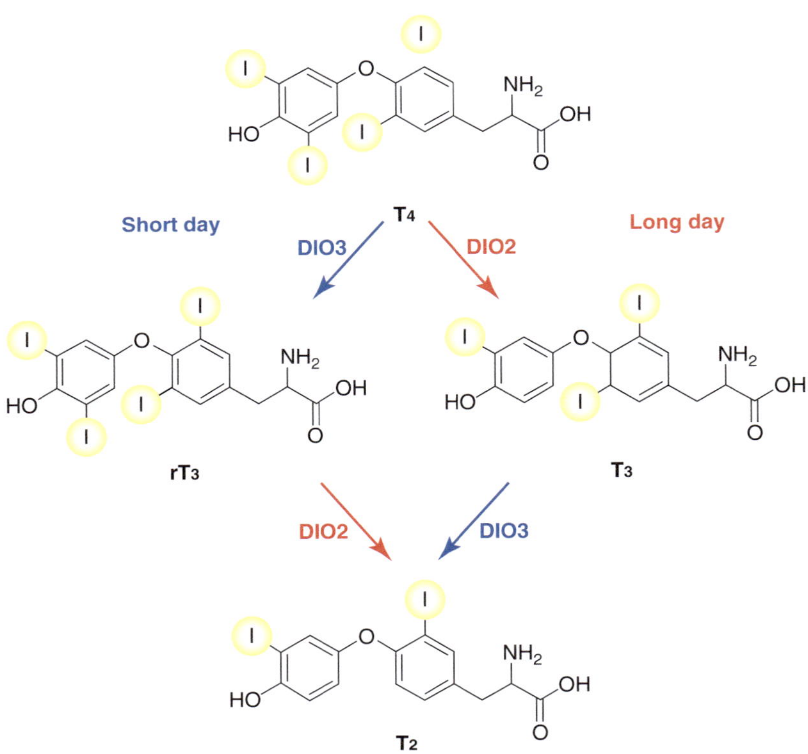

Fig. 8.3 Metabolic pathway of thyroxine (T_4). Type 2 deiodinase (DIO2) catalyzes the removal of an outer-ring iodine, while type 3 deiodinase (DIO3) catalyzes the removal of an inner-ring iodine

trolateral walls of the third ventricle within the MBH (Yoshimura et al. 2003). We later discovered photoperiodic regulation of the TH-inactivating enzyme DIO3 that converts T_4 and T_3 to inactive reverse T_3 (rT_3) and 3,3′-diiodothyronine (T_2), respectively (Fig. 8.3). In contrast to *DIO2*, *DIO3* expression is upregulated under SD conditions and downregulated under LD conditions (Yasuo et al. 2005). Reciprocal switching of *DIO2* and *DIO3* appears to fine-tune the local TH concentration within the MBH. Indeed, T_3 concentration within the MBH is about tenfold higher under LD than SD conditions, even though plasma concentrations are similar for both photoperiods (Yoshimura et al. 2003). The functional significance of this locally-activated TH has been demonstrated by intracerebroventricular (ICV) infusion of T_3 under SD conditions and a DIO2 inhibitor under LD conditions. As expected, T_3 administration induced testicular growth in a dose-dependent manner, while the administration of a DIO2 inhibitor blocked testicular growth (Yoshimura et al. 2003). Photoperiodic regulation of *DIO2* and/or *DIO3* has also been confirmed in other avian species, such as the tree sparrow (Watanabe et al. 2007), chicken (Ono et al. 2009), great tit (Perfito et al. 2012), and canary (Stevenson and Ball 2012).

8.6 LD-Induced Local Thyroid Hormone Regulates Neuro-Glial Plasticity

TH is primarily responsible for the regulation of metabolism. However, it is also important for various physiological events, such as development, bone growth, and thermogenesis (Pearce 2012; Silva 2006; Waung et al. 2012). In addition, TH regulates the development and plasticity of the central nervous system (Bernal 2002). A target site of action for the photo-induced increase in T_3 in the quail MBH is suggested by expression of TH receptors (*THRα*, *THRβ*, and *RXRα*) in the median eminence (Yoshimura et al. 2003). In order to understand the functional significance of these receptors, the ultrastructure of the quail median eminence was examined under SD and LD conditions using electron microscopy. Dynamic morphological changes were observed between the GnRH nerve terminals and glial endfeet (Yamamura et al. 2004). It is thought that the nerve terminals of hypothalamic neurons are required to directly contact the pericapillary space (i.e., the basal lamina) in order to secrete the hypothalamic neurohormone from the hypothalamus into the portal capillary (Prevot et al. 1999). Under SD conditions, many GnRH nerve terminals are encased by the endfeet of glial processes and do not contact the basal lamina, while many GnRH nerve terminals are in close proximity to the basal lamina under LD conditions. It is noteworthy that local T_3 administration to the MBH mimics these morphological changes and results in testicular growth under SD conditions (Yamamura et al. 2006). In quail, seasonal reproduction is governed primarily at the level of GnRH release. In contrast, European starlings exhibit marked plasticity in GnRH that includes regulation of GnRH production and release (MacDougall-Shackleton et al. 2009; Stevenson et al. 2012).

8.7 Uptake of Thyroid Hormones into the Hypothalamus

THs were long believed to traverse the plasma membrane by passive diffusion due to their lipophilic nature. However, it has now been established that a membrane transport system for THs exists. If a local increase in the metabolism of T_4 to T_3 in the MBH plays a key role in avian seasonal reproduction, then there must be a mechanism to transport the TH into the MBH. Some members of the organic anion transporting polypeptide (Oatp) family have been shown to transport THs in mammals (Abe et al. 2002; Hagenbuch and Meier 2004) and the possibility that a member of this family might be involved in transporting T_4 into the quail brain has been investigated (Nakao et al. 2006). The gene encoding *Oatp1c1* is highly expressed in the ECs within the MBH and the choroid plexus. Functional expression of chicken *Oatp1c1* in Chinese hamster ovary cells revealed that

Oatp1c1 is a highly specific transporter for T_4. These observations suggest that Oatp1c1 could be involved in transporting T_4 from the general circulation to CSF and from CSF to the ECs lining the ventro-lateral walls of the third ventricle in the MBH, where the *DIO2* and *DIO3* genes are expressed.

8.8 LD-Induced Thyrotropin in the Pars Tuberalis Is the Springtime Hormone

When SD quail are photostimulated, an increase in plasma LH is observed before the end of the first LD (Follett et al. 1998; Nicholls et al. 1983). This photoperiodic response is the core feature of the so-called "first day release model" of photoperiodic induction and has been used to unravel the underlying neural and molecular mechanisms of photoperiodism. As mentioned above, reciprocal switching of *DIO2* and *DIO3* plays a critical role in the regulation of seasonal reproduction. In quail, this switching precedes photoperiodic induction of LH release from the anterior pituitary gland by a couple of hours (Yasuo et al. 2005). To clarify the mechanisms regulating this photoperiodic switching, genome-wide gene expression analysis was performed during the transition from SD to LD conditions in Japanese quail (Nakao et al. 2008). Using microarrays to analyze more than 38,000 probes, induction of only two genes was observed in the pars tuberalis (PT) of the pituitary gland 14 h after dawn of the first LD. One of these genes encoded the thyroid-stimulating hormone β subunit (*TSHB*) and the other encoded the transcriptional co-activator eyes absent 3 (*EYA3*). The PT consists of thin cell layers surrounding the median eminence (Fig. 8.2). Reciprocal switching of *DIO2* and *DIO3* is observed 18 h after dawn (i.e., 4 h after *EYA3* and *TSHB* induction) in the ECs. Although EYA3 is a transcriptional co-activator, the expression sites of *EYA3* and *DIO2*/*DIO3* are different (i.e., *EYA3* in the PT and *DIO2/DIO3* in the ECs), indicating that EYA3 is unlikely to regulate *DIO2* and *DIO3* expression directly. Therefore, the possibility that PT-derived TSH may regulate *DIO2*/*DIO3* switching was examined. Expression of the TSH receptor (TSHR) and binding of ^{125}I-labeled TSH were observed in the ECs. Furthermore, ICV TSH administration induced *DIO2* expression in the ECs and testicular growth even under SD conditions, while passive immunization against TSH attenuated LD induction of *DIO2* expression. Furthermore, promoter analysis demonstrated that the TSHR-Gsα-cAMP signaling pathway is involved in this TSH regulation of *DIO2* expression. Importantly, the magnitude of testicular growth induced by ICV TSH infusion was nearly identical to that observed in birds exposed to LD, indicating that LD-induced PT TSH is the master regulator of seasonal reproduction in birds. As OPN5-positive neurons project to the external layer of the median eminence adjacent to the PT, light information received by the OPN5-positive CSF-contacting neurons appears to be transmitted to the PT and induces TSH in the PT (Fig. 8.2).

8.9 Mechanism of Seasonal Testicular Regression

As described above, the photoperiodic signal transduction pathway that activates avian seasonal reproduction has been revealed. In contrast, the underlying mechanism of seasonal testicular regression has remained a mystery. Gonadotropin-inhibitory hormone (GnIH) is a neuropeptide synthesized by the neurons of the paraventricular nucleus (PVN), and it has been shown to inhibit gonadotropin secretion (Tsutsui et al. 2000; Ubuka et al. 2006). Numerous studies reported that melatonin has a minor role in avian seasonal reproduction (Gwinner et al. 1997). However, Tsutsui and his colleagues demonstrated that melatonin suppresses reproduction by activating GnIH (Ubuka et al. 2005).

In quail, exposure to low temperatures accelerates SD-induced testicular regression (Oishi and Konishi 1978; Wada 1993). Low temperature increases the amount of circulating T_3 to support adaptive thermogenesis. In addition, circulating T_3 induced by low temperature triggers the

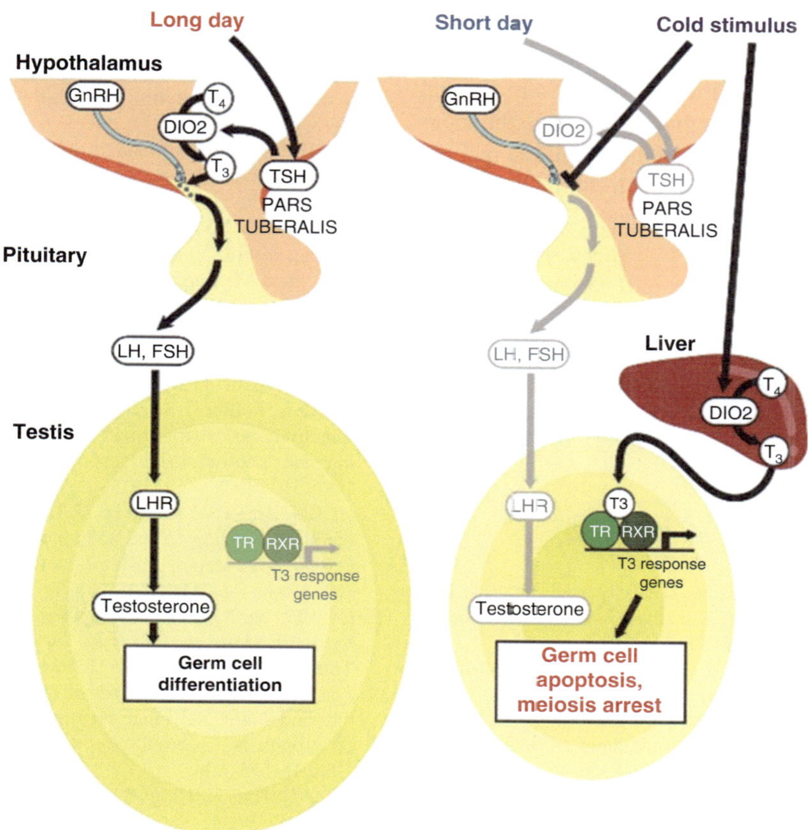

Fig. 8.4 Dual role of thyroid hormone in the regulation of seasonal testicular development and regression. Under LD conditions, centrally activated thyroid hormone triggers GnRH secretion and activates the HPG axis. In contrast, low temperature attenuates GnRH secretion and acts in concert with SD conditions to shut down the HPG axis. Low temperature increases food consumption and expression of DIO2 in the liver. Low temperature-induced circulating T_3 acts on the testis to regulate T_3 responsive genes involved in metamorphosis and in promoting apoptosis. Modified from Ikegami et al. (2015)

apoptosis of germ cells by activating genes involved in metamorphosis (Fig. 8.4; Ikegami et al. 2015). DIO2 produced in the liver through increased food intake appears to be responsible for this increase in circulating T_3 (Ikegami et al. 2015).

A role for TH in seasonal reproduction has been implicated for several decades. However, its effects were often contradictory. Some studies reported stimulatory effects, while others reported inhibitory effects (Dawson et al. 2001). Therefore, TH was suggested to play a permissive role in seasonal reproduction. A recent report explains these contradictory effects (Fig. 8.4). Central activation of TH results in seasonal testicular growth; however, peripheral activation of TH regulates adaptive thermogenesis and seasonal testicular regression. Therefore, TH has a dual role in the regulation of seasonal reproduction. Interestingly, TH also has opposite effects during the metamorphosis of the tadpole; TH induces both the outgrowth of the limb and resorption of the tail (Furlow and Neff 2006).

Conclusions

Aristotle described the seasonal changes in physiology and behavior of various organisms in his “de historia animalium,” and this phenomenon has attracted tremendous general interest. Over the past several decades, significant progress has been made in understanding the signal transduction cascade that regulates seasonal reproduction in birds, and discovery of this pathway has led to the identification of similar, but not identical mechanisms in mammals and fish (Ono et al. 2008; Yasuo et al. 2009; Nakane et al. 2013; Ikegami et al. 2014). These comparative studies have uncovered universality and diversity in the regulation of vertebrate seasonal reproduction and revealed the powerful nature of comparative analysis in understanding fundamental biological processes.

References

Abe T, Suzuki T, Unno M, Tokui T, Ito S. Thyroid hormone transporters: recent advances. Trends Endocrinol Metab. 2002;13:215–20.

Bailey MJ, Cassone VM. Melanopsin expression in the chick retina and pineal gland. Brain Res Mol Brain Res. 2005;134:345–8.

Balsalobre A. Clock genes in mammalian peripheral tissues. Cell Tissue Res. 2002;309:193–9.

Benoit J. Le role des yeux dans l'action stimulante de la lumière sur le developpement testiculaire chez le canard. C R Soc Biol (Paris). 1935;118:669–71.

Bernal J. Action of thyroid hormone in brain. J Endocrinol Invest. 2002;25:268–88.

Berson DM, Dunn FA, Takao M. Phototransduction by retinal ganglion cells that set the circadian clock. Science. 2002;295:1070–3.

Chaurasia SS, Rollag MD, Jiang G, Hayes WP, Haque R, Natesan A, Zatz M, Tosini G, Liu C, Korf HW, Iuvone PM, Provencio I. Molecular cloning, localization and circadian expression of chicken melanopsin (Opn4): differential regulation of expression in pineal and retinal cell types. J Neurochem. 2005;92:158–70.

Davies DT, Follett BK. The neuroendocrine control of gonadotrophin release in the Japanese quail. II. The role of the anterior hypothalamus. Proc R Soc Lond B. 1975;191:303–15.

Davies WI, Turton M, Peirson SN, Follett BK, Halford S, Garcia-Fernandez JM, Sharp PJ, Hankins MW, Foster RG. Vertebrate ancient opsin photopigment spectra and the avian photoperiodic response. Biol Lett. 2012;8:291–4.

Dawson A, King VM, Bentley GE, Ball GF. Photoperiodic control of seasonality in birds. J Biol Rhythms. 2001;16:365–80.

Ebihara S, Kawamura H. The role of the pineal organ and the suprachiasmatic nucleus in the control of circadian locomotor rhythms in the Java sparrow, *Padda oryzivora*. J Comp Physiol A. 1981;141:207–14.

Follett BK, Maung SL. Rate of testicular maturation, in relation to gonadotrophin and testosterone levels, in quail exposed to various artificial photoperiods and to natural daylengths. J Endocrinol. 1978;78:267–80.

Follett BK, Sharp PJ. Circadian rhythmicity in photoperiodically induced gonadotrophin release and gonadal growth in the quail. Nature. 1969;223:968–71.

Follett BK, King VM, Meddle SL. Rhythms and photoperiodism in birds. In: Lumsden PJ, Miller AJ, editors. Biological rhythms and photoperiodism in plants. Oxford: Biostatistics Scientific; 1998. p. 231–42.

Foster RG, Follett BK. The involvement of a rhodopsin-like photopigment in the photoperiodic response of the Japanese quail. J Comp Physiol A. 1985;157:519–28.

Foster RG, Follett BK, Lythgoe JN. Rhodopsin-like sensitivity of extra-retinal photoreceptors mediating the photoperiodic response in quail. Nature. 1985;313:50–2.

Foster RG, Korf HW, Schalken JJ. Immunocytochemical markers revealing retinal and pineal but not hypothalamic photoreceptor systems in the Japanese quail. Cell Tissue Res. 1987;248:161–7.

Furlow JD, Neff ES. A developmental switch induced by thyroid hormone: *Xenopus laevis* metamorphosis. Trends Endocrionol Metab. 2006;17:40–7.

Garner WW, Allard HA. Effect of the relative length of day and night and other factors of the environment on growth and reproduction in plants. J Agric Res. 1920;18:553–606.

Gwinner E, Hau H, Heigl S. Melatonin: generation and modification of avian circadian rhythms. Brain Res Bull. 1997;44:439–44.

Hagenbuch B, Meier PJ. Organic anion transporting polypeptides of the OATP/SLC21 family: phylogenetic classification as OATP/SLCO superfamily, new nomenclature and molecular/functional properties. Eur J Physiol. 2004;447:653–65.

Hahn TP, MacDougall-Shackleton SA. Adaptive specialization, conditional plasticity and phylogenetic history in the reproductive cue response systems of birds. Philos Trans R Soc B. 2008;363:267–86.

Halford S, Pires SS, Turton M, Zheng L, Gonzalez-Menendez I, Davies WL, Peirson SN, Garcia-Fernandez JM, Hankins MW, Foster RG. VA opsin-based photoreceptors in the hypothalamus of birds. Curr Biol. 2009;19:1396–402.

Homma K, Ohta M, Sakakibara Y. Photoinducible phase of the Japanese quail detected by direct stimulation of the brain. In: Suda M, Hayaishi O, Nakagawa H, editors. Biological rhythms and their central mechanism. Amsterdam: Elsevier; 1979. p. 85–94.

Ikegami K, Katou Y, Higashi K, Yoshimura T. Localization of circadian clock protein BMAL1 in the photoperiodic signal transduction machinery in Japanese quail. J Comp Neurol. 2009;517:397–404.

Ikegami K, Liao XH, Hoshino Y, Ono H, Ota W, Ito Y, Nishiwaki-Ohkawa T, Sato C, Kitajima K, Iigo M, Shigeyoshi Y, Yamada M, Murata Y, Refetoff S, Yoshimura T. Tissue-specific post-translational modification allows functional targeting of thyrotropin. Cell Rep. 2014;9:1–9.

Ikegami K, Atsumi Y, Yorinaga E, Ono H, Murayama I, Nakane Y, Ota W, Arai N, Tega A, Iigo M, Darras VM, Tsutsui K, Hayashi Y, Yoshida S, Yoshimura T. Low temperature-induced circulating triiodothyronine accelerates seasonal testicular regression. Endocrinology. 2015;156:647–59.

Juss TS, Meddle SL, Servant RS, King VM. Melatonin and photoperiodic time measurement in Japanese quail (*Coturnix coturnix japonica*). Proc R Soc Lond B Biol Sci. 1993;254:21–8.

Kang SW, Leclerc B, Kosonsiriluk S, Mauro LJ, Iwasawa A, El Halawani ME. Melanopsin expression in dopamine–melatonin neurons of the premammillary nucleus of the hypothalamus and seasonal reproduction in birds. Neuroscience. 2010;170:200–13.

Konishi H, Foster RG, Follett BK. Evidence for a daily rhythmicity in the acute release of LH in response to electrical stimulation in the Japanese quail. J Comp Physiol A Sens Neural Behav Physiol. 1987;161:315–9.

Lamb TD. Evolution of vertebrate retinal photoreception. Phil Trans R Soc B. 2009;364:2911–24.

Lofts B, Murton RK, Westwood NJ. Photoresponses of the Woodpigeon *Columba palumbus* in relation to the breeding season. Ibis. 1967;109:338–51.

MacDougall-Shackleton SA, Stevenson TJ, Watts HE, Pereyra ME, Hahn TP. The evolution of photoperiod response systems and seasonal GnRH plasticity in birds. Integr Comp Biol. 2009;49:580–9.

Marcovitch S. Plant lice and light exposure. Science. 1923;58:537–8.

Max M, McKinnon PJ, Seidenman KJ, Barrett RK, Applebury ML, Takahashi JS, Margolskee RF. Pineal opsin: a nonvisual opsin expressed in chick pineal. Science. 1995;267:1502–6.

Meddle SL, Follett BK. Photoperiodically driven changes in Fos expression within the basal tuberal hypothalamus and median eminence of Japanese quail. J Neurosci. 1997;17:8909–18.

Menaker M. Extraretinal light perception in the sparrow. I. Entrainment of the biological clock. Proc Natl Acad Sci U S A. 1968;59:414–21.

Menaker M, Roberts R, Elliott J, Underwood H. Extraretinal light perception in the sparrow. III. The eyes do not participate in photoperiodic photoreception. Proc Natl Acad Sci U S A. 1970;67:320–5.

Nakane Y, Yoshimura T. Universality and diversity in the signal transduction pathway that regulates seasonal reproduction in vertebrates. Front Neurosci. 2014;8:115.

Nakane Y, Ikegami K, Ono H, Yamamoto N, Yoshida S, Hirunagi K, Ebihara S, Kubo Y, Yoshimura T. A mammalian neural tissue opsin (Opsin 5) is a deep brain photoreceptor in birds. Proc Natl Acad Sci U S A. 2010;107:15264–8.

Nakane Y, Ikegami K, Iigo M, Ono H, Takeda K, Takahashi D, Uesaka M, Kimijima M, Hashimoto R, Arai N, Suga T, Kosuge K, Abe T, Maeda R, Senga T, Amiya N, Azuma T, Amano M, Abe H, Yamamoto N, Yoshimura T. The saccus vasculosus of fish is a sensor of seasonal changes in day length. Nat Commun. 2013;4:2108.

Nakane Y Shimmura T, Abe H, Yoshimura T. Intrisic photosensitivity of a deep brain photoreceptor. Curr Biol. 2014;24:R596–7.

Nakao N, Takagi T, Iigo M, Tsukamoto T, Yasuo S, Masuda T, Yanagisawa T, Ebihara S, Yoshimura T. Possible involvement of organic anion transporting polypeptide 1c1 in the photoperiodic response of gonads in birds. Endocrinology. 2006;147:1067–73.

Nakao N, Ono H, Yamamura T, Anraku T, Takagi T, Higashi K, Yasuo S, Katou Y, Kageyama S, Uno Y, Kasukawa T, Iigo M, Sharp PJ, Iwasawa A, Suzuki Y, Sugano S, Niimi T, Mizutani M, Namikawa T, Ebihara S, Ueda HR, Yoshimura T. Thyrotrophin in the pars tuberalis triggers photoperiodic response. Nature. 2008;452:317–22.

Nicholls TJ, Follett BK, Robinson JE. A photoperiodic response in gonadectomized Japanese quail exposed to a single long day. J Endocrinol. 1983;97:121–6.

Nicholls TJ, Goldsmith AR, Dawson A. Photorefractoriness in birds and comparison with mammals. Physiol Rev. 1988;68:133–76.

Oishi T, Konishi T. Effects of photoperiod and temperature on testicular and thyroid activity of the Japanese quail. Gen Comp Endocrinol. 1978;36:250–4.

Okano T, Yoshizawa T, Fukada Y. Pinopsin is a chicken pineal photoreceptive molecule. Nature. 1994;372:94–7.

Oliver J, Bayle JD. Brain photoreceptors for the photoinduced testicular response in birds. Experientia. 1982;38:1020–9.

Ono H, Hoshino Y, Yasuo S, Watanabe M, Nakane Y, Murai A, Ebihara S, Korf HW, Yoshimura T. Involvement of thyrotropin in photoperiodic signal transduction in mice. Proc Natl Acad Sci U S A. 2008;105:18238–42.

Ono H, Nakao N, Yamamura T, Kinoshita K, Mizutami M, Namikawa T, Iigo M, Ebihara S, Yoshimura T. Red jungle fowl (*Gallus gallus*) as a model for studying the molecular mechanism of seasonal reproduction. Anim Sci J. 2009;80:328–32.

Pearce EN. Thyroid hormone and obesity. Curr Opin Endocrinol Diabetes Obes. 2012;19:408–13.

Perfito N, Jeong SY, Silverin B, Calisi RM, Bentley GE, Hau M. Anticipating spring: wild populations of great tits (*Parus major*) differ in expression of key genes for photoperiodic time measurement. PLoS One. 2012;7:e34997.

Prevot V, Croix D, Bouret S, Dutoit S, Tramu G, Stefano GB, Beauvillain JC. Definitive evidence for the existence of morphological plasticity in the external zone of the median eminence during the rat estrous cycle: implication of neuro-glio-endothelial interactions in gonadotropin-releasing hormone release. Neuroscience. 1999;94:809–19.

Reppert SM, Weaver DR. Coordination of circadian timing in mammals. Nature. 2002;418:935–41.

Rowan W. Relation of light to bird migration and developmental changes. Nature. 1925;115:494–5.

Sharp PJ, Follett BK. The effect of hypothalamic lesions on gonadotrophin release in Japanese quail (*Coturnix coturnix japonica*). Neuroendocrinol. 1969;5:205–18.

Silva JE. Thermogenic mechanisms and their hormonal regulation. Physiol Rev. 2006;86:435–64.

Silver R, Witkovsky P, Horvath P, Alones V, Barnstable CJ, Lehman MN. Coexpression of opsin- and VIP-like-immunoreactivity in CSF-contacting neurons of the avian brain. Cell Tissue Res. 1988;253:189–98.

Siopes TD, Wilson WO. Extraocular modification of photoreception in intact and pinealectomized coturnix. Poult Sci. 1974;53:2035–41.

Steele CT, Zivkovic BD, Siopes T, Underwood H. Ocular clocks are tightly coupled and act as pacemakers in the circadian system of Japanese quail. Am J Physiol Regul Integr Comp Physiol. 2003;284:R208–18.

Stevenson TJ, Ball GF. Disruption of neuropsin mRNA expression via RNA interference facilitates the photo-induced increase in thyrotropin-stimulating subunit β in birds. Eur J Neurosci. 2012;36:2859–65.

Stevenson TJ, Hahn TP, MacDougall-Shackleton SA, Ball GF. Gonadotropin-releasing hormone plasticity: a comparative perspective. Front Neuroendocrionol. 2012;33:287–300.

Takahashi JS, Menaker M. Role of the suprachiasmatic nucleus in the circadian system of the house sparrow. J Neurosci. 1982;2:815–28.

Tarttelin EE, Bellingham J, Hankins MW, Foster RG, Lucas RJ. Neuropsin (Opn5): a novel opsin identified in mammalian neural tissue. FEBS Lett. 2003;554:410–6.

Tomonari S, Takagi A, Akamatsu S, Noji S, Ohuchi H. A non-canonical photopigment, melanopsin, is expressed in the differentiating ganglion, horizontal, and bipolar cells of the chicken retina. Dev Dyn. 2005;234:783–90.

Tomonari S, Takagi A, Noji S, Ohuchi H. Expression pattern of the melanopsin-like (cOpn4m) and VA opsin-like genes in the developing chicken retina and neural tissues. Gene Expr Patterns. 2007;7:746–53.

Tsutsui K, Saigoh E, Ukena K, Teranishi H, Fujisawa Y, Kikuchi M, Ishii S, Sharp PJ. A novel avian hypothalamic peptide inhibiting gonadotropin release. Biochem Biophys Res Commun. 2000;275:661–7.

Ubuka T, Bentley GE, Ukena K, Wingfield JC, Tsutsui K. Melatonin induces the expression of gonadotropin-inhibitory hormone in the avian brain. Proc Natl Acad Sci U S A. 2005;102:3052–7.

Ubuka T, Ukena K, Sharp PJ, Bentley GE, Tsutsui K. Gonadotropin-inhibitory hormone inhibits gonadal development and maintenance by decreasing gonadotropin synthesis and release in male quail. Endocrinology. 2006;147:1187–94.

Vigh B, Vigh-Teichmann I. Actual problems of the cerebrospinal fluid-contacting neurons. Microsc Res Tech. 1998;41:57–83.

von Frisch K. Beitrage zur Physiologie der Pigmentzellen in der Fischhaut. Pfluger's Archiv für die Gesamte Physiologie des Menschen und der Tiere. 1911;138:319–87.

Wada M. Low temperature and short days together induce thyroid activation and suppression of LH release in Japanese quail. Gen Comp Endocrinol. 1993;90:355–63.

Wada Y, Okano T, Adachi A, Ebihara S, Fukada Y. Identification of rhodopsin in the pigeon deep brain. FEBS Lett. 1998;424:53–6.

Watanabe T, Yamamura T, Watanabe M, Yasuo S, Nakao N, Dawson A, Ebihara S, Yoshimura T. Hypothalamic expression of thyroid hormone-activating and -inactivating enzyme genes in relation to photorefractoriness in birds and mammals. Am J Physiol Regul Integr Comp Physiol. 2007;292:R568–72.

Waung JA, Bassett JH, Williams GR. Thyroid hormone metabolism in skeletal development and adult bone maintenance. Trends Endocrinol Metab. 2012;23:155–62.

Yamamura T, Hirunagi K, Ebihara S, Yoshimura T. Seasonal morphological changes in the neuro-glial interaction between gonadotropin-releasing hormone nerve terminals and glial endfeet in Japanese quail. Endocrinology. 2004;145:4264–7.

Yamamura T, Yasuo S, Hirunagi K, Ebihara S, Yoshimura T. T_3 implantation mimics photoperiodically reduced encasement of nerve terminals by glial processes in the median eminence of Japanese quail. Cell Tissue Res. 2006;324:175–9.

Yamashita T, Ohuchi H, Tomonari S, Ikeda K, Sakai K, Shichida Y. Opn5 is a UV-sensitive bistable pigment that couples with Gi subtype of G protein. Proc Natl Acad Sci U S A. 2010;107:22084–9.

Yamazaki S, Numano R, Abe M, Hida A, Takahashi R, Ueda M, Block GD, Sakaki Y, Menaker M, Tei H. Resetting central and peripheral circadian oscillators in transgenic rats. Science. 2000;288:682–5.

Yasuo S, Watanabe M, Okabayashi N, Ebihara S, Yoshimura T. Circadian clock genes and photoperiodism: comprehensive analysis of clock gene expression in the mediobasal hypothalamus, the suprachiasmatic nucleus, and the pineal gland of Japanese quail under various light schedules. Endocrinology. 2003;144:3742–8.

Yasuo S, Watanabe M, Nakao N, Takagi T, Follett BK, Ebihara S, Yoshimura T. The reciprocal switching of two thyroid hormone-activating and -inactivating enzyme genes is involved in the photoperiodic gonadal response of Japanese quail. Endocrinology. 2005;146:2551–4.

Yasuo S, Yoshimura T, Ebihara S, Kolf HW. Melatonin transmits photoperiodic signals through the MT1 melatonin receptor. J Neurosci. 2009;29:2885–9.

Yoshimura T, Yasuo S, Suzuki Y, Makino E, Yokota Y, Ebihara S. Identification of the suprachiasmatic nucleus in birds. Am J Physiol Regul Integr Comp Physiol. 2001;280:R1185–9.

Yoshimura T, Yasuo S, Watanabe M, Iigo M, Yamamura T, Hirunagi K, Ebihara S. Light-induced hormone conversion of T_4 to T_3 regulates photoperiodic response of gonads in birds. Nature. 2003;426:178–81.

Young MW, Kay SA. Time zones: a comparative genetics of circadian clocks. Nat Rev Genet. 2001;2:702–15.

Hormonal Responses to a Potential Mate in Male Birds

9

Yasuko Tobari, Yoshimi Sato, and Kazuo Okanoya

Abstract

Social interactions rapidly modulate circulating hormone levels and behavioral patterns in most male animals. In male birds, sexual interaction or visual exposure to a conspecific female usually causes an increase in the levels of peripheral reproductive hormones, such as gonadotropins and androgens. Although the perception of a female presence is processed in the brain and peripheral hormonal levels are regulated by the hypothalamus–pituitary–gonadal (HPG) axis, the specific neural circuitry and neurochemical systems that translate social signals into reproductive physiology in male birds were not well understood until 2008. Today, there is growing evidence that two neuropeptides localized in the hypothalamus, gonadotropin-releasing hormone and gonadotropin-inhibitory hormone, are responsive to social information. These two neuropeptides have thus begun to be regarded as modulators translating social stimuli into changes in the levels of peripheral reproductive hormones. Here, we review previous studies that investigated the male responses of the HPG axis to the mere presence of a female or to sexual interaction and describe the neurochemical pathways linking visual perception of a potential mate to rapid peripheral hormonal changes via the brain–pituitary endocrine system in sexually mature male Japanese quail.

Y. Tobari (✉) • Y. Sato • K. Okanoya
Laboratory of Animal Genetics and Breeding,
Department of Animal Science and Biotechnology,
School of Veterinary Medicine, Azabu University,
Fuchinobe 1-17-71, Chuo-ku, Sagamihara 252-5201,
Japan

Graduate School of Arts and Sciences,
The University of Tokyo,
Room 215 bldg.3 Komaba campus, 3-8-1 Komaba,
Meguro-ku, Tokyo 153-8902, Japan
e-mail: tobari@azabu-u.ac.jp

T. Sasanami (ed.), *Avian Reproduction*, Advances in Experimental Medicine and Biology 1001,
DOI 10.1007/978-981-10-3975-1_9

Keywords
Call • Courtship behavior • Female presence • Luteinizing hormone • Monoamine • Noradrenaline • Norepinephrine • Social environment • Song • Testosterone

9.1 Introduction

Many environmental factors can influence the reproductive physiology and behavior of birds. While a change in photoperiod induces or suppresses reproductive function in some species, additional environmental cues, such as food availability, temperature, humidity, competition, and the presence of the opposite sex, may also be important in fine-tuning the timing of breeding (Wingfield 1983). Social birds receive multiple forms of sensory information from conspecifics, and use specific signals to make appropriate physiological and behavioral responses for reproduction. The courtship behavior of male birds has served as a good model for studying the rapid neural processes that regulate hormonal and behavioral responses to social cues. In many animals, when a male recognizes, as well as when he copulates with, a female, the male shows acute changes in the levels of both blood androgens and blood gonadotropins. This social modulation of the levels of blood androgens is thought to be a mechanism for adjusting androgen-dependent behaviors to the current social environment (Oliveira 2009). In more detail, in his review, Oliveira (2009) stated the following: "the social interactions in which an individual participates or to which he is exposed, influence its androgen levels, which in turn will modulate perceptive, motivational, and cognitive mechanisms that will affect its subsequent behavior in social interaction." In this context, the following question arises: what are the proximate mechanisms that translate social signals into reproductive physiology to modulate plasma hormone levels? Social animals receive multiple forms of sensory information, and the perception/cognition of social information is processed in the brain. The peripheral production and release of androgens is under the control of the hypothalamus–pituitary–gonadal (HPG) axis (Fig. 9.1). There is growing evidence that two neuropeptides localized in the hypothalamus, gonadotropin-releasing hormone (GnRH) and gonadotropin-inhibitory hormone (GnIH), are responsive to social information. Therefore, these two hypothalamic neuropeptides are viewed as integrators that translate social environmental stimuli into reproductive physiology and its associated behaviors (Stevenson et al. 2012; Kriegsfeld et al. 2015). In this chapter, we review the social modulation of courtship behaviors and the HPG axis at each level in avian species, particularly with reference to male behavioral and hormonal responses to female conspecifics.

9.2 Male Behavioral Responses to Female Conspecifics

In the males of many species, exposure to a female results in rapid changes in behavioral frequency and pattern. Sexually mature males exposed to the stimulus of a receptive female show increases in the activation of courtship behavior. Courtship behavior appears to reflect male sexual arousal, and most courtship behavior of males is androgen-dependent. Until 1980, the effects of social factors on male endocrine activity in birds were measured by assessing changes in courtship behavior. Both observations and experiments indicated that the presence of a female rapidly changes the frequency and pattern of male courtship behavior in some avian species. This section provides a brief review of this subject.

In songbirds, such as the zebra finch (*Taeniopygia guttata*) and Bengalese finch (*Lonchura striata* var. *domestica*), the most characteristic courtship behavior is a song that each male learns from his father during development. Following on from male song and a suite of asso-

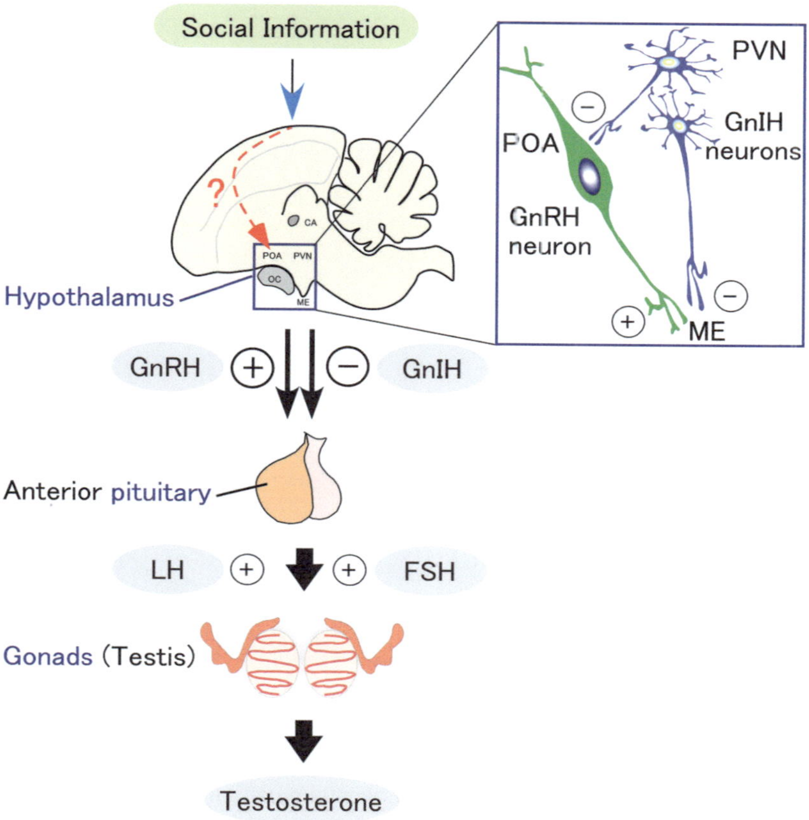

Fig. 9.1 Schematic representation of the avian hypothalamus–pituitary–gonadal (HPG) axis and hypothetical mechanism for social regulation of the levels of blood androgens, such as testosterone, in male birds. Under stimulatory conditions, gonadotropin-releasing hormone (GnRH) is released from the hypothalamus, stimulating the production and release of luteinizing hormone (LH) and follicle-stimulating hormone (FSH) from the pituitary gland, which in turn regulate sperm production and testosterone synthesis and release. Gonadotropin-inhibitory hormone (GnIH) can downregulate the HPG axis at the level of GnRH neurons and/or the level of the pituitary. In this model, information from social contexts is integrated by GnRH and/or GnIH neurons in the hypothalamus to regulate blood testosterone levels. *CA* commissura anterior, *ME* median eminence, *POA* preoptic area, *PVN* paraventricular nucleus of the hypothalamus, *OC* optic chiasma

ciated masculine courtship behaviors, including dancing back and forth on a perch, the sequence of reproductive behaviors proceeds to copulation (Fig. 9.2a) (Wade 2001; Ullrich et al. 2016). Ernst and Bentley (2016) experimentally showed that male zebra finches receiving a female stimulus exhibit higher levels of activity, song behavior, and copulation attempts than isolates or those receiving a male stimulus. The European starling (*Sturnus vulgaris*) is a hole-nesting songbird. Male starlings sing very long and complex courtship songs during the breeding season (Eens et al. 1991). Eens et al. (1991) showed that male starlings exhibit a significant increase in the time spent at the nest box, as well as in the total time spent singing, after a female was introduced. Sitting and singing in the nest box also increased significantly. Wing-waving is a visual display associated with singing in which the tail is tucked downward, the beak pointed upward, the throated feathers ruffled, and the wings rotated (Fig. 9.2b) (Eens et al. 1990). Wing-waving and carrying nest material occurred only after the introduction of a female, and not with introduction of a male

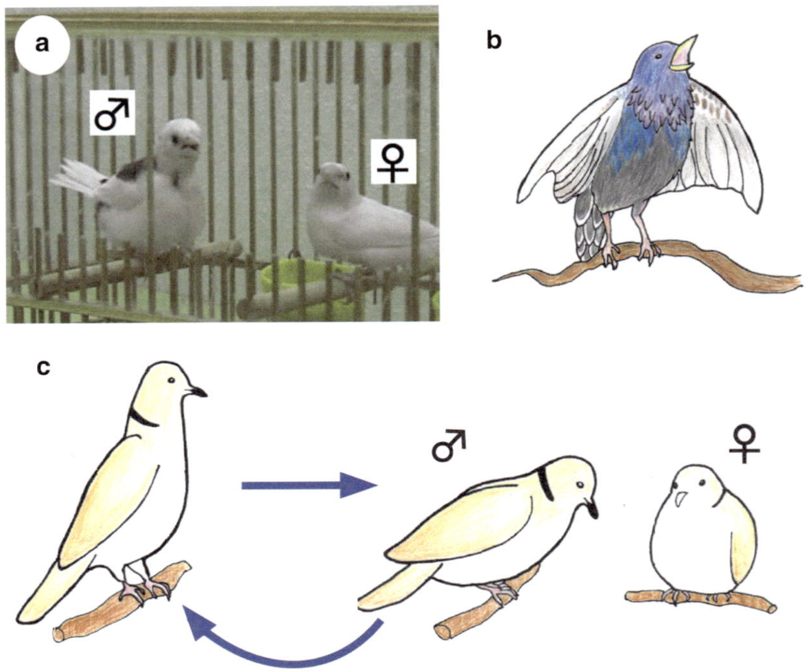

Fig. 9.2 Photograph of Bengalese finch courtship song behavior. A male (with ruffled feathers) singing toward a female (**a**). The wing-waving display. Male European starlings spread their wings and move them in a rotating manner with vertically flexed tail (**b**). The "bow-coo" display. In the ring dove, this coo is always made together with a bow. It is always directed toward a particular bird and repeated quickly with short intervals (**c**)

(Eens et al. 1990; Pinxten et al. 2003). These observations suggest that the primary function of wing-waving is to attract a female. Male ring doves (*Streptopelia risoria*) exposed to female conspecifics also exhibit a variety of courtship behaviors. Courtship begins soon after a male and female ring dove are paired in a cage containing nesting material and a nest bowl. They then show a regular series of behaviors from courtship and mating to nest building. Male-typical courtship behavior is characterized by the "bow-coo" display (Fig. 9.2c) (Lehrman 1964). If a male ring dove taken from isolation is presented with a female conspecific, the male will immediately approach her and show the "bow-coo" display many times at once, in which the cooing is accompanied by a bowing directed toward her. In addition, Japanese quail (*Coturnix japonica*) display a variety of reproductive behaviors. In the reproductive courtship of quail, one of the most characteristic behaviors is crowing vocalization, which is loud, has a broad band, and consists of two to three syllables (Fig. 9.3a) (Goodson and Adkins-Regan 1997). To examine whether the presence of a female changes the behavioral patterns in male quail, we analyzed male vocalizations and positioning in the behavioral test cage by the presence of a female bird. The basic experimental paradigm used in this study to assess the effects of female presence is shown in Fig. 9.3b–d. In detail, the test cage (28 cm high, 60 cm long, and 45 cm wide) was located in an empty room. The cage was divided in the middle with a transparent acrylic glass wall to physically separate the birds on both sides but which allows the birds to view each other freely. Five minutes after placing an experimental male on one side of the cage, a stimulus bird was added to the other side of the cage, except in the isolated group (control), and kept in place for 60 min (Fig. 9.3d). The stimulus bird was unfamiliar to the experimental male in all cases. The behavior of the experimental male was recorded with a digital video camera as soon as the stimulus bird was introduced into the cage. These tests were conducted using a within-subjects design, spaced 24 h apart. The mean frequency of crowing was significantly reduced by the presence of a female (t = 6.108, P = 0.0036, paired t test; Fig. 9.3e). When a female quail was visible in the adjacent compartment, all males spent significantly more time close to the acrylic glass wall (t = 7.889, P = 0.0014, paired t test;

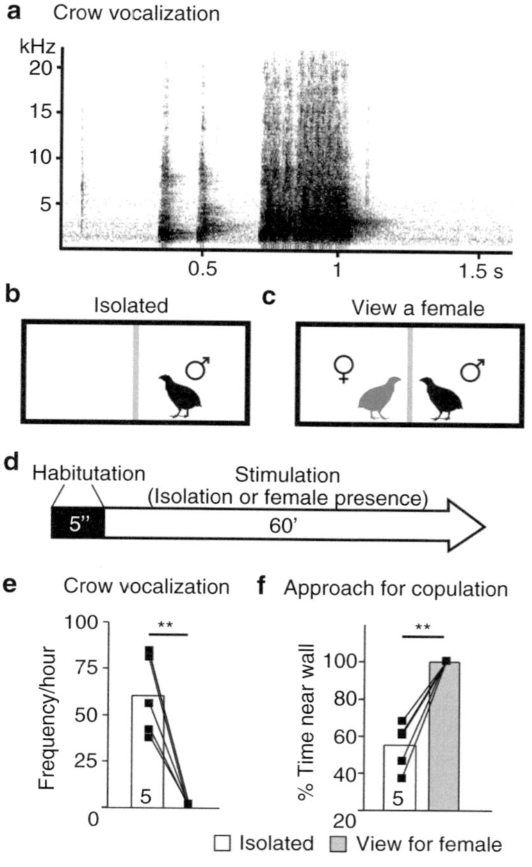

Fig. 9.3 Female presence changes the behavioral patterns in males. Sonogram of a typical crow of a male Japanese quail (**a**). The experimental cages were divided in the middle with an acrylic glass wall, indicated by the *gray lines* (**b**, **c**). Schematic representation of the experimental design depicting the habituation period (5 min, *black bar*) and the stimulation period (60 min, *white arrow*) (**d**). Effects of the presence of a female on frequency of crow vocalization (**e**), and social proximity behavior of male birds as indicated by the percentage of time that the subject was positioned within 10 cm to the acrylic glass wall (**f**). *Numbers in bars* indicate the number of birds analyzed. $^{**}P < 0.01$ vs isolated group, paired *t* test. Data are presented as means ± SEM. Tobari et al. (unpublished)

Fig. 9.3f). Our results confirmed that sexually motivated male quail produce crows in the absence of a female (Domjan and Nash 1988; Meddle et al. 1997; Charlier et al. 2005). The functions of crowing are mate attraction (Goodson and Adkins-Regan 1997) and territorial advertisement (Deregnaucourt 2010); thus, they do not require the presence of a target female. Once males are in the presence of a female, they suppress this crowing and actively approach her for copulation (Fig. 9.3f) (Domjan and Nash 1988; Meddle et al. 1997; Charlier et al. 2005).

9.3 Male Peripheral Hormonal Responses to Female Conspecifics

The role of social cues in reproductive physiology has been a topic of interest in a wide variety of vertebrate taxa; a cue from the opposite sex has been found to up- and down-regulate reproductive function. Many reports have indicated that cues from the opposite sex can change the morphology of gonadal and accessory sex tissue and peripheral hormonal levels in male birds. Burger (1953) showed that male starlings exposed to females exhibited larger testicular volumes than males in the absence of females. Haase et al. (1976) also showed that male domestic pigeons (*Columba livia*) paired with a female for 2 months had a higher level of plasma luteinizing hormone (LH) and larger testes than unpaired males. In male ring doves, it was reported that the concentration of plasma LH increased after 24 h of pairing with a female (Silver et al. 1980) and that, within 4 h after pairing with a female, male androgen levels rose (Feder et al. 1977) independently. In another study, Dufty and Wingfield (1986) exposed captive male brown-headed cowbirds (*Molothrus ater*) to long days. These male birds exhibited gonadal growth and had elevated plasma testosterone. They showed that this photoperiodic response was enhanced by the presence of females. Photostimulated male brown-headed cowbirds housed with females exhibited an increase in plasma testosterone levels sooner and maintained peak testosterone concentrations longer than individually housed males (Dufty and Wingfield 1986). Additionally, Pinxten et al. (2003) examined the short-term effects of a female conspecific on plasma testosterone concentration in male birds. They exposed individually housed male European starlings to a female conspecific for

60 min. The male starlings exposed to females exhibited a higher level of plasma testosterone than the males without females. Their results suggested that a stimulation time of 1 h was sufficient to increase the plasma testosterone level in male starlings (Pinxten et al. 2003).

It was also reported that, when a male recognizes the presence of, as well as when he copulates with, a female he shows changes in plasma androgen concentrations. Male ring doves exposed to females through glass partitions for 24 h had androgen levels comparable to those of males given free access to females, and higher androgen levels than males in isolation (O'Connell et al. 1981). This indicates that contact cues from the mating situation do not necessarily influence female-induced elevations in male plasma androgens of ring doves (O'Connell et al. 1981). In male Japanese quail, visual cues appear to be of primary importance in the perception of social context. Interestingly, it has been reported that the presence of a female bird, as well as copulation, actually rapidly decreases the plasma testosterone levels in male quail (Delville et al. 1984; Cornil et al. 2009). This is unusual among vertebrate species, in which increases in plasma LH and testosterone generally follow interaction with a female. We do not have any explanation for this seemingly unusual phenomenon at present.

9.4 Neuroestrogen Activates Male Sociosexual Behaviors and Social Regulation of Brain Aromatase Activity in Birds

Estrogens, such as 17β-estradiol (E_2), profoundly alter physiological and behavioral processes, including gonadotropin secretion and social behavior. Estrogens are synthesized from androgens by the enzyme aromatase. This enzyme is mainly expressed in the ovaries, but is also found in many other tissues, including the brain. In birds, brain aromatase expression is localized in specific neuronal regions, such as the hypothalamic preoptic area (HPOA) (Steimer and Hutchison 1980; Balthazart et al. 1996, 1990b; Schlinger and Arnold 1991). In Japanese quail and ring dove, the expression of male sexual behaviors is dependent on and modulated rapidly by the local production of E_2 in the POA via aromatization of circulating testosterone (Balthazart et al. 1990a; Watson and Adkins-Regan 1989; Cornil et al. 2006; Fusani et al. 2001). It was recently demonstrated that aromatase activity (AA) can also be rapidly modulated by the glutamate or calcium-dependent phosphorylation resulting from neuronal depolarization (Balthazart et al. 2003, 2006; Charlier et al. 2011). Moreover, rapid AA fluctuations occur in vivo in response to social interactions (Cornil et al. 2005; Remage-Healey et al. 2008). Cornil et al. (2005) provided the first demonstration of rapid changes in AA occurring in the HPOA of male quail following visual access to or copulation with a female. Both social interactions produced a significant decrease in AA in the HPOA, with the maximal effect seen after 5 min. Furthermore, de Bournonville et al. (2013) investigated the time course of the effect of sexual interaction on AA measured in different populations of aromatase-expressing cells microdissected from the brains of male quail that experienced varying durations of visual exposure to, or copulation with, a female. Copulation produced rapid AA inhibition in the medial preoptic nucleus of the HPOA that was detectable after 5 min and reached a maximum after 15 min, and in the tuberal hypothalamus that was detectable after 2 min and reached a maximum after 15 min. Viewing a female induced enzymatic fluctuations in the HPOA, but not in the tuberal hypothalamus. Although the functional significance of such decreases in brain AA is unclear, de Bournonville et al. (2013) suggested that rapid estrogen effects are important during the motivational phase of the behavior to trigger physiological events essential to activate mate search and copulation.

9.5 Avian Neuropeptide-Related Reproduction

Avian reproductive physiology and peripheral androgen levels are regulated by the HPG axis (Fig. 9.1) (Ottinger and Bakst 1995). In stimulatory conditions, GnRH is released from the hypothalamus, stimulating the production and release of LH and follicle-stimulating hormone from the pituitary gland, which, in turn, regulate testicular and ovarian function. It is now generally accepted that GnRH is the only hypothalamic regulator of the release of pituitary gonadotropins, and that no other neuropeptide has a direct influence on the reproductive axis.

The recent discovery of two hypothalamic RFamide peptides (peptides with a common Arg-Phe-NH_2 C-terminus), kisspeptins and GnIH, revealed a new pathway for reproductive control (Kotani et al. 2001; Muir et al. 2001; Ohtaki et al. 2001; Tsutsui et al. 2000). Kisspeptins are a family of neuropeptides encoded by the KiSS1 gene; they are the endogenous ligands for the G protein-coupled receptor GPR54 (Kotani et al. 2001; Muir et al. 2001; Ohtaki et al. 2001). It is now known that kisspeptins play a key role in the maintenance of the reproductive axis and may be critical for the onset of puberty (Decourt et al. 2008; Seminara et al. 2003). Kisspeptins stimulate gonadotropin secretion in mammals (Gottsch et al. 2004; Messager et al. 2005), with an effect that appears to be mediated exclusively by GnRH neurons that express GPR54 (Irwig et al. 2004). On the other hand, GnIH was first identified in the quail hypothalamo–hypophysial system and named for its ability to inhibit gonadotropin secretion (Tsutsui et al. 2000). GnIH neurons project to the hypothalamus as well as the median eminence (Tsutsui et al. 2000; Ubuka et al. 2003; Ukena et al. 2003). GnIH may also decrease GnRH release, because GnIH neurons make axosomatic contact with GnRH neurons (Bentley et al. 2003; Ubuka et al. 2008). GnIH receptors have been identified on the pituitary and in several brain regions, including the hypothalamus (Yin et al. 2005), and are also expressed in the GnRH neurons (Ubuka et al. 2008). These findings suggest that, as well as acting on the pituitary gland, GnIH acts on the GnRH neurons to inhibit gonadotropin release.

The hypothalamic neuropeptides GnRH, kisspeptin, and GnIH are considered to be integrators of environmental stimuli for reproduction (Parhar et al. 2012; Stevenson et al. 2012). Meanwhile, the KiSS1 gene is absent in avian species (Scanes 2015; Um et al. 2010) and, to date, kisspeptin-like peptides have not been identified in avian brains (Tobari et al. 2010). Therefore, recent studies have investigated whether avian GnRH and GnIH in the hypothalamus are influenced by various factors in the social environment. In the next section, we review the social regulation and modulation of GnRH and GnIH neuronal plasticities in avian species, particularly with regard to the effects of a female presence on male hypothalamic function.

9.6 Social Environmental Regulation of GnRH and GnIH in Birds

A longstanding goal has been to clarify how social stimuli from individuals of the opposite sex might increase gonadotropin and androgen secretion in birds. Social information is perceived by the brain via sensory systems. The information seems to converge on the hypothalamus and is then transduced into an endocrine response through the portal vessels to the adenohypophysis (Ball 1993). In the 1990s, researchers began to examine changes in the activity of GnRH neurons in the hypothalamus of many species, from fish to mammals, following various social interactions (Burmeister et al. 2005; Burmeister and Wilczynski 2005; Dellovade and Rissman 1994; Lambert et al. 1992). In avian species, Mantei et al. (2008) showed that the number of cells expressing GnRH mRNA and protein increases significantly in male ring doves after short-term exposure to a receptive female for 1 or 2 hours. In starlings, photostimulated male starlings housed with a females for 2 weeks facilitated a photoperiodic increase in the number of GnRH cells in the rostral region in the preoptic area (Stevenson and Ball 2009). On the other hand, no significant effect of a female presence was shown in male quail with regard to GnRH mRNA expression after visual exposure to a receptive female for 1 h (Fig. 9.4d) (Tobari et al. 2014). This lack of a change in GnRH mRNA expression is consistent with previous reports indicating that quail maintain constant high GnRH cell numbers and that the plasticity in GnRH expression in quail is negligible (MacDougall-Shackleton et al. 2009; Teruyama and Beck 2000; Stevenson et al. 2012). Male zebra finch also did not exhibit an increase in GnRH expression after a 30-min interaction with a female (Ernst and Bentley 2016). Taken together, these data suggest that the GnRH neurons in particular bird species respond to social stimuli sensitively and that social interaction with female conspecifics increases GnRH synthesis and GnRH cell numbers in male birds that exhibit GnRH plasticity.

Maney et al. (2007) hypothesized that a rapid neuroendocrine response to social cues will resemble the neuroendocrine response to photic cues (long days), during which GnRH release from the brain is induced (Shinomiya et al. 2014). To test this hypothesis, they played recordings of conspecific male song to female white-throated sparrows (*Zonotrichia albicollis*) and quantified the resulting rapid changes in plasma LH, as well as the expression of the immediate early gene, Egr-1, in the mediobasal hypothalamus (MBH), which is considered to be

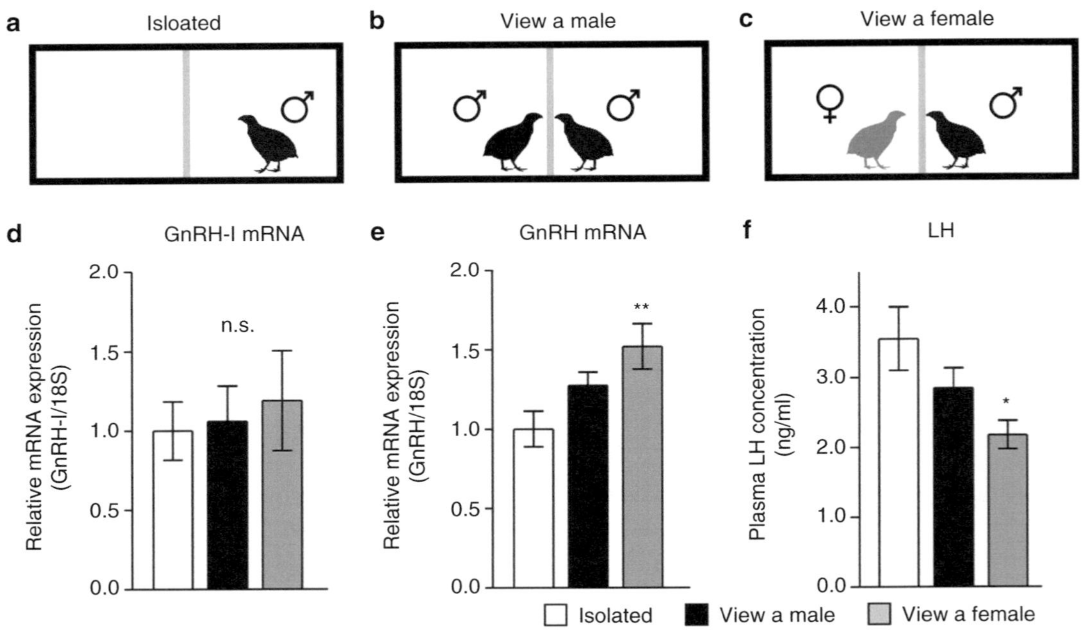

Fig. 9.4 Effects of male or female presence on hypothalamic GnRH and GnIH mRNA expressions and plasma LH concentration in male quail. The experimental cages were divided in the middle with an acrylic glass wall, indicated by the *gray lines* (**a–c**). Effects of the presence of a female on hypothalamic GnRH and GnIH mRNA expressions (**d**, **e**), and circulating LH level (**f**). One-way ANOVA followed by Tukey's multiple comparisons test, $*P < 0.05$, $**P < 0.01$ versus isolated group. Data are presented as means ± SEM. Modified from Tobari et al. (2014)

the avian brain region involved in the control of GnRH release. They showed that hearing courtship songs for 42 min induces a rise in plasma LH and an increase in Egr-1 expression in the MBH. The time course of LH release and the hypothalamic distribution of Egr-1 induction suggest that auditory courtship signals act as a trigger to initiate rapid release of GnRH in female sparrows. Thus, female presence may also modify GnRH release in male quail, resulting in a decrease in plasma LH concentration in a manner resembling auditory regulation of GnRH secretion in female sparrows.

Recent studies have suggested that GnIH expression is also influenced by social status and breeding conditions (Calisi et al. 2011; Ubuka et al. 2005). Therefore, GnIH has also begun to be regarded as a modulator of reproductive function in response to the social environment (Maruska and Fernald 2011; Tsutsui et al. 2015). In Japanese quail, sexually active male birds exhibit a transient decline in testosterone as a result of viewing a female (Cornil et al. 2009; Delville et al. 1984), as described above. Based on these earlier observations, Tobari et al. (2014) investigated the neural pathways from the visual perception of the presence of a female that leads to a change in plasma testosterone level in the male quail. Tobari et al. (2014) first conducted experiments to determine whether female presence modifies hypothalamic GnRH or GnIH precursor mRNA expression in male quail. The basic experimental paradigm used in the study to assess the effects of male or female presence is shown in Figs. 9.3d and 9.4a–c. Viewing a female increased the level of GnIH mRNA expression in the paraventricular nucleus (PVN) of the hypothalamus (Fig. 9.4e), which was associated with a decrease in the plasma levels of LH (Fig. 9.4f). Furthermore, norepinephrine (NE) release increased rapidly in the male PVN just after viewing a female. NE application stimulated GnIH release from the diencephalic tissue blocks, and the intracerebroventricular injection of NE reduced the plasma levels of LH. Fluorescent double-labeling immunohistochemistry revealed that GnIH neurons are innervated by noradrenergic fibers, and immunohistochemistry combined with in situ hybridization demonstrated that GnIH neurons expressed α2A-adrenergic receptor (α2A-ADR) mRNA. These findings suggest that viewing a female causes an increase in NE release in the PVN, which directly stimulates the secretion of GnIH, inhibiting LH secretion from the pituitary (Fig. 9.5). This study outlined the molecular pathways linking social encounters to

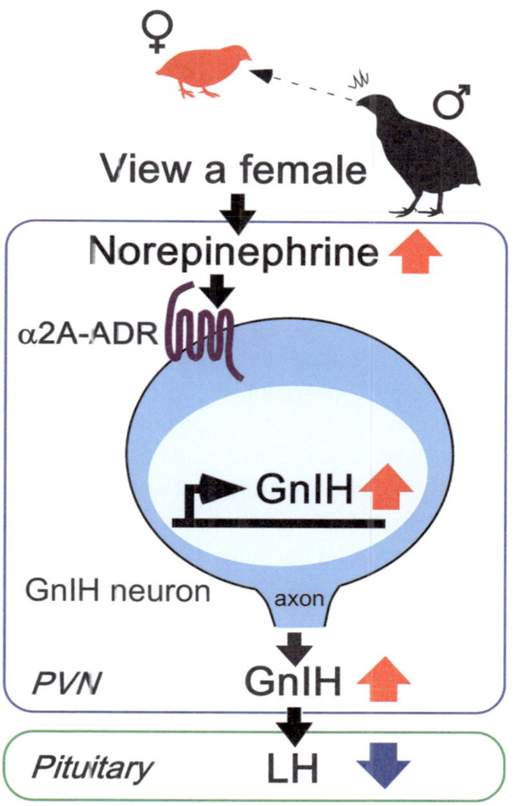

Fig. 9.5 Schematic model of the mechanism of social regulation of reproductive physiology through activation of GnIH neurons. The α2A-adrenergic receptor (α2A-ADR) is expressed on gonadotropin-inhibitory hormone (GnIH) neurons in the hypothalamus. Female presence rapidly increases norepinephrine (NE) release in the paraventricular nucleus (PVN) of the hypothalamus, where GnIH neuronal cell bodies are present, and stimulates GnIH release, resulting in the suppression of luteinizing hormone (LH) release in sexually active male quail

peripheral hormonal changes via the brain–pituitary endocrine system in sexually mature male birds.

9.7 Summary and Future Directions

The presence of an individual of the opposite sex often triggers changes in animals' reproductive hormones and behaviors. These hormonal responses to a potential mate are thought to act on brain regions involved in memory, cognition, emotion, genital reflexes, behavior in subsequent social interactions (Oliveira 2004), and future parental behaviors, and to contribute to the optimization of reproductive success (Gleason and Marler 2010). In this chapter, we focused on data from studies of birds suggesting that sexually mature male birds show dynamic changes in behavioral pattern, brain AA, and the HPG axis in response to social interaction. We also presented findings revealing a new pathway mediating social effects on the endocrine system: the noradrenergic innervation of GnIH neurons that leads to changes in the secretion of reproductive hormones in response to a potential mate. It is gradually being clarified how social information influences the HPG axis. However, unanswered questions remain. For example, is the noradrenergically induced hormonal response to a female conserved in animal species other than quail? What are the neural pathways from reception/perception of a female to noradrenergic neurons in male birds? Does social information influence GnRH and/or GnIH release into hypophysial portal vessel blood? Further studies are needed to answer these questions. The development of hypophysial portal blood collection techniques (Youngren et al. 1996) for application to social birds and the use of recently developed techniques in birds, such as transgenic methods (Agate et al. 2009; Abe et al. 2015; Liu et al. 2015) and optogenetics (Roberts et al. 2012), should increase our understanding of how the social environment influences reproductive physiology.

References

Abe K, Matsui S, Watanabe D. Transgenic songbirds with suppressed or enhanced activity of CREB transcription factor. Proc Natl Acad Sci U S A. 2015;112(24):7599–604. doi:10.1073/pnas.1413484112.

Agate RJ, Scott BB, Haripal B, Lois C, Nottebohm F. Transgenic songbirds offer an opportunity to develop a genetic model for vocal learning. Proc Natl Acad Sci U S A. 2009;106(42):17963–7. doi:10.1073/pnas.0909139106.

Ball GF. The neural integration of environmental information by seasonally breeding birds. Am Zool. 1993;33:185–99.

Balthazart J, Evrard L, Surlemont C. Effects of the non-steroidal inhibitor R76713 on testosterone-induced sexual behavior in the Japanese quail (*Coturnix coturnix japonica*). Horm Behav. 1990a;24(4):510–31.

Balthazart J, Foidart A, Surlemont C, Vockel A, Harada N. Distribution of aromatase in the brain of the Japanese quail, ring dove, and zebra finch: an immunocytochemical study. J Comp Neurol. 1990b;301(2):276–88.

Balthazart J, Absil P, Foidart A, Houbart M, Harada N, Ball GF. Distribution of aromatase-immunoreactive cells in the forebrain of zebra finches (Taeniopygia guttata): implications for the neural action of steroids and nuclear definition in the avian hypothalamus. J Neurobiol. 1996;31(2):129–48.

Balthazart J, Baillien M, Charlier TD, Ball GF. Calcium-dependent phosphorylation processes control brain aromatase in quail. Eur J Neurosci. 2003;17(8):1591–606. doi:10.1046/j.1460-9568.2003.02598.x.

Balthazart J, Baillien M, Ball GF. Rapid control of brain aromatase activity by glutamatergic inputs. Endocrinology. 2006;147(1):359–66. doi:10.1210/en.2005-0845.

Bentley GE, Perfito N, Ukena K, Tsutsui K, Wingfield JC. Gonadotropin-inhibitory peptide in song sparrows (*Melospiza melodia*) in different reproductive conditions, and in house sparrows (*Passer domesticus*) relative to chicken-gonadotropin-releasing hormone. J Neuroendocrinol. 2003;15(8):794–802.

Burger JW. The effect of photic stimuli on the reproductive cycle of the male starling, *Sturnus vulgaris*. J Exp Zool. 1953;124:227–39.

Burmeister SS, Wilczynski W. Social signals regulate gonadotropin-releasing hormone neurons in the green treefrog. Brain Behav Evol. 2005;65(1):26–32. doi:10.1159/000081108.

Burmeister SS, Jarvis ED, Fernald RD. Rapid behavioral and genomic responses to social opportunity. PLoS Biol. 2005;3(11):e363. doi:10.1371/journal.pbio.0030363.

Calisi RM, Diaz-Munoz SL, Wingfield JC, Bentley GE. Social and breeding status are associated with the expression of GnIH. Genes Brain Behav. 2011;10(5):557–64. doi:10.1111/j.1601-183X.2011.00693.x.

Charlier TD, Ball GF, Balthazart J. Sexual behavior activates the expression of the immediate early genes c-fos and Zenk (egr-1) in catecholaminergic neurons of male Japanese quail. Neuroscience. 2005;131(1):13–30. doi:10.1016/j.neuroscience.2004.09.068.

Charlier TD, Harada N, Balthazart J, Cornil CA. Human and quail aromatase activity is rapidly and reversibly inhibited by phosphorylating conditions. Endocrinology. 2011;152(11):4199–210. doi:10.1210/en.2011-0119.

Cornil CA, Dalla C, Papadopoulou-Daifoti Z, Baillien M, Dejace C, Ball GF, Balthazart J. Rapid decreases in preoptic aromatase activity and brain monoamine concentrations after engaging in male sexual behavior. Endocrinology. 2005;146(9):3809–20. doi:10.1210/en.2005-0441.

Cornil CA, Dalla C, Papadopoulou-Daifoti Z, Baillien M, Balthazart J. Estradiol rapidly activates male sexual behavior and affects brain monoamine levels in the quail brain. Behav Brain Res. 2006;166(1):110–23.

Cornil CA, Stevenson TJ, Ball GF. Are rapid changes in gonadal testosterone release involved in the fast modulation of brain estrogen effects? Gen Comp Endocrinol. 2009;163(3):298–305. doi:10.1016/j.ygcen.2009.04.029.

de Bournonville C, Dickens MJ, Ball GF, Balthazart J, Cornil CA. Dynamic changes in brain aromatase activity following sexual interactions in males: where, when and why? Psychoneuroendocrinology. 2013;38(6):789–99. doi:10.1016/j.psyneuen.2012.09.001.

Decourt C, Tillet Y, Caraty A, Franceschini I, Briant C. Kisspeptin immunoreactive neurons in the equine hypothalamus interactions with GnRH neuronal system. J Chem Neuroanat. 2008;36(3-4):131–7.

Dellovade TL, Rissman EF. Gonadotropin-releasing hormone-immunoreactive cell numbers change in response to social interactions. Endocrinology. 1994;134(5):2189–97.

Delville Y, Sulon J, Hendrick JC, Balthazart J. Effect of the presence of females on the pituitary-testicular activity in male Japanese quail (*Coturnix coturnix japonica*). Gen Comp Endocrinol. 1984;55:295–305.

Deregnaucourt S. Interspecific hybridization as a tool to understand vocal divergence: the example of crowing in quail (*Genus Coturnix*). PLoS One. 2010;5(2):e9451. doi:10.1371/journal.pone.0009451.

Domjan M, Nash S. Stimulus control of social behaviour in male Japanese quail, *Coturnix coturnix japonica*. Anim Behav. 1988;36:1006–15.

Dufty AM, Wingfield JC. The influence of social cues on the reproductive endocrinology of male brown-headed cowbirds: field and laboratory studies. Horm Behav. 1986;20:222–34.

Eens M, Pinxten R, Verheyen RF. On the function of singing and wing-waving in the European starling *Sturnus vulgaris*. Bird Study. 1990;37(1):48–52. doi:10.1080/00063659009477038.

Eens M, Pinxten R, Verheyen RF. Male song as a cue for mate choice in the European starling. Behaviour. 1991;116:210–38.

Ernst DK, Bentley GE. Neural and neuroendocrine processing of a non-photic cue in an opportunistically-breeding songbird. J Exp Biol. 2016; doi:10.1242/jeb.126987.

Feder HH, Storey A, Goodwin SD, Reboulleau C, Silver R. Testosterone and "5α-dihydrotestosterone" levels in peripheral plasma of male and female ring Doves (*Streptopelia risoria*) during the reproductive cycle. Biol Reprod. 1977;16:666–77.

Fusani L, Gahr M, Hutchison JB. Aromatase inhibition reduces specifically one display of the ring dove courtship behavior. Gen Comp Endocrinol. 2001;122(1):23–30.

Gleason ED, Marler CA. Testosterone response to courtship predicts future paternal behavior in the California mouse, *Peromyscus californicus*. Horm Behav. 2010;57(2):147–54. doi:10.1016/j.yhbeh.2009.10.006.

Goodson JL, Adkins-Regan E. Playback of crows of male Japanese quail elicits female phonotaxis. Condor. 1997;99:990–3.

Gottsch ML, Cunningham MJ, Smith JT, Popa SM, Acohido BV, Crowley WF, Seminara S, Clifton DK, Steiner RA. A role for kisspeptins in the regulation of gonadotropin secretion in the mouse. Endocrinology. 2004;145(9):4073–7.

Haase B, Paulke E, Sharp PJ. Effects of seasonal and social factors on testicular activity and hormone levels in domestic pigeons. J Exp Zool. 1976;197:81–8.

Irwig MS, Fraley GS, Smith JT, Acohido BV, Popa SM, Cunningham MJ, Gottsch ML, Clifton DK, Steiner RA. Kisspeptin activation of gonadotropin releasing hormone neurons and regulation of KiSS-1 mRNA in the male rat. Neuroendocrinology. 2004;80(4):264–72.

Kotani M, Detheux M, Vandenbogaerde A, Communi D, Vanderwinden JM, Le Poul E, Brezillon S, Tyldesley R, Suarez-Huerta N, Vandeput F, Blanpain C, Schiffmann SN, Vassart G, Parmentier M. The metastasis suppressor gene KiSS-1 encodes kisspeptins, the natural ligands of the orphan G protein-coupled receptor GPR54. J Biol Chem. 2001;276(37):34631–6.

Kriegsfeld LJ, Ubuka T, Bentley GE, Tsutsui K. Seasonal control of gonadotropin-inhibitory hormone (GnIH) in birds and mammals. Front Neuroendocrinol. 2015;37:65–75. doi:10.1016/j.yfrne.2014.12.001.

Lambert GM, Rubin BS, Baum MJ. Sexual dimorphism in the effects of mating on the in vitro release of LHRH from the ferret mediobasal hypothalamus. Physiol Behav. 1992;52(4):809–13.

Lehrman DS. The reproductive behavior of ring doves. Sci Am. 1964;211:48–54.

Liu WC, Kohn J, Szwed SK, Pariser E, Sepe S, Haripal B, Oshimori N, Marsala M, Miyanohara A, Lee R. Human mutant huntingtin disrupts vocal learning in transgenic songbirds. Nat Neurosci. 2015;18(11):1617–22. doi:10.1038/nn.4133.

MacDougall-Shackleton SA, Stevenson TJ, Watts HE, Pereyra ME, Hahn TP. The evolution of photoperiod response systems and seasonal GnRH plasticity in birds. Integr Comp Biol. 2009;49(5):580–9. doi:10.1093/icb/icp048.

Maney DL, Goode CT, Lake JI, Lange HS, O'Brien S. Rapid neuroendocrine responses to auditory courtship signals. Endocrinology. 2007;148(12):5614–23. doi:10.1210/en.2007-0879.

Mantei KE, Ramakrishnan S, Sharp PJ, Buntin JD. Courtship interactions stimulate rapid changes in GnRH synthesis in male ring doves. Horm Behav. 2008;54(5):669–75. doi:10.1016/j.yhbeh.2008.07.005.

Maruska KP, Fernald RD. Social regulation of gene expression in the hypothalamic–pituitary–gonadal axis. Physiology (Bethesda). 2011;26(6):412–23. doi:10.1152/physiol.00032.2011.

Meddle SL, King VM, Follett BK, Wingfield JC, Ramenofsky M, Foidart A, Balthazart J. Copulation activates Fos-like immunoreactivity in the male quail forebrain. Behav Brain Res. 1997;85(2):143–59.

Messager S, Chatzidaki EE, Ma D, Hendrick AG, Zahn D, Dixon J, Thresher RR, Malinge I, Lomet D, Carlton MBL, Colledge WH, Caraty A, Aparicio SAJR. Kisspeptin directly stimulates gonadotropin-releasing hormone release via G protein-coupled receptor 54. Proc Natl Acad Sci U S A. 2005;102(5):1761–6.

Muir AI, Chamberlain L, Elshourbagy NA, Michalovich D, Moore DJ, Calamari A, Szekeres PG, Sarau HM, Chambers JK, Murdock P, Steplewski K, Shabon U, Miller JE, Middleton SE, Darker JG, Larminie CG, Wilson S, Bergsma DJ, Emson P, Faull R, Philpott KL, Harrison DC. AXOR12, a novel human G protein-coupled receptor, activated by the peptide KiSS-1. J Biol Chem. 2001;276(31):28969–75.

O'Connell M, Reboulleau C, Feder H, Silvers R. Social interactions and androgen levels in birds I. Female characteristics associated with increased plasma androgen levels in the male ring dove (*Streptopelia risoria*). Gen Comp Endocrinol. 1981;44:454–63.

Ohtaki T, Shintani Y, Honda S, Matsumoto H, Hori A, Kanehashi K, Terao Y, Kumano S, Takatsu Y, Masuda Y, Ishibashi Y, Watanabe T, Asada M, Yamada T, Suenaga M, Kitada C, Usuki S, Kurokawa T, Onda H, Nishimura O, Fujino M. Metastasis suppressor gene KiSS-1 encodes peptide ligand of a G-protein-coupled receptor. Nature. 2001;411(6837):613–7.

Oliveira RF. Social modulation of androgens in vertebrates: mechanisms and function. Adv Study Behav. 2004;34:165–239.

Oliveira RF. Social behavior in context: hormonal modulation of behavioral plasticity and social competence. Integr Comp Biol. 2009;49(4):423–40. doi:10.1093/icb/icp055.

Ottinger MA, Bakst MR. Endocrinology of the avian reproductive system. J Avian Med Surg. 1995;9(4):242–50.

Parhar I, Ogawa S, Kitahashi T. RFamide peptides as mediators in environmental control of GnRH neurons. Prog Neurobiol. 2012;98(2):176–96.

Pinxten R, de Ridder E, Eens M. Female presence affects male behavior and testosterone levels in the European starling (*Sturnus vulgaris*). Horm Behav. 2003;44(2):103–9.

Remage-Healey L, Maidment NT, Schlinger BA. Forebrain steroid levels fluctuate rapidly during social interactions. Nat Neurosci. 2008;11(11):1327–34.

Roberts TF, Gobes SM, Murugan M, Olveczky BP, Mooney R. Motor circuits are required to encode a sensory model for imitative learning. Nat Neurosci. 2012;15(10):1454–9. doi:10.1038/nn.3206.

Scanes CG. Avian endocrine system. In: Scanes C, editor. Sturkie's avian physiology. 6th ed. Amsterdam: Academic Press; 2015. p. 489–96.

Schlinger BA, Arnold AP. Brain is the major site of estrogen synthesis in a male songbird. Proc Natl Acad Sci U S A. 1991;88:4191–4.

Seminara SB, Messager S, Chatzidaki EE, Thresher RR, Acierno JSJ, Shagoury JK, Bo-Abbas Y, Kuohung W, Schwinof KM, Hendrick AG, Zahn D, Dixon J, Kaiser UB, Slaugenhaupt SA, Gusella JF, O'Rahilly S, Carlton MBL, Crowley WFJ, Aparicio SAJR, Colledge WH. The GPR54 gene as a regulator of puberty. N Engl J Med. 2003;349(17):1614–27.

Shinomiya A, Shimmura T, Nishiwaki-Ohkawa T, Yoshimura T. Regulation of seasonal reproduction by hypothalamic activation of thyroid hormone. Front Endocrinol (Lausanne). 2014;5:12. doi:10.3389/fendo.2014.00012.

Silver R, Goldsmith AR, Follett BK. Plasma luteinizing hormone in male ring doves during the breeding cycle. Gen Comp Endocrinol. 1980;42:19–24.

Steimer T, Hutchison JB. Aromatization of testosterone within a discrete hypothalamic area associated with the behavioral action of androgen in the male dove. Brain Res Bull. 1980;192:586–91.

Stevenson TJ, Ball GF. Anatomical localization of the effects of reproductive state, castration, and social milieu on cells immunoreactive for gonadotropin-releasing hormone-I in male European starlings (*Sturnus vulgaris*). J Comp Neurol. 2009;517:146–55.

Stevenson TJ, Hahn TP, MacDougall-Shackleton SA, Ball GF. Gonadotropin-releasing hormone plasticity: A comparative perspective. Front Neuroendocrinol. 2012;33(3):287–300. doi:10.1016/j.yfrne.2012.09.001.

Teruyama R, Beck MM. Changes in immunoreactivity to anti-cGnRH-I and -II are associated with photostimulated sexual status in male quail. Cell Tissue Res. 2000;300(3):413–26.

Tobari Y, Iijima N, Tsunekawa K, Osugi T, Okanoya K, Tsutsui K, Ozawa H. Identification of gonadotropin-inhibitory hormone in the zebra finch (*Taeniopygia guttata*): Peptide isolation, cDNA cloning and brain distribution. Peptides. 2010;31:816–26.

Tobari Y, Son YL, Ubuka T, Hasegawa Y, Tsutsui K. A new pathway mediating social effects on the endocrine system: female presence acting via norepinephrine release stimulates gonadotropin-inhibitory hormone in the paraventricular nucleus and suppresses luteinizing hormone in quail. J Neurosci. 2014;34(29):9803–11. doi:10.1523/JNEUROSCI.3706-13.2014.

Tsutsui K, Saigoh E, Ukena K, Teranishi H, Fujisawa Y, Kikuchi M, Ishii S, Sharp PJ. A novel avian hypothalamic peptide inhibiting gonadotropin release. Biochem Biophys Res Commun. 2000;275(2):661–7.

Tsutsui K, Ubuka T, Son YL, Bentley GE, Kriegsfeld LJ. Contribution of GnIH research to the progress of reproductive neuroendocrinology. Front Endocrinol (Lausanne). 2015;6:179. doi:10.3389/fendo.2015.00179.

Ubuka T, Ueno M, Ukena K, Tsutsui K. Developmental changes in gonadotropin-inhibitory hormone in the Japanese quail (*Coturnix japonica*) hypothalamo–hypophysial system. J Endocrinol. 2003;178(2):311–8.

Ubuka T, Bentley GE, Ukena K, Wingfield JC, Tsutsui K. Melatonin induces the expression of gonadotropin-inhibitory hormone in the avian brain. Proc Natl Acad Sci U S A. 2005;102(8):3052–7. doi:10.1073/pnas.0403840102.

Ubuka T, Kim S, Huang YC, Reid J, Jiang J, Osugi T, Chowdhury VS, Tsutsui K, Bentley GE. Gonadotropin-inhibitory hormone neurons interact directly with gonadotropin-releasing hormone-I and -II neurons in European starling brain. Endocrinology. 2008;149(1):268–78.

Ukena K, Ubuka T, Tsutsui K. Distribution of a novel avian gonadotropin-inhibitory hormone in the quail brain. Cell Tissue Res. 2003;312(1):73–9.

Ullrich R, Norton P, Scharff C. Waltzing Taeniopygia: integration of courtship song and dance in the domesticated Australian zebra finch. Anim Behav. 2016;112:285–300.

Um HN, Han JM, Hwang JI, Hong SI, Vaudry H, Seong JY. Molecular coevolution of kisspeptins and their receptors from fish to mammals. Ann N Y Acad Sci. 2010;1200:67–74. doi:10.1111/j.1749-6632.2010.05508.x.

Wade J. Zebra finch sexual differentiation: the aromatization hypothesis revisited. Microsc Res Tech. 2001;54:354–63.

Watson JT, Adkins-Regan E. Testosterone implanted in the preoptic area of male Japanese quail must be aromatized to activate copulation. Horm Behav. 1989;23(3):432–47.

Wingfield JC. Environmental and endocrine control of reproduction: an ecological approach. In: Mikami S, Homma K, Wada M, editors. Avian endocrinology. Tokyo: Japan Scientific Societies Press; 1983. p. 265–88.

Yin H, Ukena K, Ubuka T, Tsutsui K. A novel G protein-coupled receptor for gonadotropin-inhibitory hormone in the Japanese quail (*Coturnix japonica*): identification, expression and binding activity. J Endocrinol. 2005;184(1):257–66. doi:10.1677/joe.1.05926.

Youngren O, Chaiseha Y, Phillips R, El Halawani M. Vasoactive intestinal peptide concentrations in turkey hypophysial portal blood differ across the reproductive cycle. Gen Comp Endocrinol. 1996;103:323–30.

Neuroendocrine Control of Broodiness

10

Takeshi Ohkubo

Abstract

In the majority of vertebrates, survival of offspring to sexual maturation is important for increasing population size, and parental investment in the young is important for reproductive success. Consequently, parental care is critical for the survival of offspring in many species, and many vertebrates have adapted this behavior to their social and ecological environments. Parental care is defined as any behavior that is performed in association with one's offspring (Rosenblatt, Mayer, Siegel. Maternal behavior among nonprimate mammals. In: Adler, Pfaff, Goy, editors. Handbook of behavioral neurobiology. New York: Plenum; 1985. p. 229–98) and is well characterized in mammals and birds. In birds (class Aves), this is due to the high level of diversity across species. Parental behavior in birds protects the young from intruders, and generally involves nest building, incubation, and broody behavior which protect their young from an intruder, and the offspring are reared to independence. Broodiness is complexly regulated by the central nervous system and is associated with multiple hormones and neurotransmitters produced by the hypothalamus and pituitary gland. The mechanism of this behavior has been extensively characterized in domestic chicken (*Gallus domesticus*), turkey (*Meleagris gallopavo*), and pigeons and doves (family Columbidae). This chapter summarizes broodiness in birds from a physiology, genetics, and molecular biology perspective.

Keywords

Broodiness • Chicken • Incubation behavior • Pituitary gland • Prolactin (PRL) • PRL receptor • Signal transduction • Transcription • Vasoactive intestinal peptide (VIP)

T. Ohkubo
College of Agriculture, Ibaraki University, Ibaraki, Japan
e-mail: takeshi.ohkubo.0533@vc.ibaraki.ac.jp

T. Sasanami (ed.), *Avian Reproduction*, Advances in Experimental Medicine and Biology 1001,
DOI 10.1007/978-981-10-3975-1_10

10.1 Broodiness

Parental care is defined as any behavior that is performed in association with one's offspring in vertebrates (Rosenblatt et al. 1985). The parental behavior is highly diversified in birds. At one end of the spectrum, species such as cuckoos (family Cuculidae) and cowbirds (*Molothrus* spp.), deposit their eggs in the nests of other birds and never engage in parental behavior. The megapodes (family Megapodiidae) also do not incubate their eggs themselves, but rather bury their eggs in a nest mound made of compost to keep them warm until hatching. In some species, such as chicken and duck, the female displays the parental behavior, while male parental care, such as is provided by the emperor penguin (*Aptenodytes forsteri*) in Antarctica for more than 2 months to protect and warm the egg, is much rarer. However, the most common pattern of parental behavior in birds is biparental care, where both males and females provide equivalent levels of parental care toward their eggs and offspring. Approximately 90% of avian species are believed to be monogamous, and both sexes exhibit this behavior to enhance their reproductive success (Nelson 1995).

The extent of broodiness depends on the developmental state of the young at hatching. In precocial birds such as chicken, turkey, and duck, the chicks are able to independently search for and eat food immediately after hatching. Therefore, maternal care in chickens involves nest building, incubation, and broody behavior, all of which protect their young from intruders. By contrast, in altricial birds, the newly hatched nestlings are usually helpless and so the parents take care of their chicks until fledging. Consequently, in these species, the parents not only carry out nest building and incubation of the eggs before hatching but also feed and brood their young.

10.1.1 Characteristics of Incubation Behavior

In domestic chicken (*Gallus domesticus*) and turkey (*Meleagris gallopavo*), incubation behavior is triggered by the accumulation of a sufficient number of eggs to incubate; for example, bantam hens lay 10–20 eggs (Lea and Sharp 1982). The onset of incubation behavior is preceded by an increased frequency of nest visits (Haller and Cherms 1961). Furthermore, in laying bantams, the time spent at the nest is gradually increased closer to the onset of incubation behavior (Lea et al. 1981). Incubation behavior is preceded by an increase in the amplitude and duration of a nocturnal surge in plasma prolactin (PRL) levels (Sharp 1980) which, together with estrogen, stimulates brood patch formation (Hutchison et al. 1967; Hutchison 1975)—a defeathered area on the belly skin that is well supplied with blood vessels to effectively transfer body heat to the eggs (Jones 1971).

When incubation begins, chicken and turkey hens spend almost all of their time at the nest warming their eggs, occasionally turning and rearranging them (El Halawani et al. 1984). During this time, nutrient and water intake is drastically reduced, which results in a dramatic decrease in body weight (Sherry et al. 1980; Zadworny et al. 1988). When hens are incubating their eggs, they display aggressive and defensive behaviors, such as clucking (Romanov et al. 2002). Hens can be removed from their nest and transferred to a different environment to stop the expression of incubation behavior (El Halawani et al. 1980; Sharp et al. 1988), but it takes several days to disrupt this. Furthermore, in the case of bantams, approximately 90% of hens never renest following 72 h of nest deprivation (Sharp et al. 1988), and turkeys never resume incubation behavior after 5 days of nest deprivation (El Halawani et al. 1980). These behavioral changes are also under the control of hormones, including PRL.

10.2 Prolactin and Broodiness

The mechanism that controls broodiness has interested avian biologists for a long time, and poultry scientists have also sought to clarify how incubation behavior is expressed in domestic birds, because broodiness causes the ovary and oviduct to regress, which results in the cessation of egg laying and thus economic losses for the poultry industry.

In 1930s, Oscar Riddle and his colleagues provided the first evidence that PRL induces broodiness in pigeons and chicken (Riddle et al. 1933, 1935), and it has since been confirmed that PRL is involved in brooding of the chicks (Yamashina 1952) and incubation behavior (Saeki and Tanabe 1955). Following the development of a radioimmunoassay for avian PRL, it has been shown that the onset of incubation behavior in chicken is correlated with an increase of PRL in the plasma (Sharp et al. 1979, 1988), and that the blood PRL level remains high throughout the incubation period in chicken and turkey (Sharp et al. 1979, 1988; Zadworny et al. 1985, 1988; Kuwayama et al. 1992). Furthermore, active immunization against recombinant-derived PRL reduces the incidence, delays the development, or prevents the occurrence of incubation behavior in birds (March et al. 1994); while administration of ovine PRL induces incubation behavior in turkey (El Halawani et al. 1986; Youngren et al. 1991). Therefore, it is well established that PRL is involved in the initiation and maintenance of incubation behavior in these strains (Fig. 10.1).

This characteristic increase in blood PRL levels during broodiness is common to both domestic and wild birds. Indeed, plasma PRL levels even increase during the breeding season in a brood parasite, the brown-headed cowbird (*Molothrus ater*) (Dufty et al. 1987), and in male emperor penguins (*Aptenodytes forsteri*) during egg incubation (Lormée et al. 1999). However, the plasma PRL levels differ between precocial and altricial birds: in precocial birds belonging to the Galliformes and Anceriformes, PRL levels rapidly decline following the hatching of chicks and can be modified by manipulating the young chicks (Kuwayama et al. 1992), while in altricial birds, plasma PRL levels decrease more slowly after hatching, presumably because brooding by the parent(s) is imperative for chicks to survive to fledging (Goldsmith 1983).

Since PRL has a specific role in controlling incubation behavior, it is expected that the

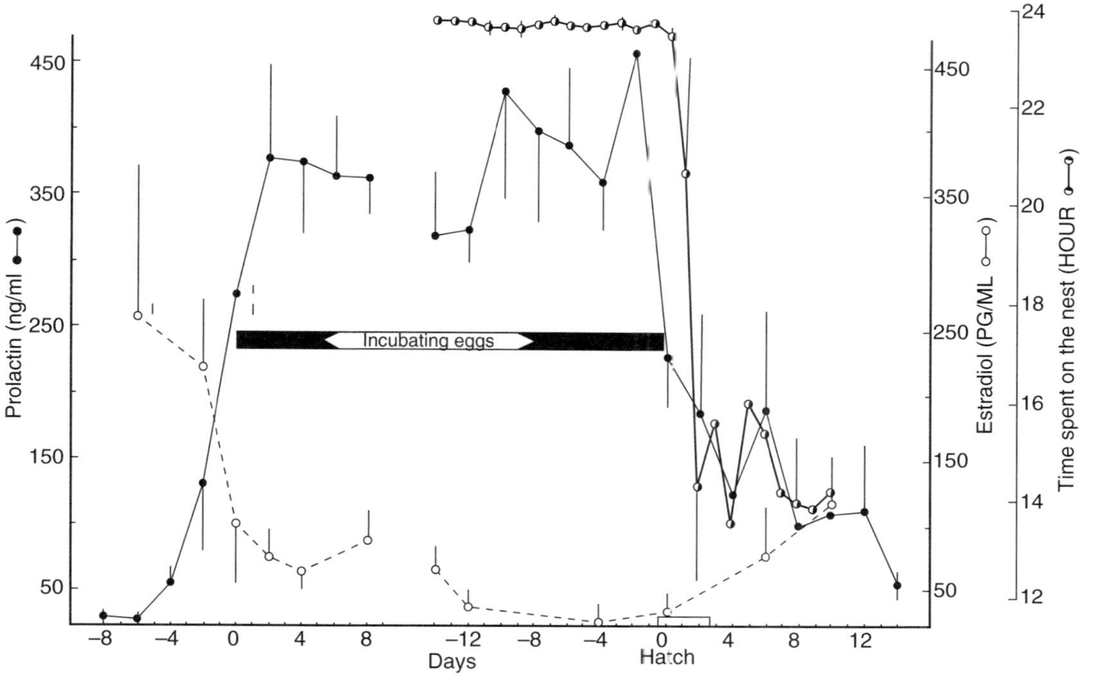

Fig. 10.1 Concentration of plasma PRL increased throughout the incubation period, and concentration of estradiol decreased at onset of incubation. In addition, incubating hens spent over 95% of the day on their nest (adapted from Zadworny et al. 1988)

hypothalamus is the regulatory center for exhibiting broodiness. This hypothesis is supported by the observation that ^{125}I-labeled PRL specifically binds with the hypothalamic membrane in ring dove (*Streptopelia risoria*) (Buntin and Ruzycki 1987), with the binding sites localized in the preoptic area, tuberal hypothalamus, and bed nucleus of the pallial commissure (Buntin et al. 1993). It has been predicted that the PRL receptor (PRLR) in the preoptic hypothalamus mediates the expression of parental behavior, as lesions in this area block the development of incubation behavior in turkey (Youngren et al. 1989) and disrupt PRL-induced parental feeding behavior in ring dove (Slawski and Buntin 1995); and injections of PRL into the preoptic hypothalamus induce incubation behavior in turkey (Opel 1971). Furthermore, the development of incubation behavior in ring dove is associated with increased c-*fos* expression and the uptake of 2-[^{14}C] deoxyglucose in the preoptic hypothalamus (Georgiou et al. 1995; Sharp et al. 1996). The PRLR receptor in either the preoptic hypothalamus or the basal hypothalamus may also mediate the suppression of gonadotropin secretion (Buntin et al. 1988), and an injection of PRL into the ventromedial hypothalamus induces hyperphagia in ring dove (Hnasko and Buntin 1993).

High amounts of PRLR mRNA expression have been found in the basal and preoptic hypothalamus (Ohkubo et al. 1998a), suggesting the presence of indispensable, but perhaps different functions of PRL in these two areas. PRLRs in the basal hypothalamus occur in the same area as vasoactive intestinal polypeptide (VIP) neurons, which have a stimulatory effect on PRL synthesis and release (Sharp et al. 1989; Talbot et al. 1991, 1995). Thus, these neurons may be the targets for PRL action, allowing circulating levels of the hormone to exert a negative feedback control over its secretion. This is supported by the observation that intracerebroventricular injection of PRL in incubating ring doves suppresses the activity of basal hypothalamic VIP neurons, as assessed by quantitative immunocytochemistry (Saldanha and Silver 1995).

Rearing behavior is essential for reproductive success. In altricial birds especially, where post-hatch care is important for the hatchlings, the parents put a lot of effort into rearing their young, including feeding, protecting, and warming them. Plasma PRL levels significantly increase during late incubation and the post-hatch stage (Buntin 1996; Sockman et al. 2006; Angelier et al. 2016). During the rearing period, the presence of chicks is transmitted to the parents as various stimuli, including tactile, visual, auditory, and olfactory, which modify the neuroendocrine mechanisms (Numan and Woodside 2010). This change in the internal environment alters the behavior of the parents, which has a significant impact on fledging success. There is evidence that plasma PRL levels are positively correlated with the frequency of feeding and protecting the offspring (Vleck et al. 1991; Khan et al. 2001; Duckworth et al. 2003; Smiley and Adkins-Regan 2016), which, in turn, increases the number of chicks that successfully fledge (Smiley and Adkins-Regan 2016). Furthermore, it has been shown that the administration of PRL to non-breeding ring doves promotes feeding and care toward the nestlings (Wang and Buntin 1999).

By contrast, avian species that hatch precocial offspring exhibit crouching behavior immediately after their chicks hatch, and this behavior can be triggered in Galliformes by introducing newly hatched chicks to them (Richard-Yris and Leboucher 1987; Opel and Proudman 1989). In precocial species, the chicks are able to walk and search for food immediately after hatching. However, hens must protect them from lower temperatures during the first 2 weeks as their thermoregulatory mechanisms are not yet mature (Mills et al. 1997). These birds also display typical maternal behaviors such as clucking and food calling, where clucking (or the "follow-me call"; Field et al. 2007) is emitted to initiate following behavior in the chicks (Collias and Joos 1953; Collias 1987; Field et al. 2007), and food-calling is emitted to attract their young to food that they have found (Wauters and Richard-Yris 2002). These rearing behaviors from incubation onward are induced by tactile stimulation from the newly hatched chicks, occasionally in combination with visual and/or auditory stimuli, which causes a decline in PRL in the plasma (Richard-Yris et al. 1998).

10.2.1 Neuroendocrine Control of Pituitary Function in Association with Incubation Behavior

Reproduction is triggered by external stimuli that act on the central nervous system. These stimuli include photoperiod, ambient temperature, and the presence of eggs and chicks, and are transmitted to the brain via the sensory organs. These signals are then transmitted to the hypothalamus, which regulates the hypothalamus–pituitary–gonadal axis through the action of a variety of neuropeptides and neurotransmitters. The major hypothalamic neuropeptides that are involved in the regulation of reproduction and incubation behavior are gonadotropin-releasing hormone (GnRH) and VIP.

Two structurally different forms of GnRH have been identified in birds: GnRH-I and GnRH-II. Both of these forms stimulate luteinizing hormone (LH) in vivo and in vitro (Sharp et al. 1987; Guémené and Williams 1992); however, the physiological significance of GnRH-II is not yet fully understood. The release of GnRH-I (but not GnRH-II) into the hypophyseal portal system leads to regression of the gonad in domestic chicken (Sharp et al. 1990). The secretion of GHRH-I is stimulated by epinephrine and norepinephrine and inhibited by dopamine (DA) and enkephalins. GnRH-I particularly stimulates the secretion of LH and follicle-stimulating hormone (FSH) from the pituitary gland, which control follicle development and the sex steroid hormones in the ovary.

10.2.1.1 The Role of VIP with Dopamine on PRL Release

VIP is a PRL-releasing hormone that acts directly on the lactotrophs of the pituitary gland in birds to increase the release of PRL (Macnamee et al. 1986; Opel and Proudman 1988; El Halawani et al. 1990), and also induces PRL gene expression in the pituitary gland (El Halawani et al. 1997). Avian VIP is a 28 amino acid peptide that was originally isolated from the intestine of chicken (Nilsson 1975). Alternative splicing generates VIP transcripts, the most abundant and shortest of which encode VIP (Talbot et al. 1995; You et al. 1995a). Active immunization against VIP prevents photoinduced PRL release (El Halawani et al. 1996) and PRL release in incubating chickens (Sharp et al. 1989), while passive immunization inhibits PRL and causes the termination of incubation behavior in chicken (Sharp et al. 1989). It has also been shown that VIP secretion is controlled by DA, and so DA also plays an important role in controlling incubation behavior. The VIP neuron is widely localized in the basal hypothalamus, projecting to the median eminence and having axons that terminate in the external layer. It has previously been found that VIP concentrations in the hypophyseal portal vein are correlated with blood PRL levels in turkey (You et al. 1995a); and the number of VIP neurons in the basal hypothalamus also increases during incubation in hens (Sharp et al. 1989; Cloues et al. 1990; Mauro et al. 1992). Furthermore, both PRL gene expression in the pituitary gland and VIP mRNA in the hypothalamus increase in response to tactile stimuli in incubating hens (Silver et al. 1988; Buntin et al. 1991; Massaro et al. 2007).

The mechanism of DAergic regulation of VIP in the hypothalamus has been extensively analyzed in domestic turkey. Although DA inhibits PRL release in mammals (Ben-Jonathan and Hnasko 2001), it stimulates VIP release in birds. Furthermore, DA has both a stimulatory and inhibitory effect on PRL release in birds, depending on the subtype of DA receptor (Youngren et al. 1995, 1996). The distribution of DA neurons in the hypothalamus has been established in a number of species, such as chicken (Knigge and Piekut 1985; Moons et al. 1995), turkey (Al-Zailaie and El Halawani 2000), Japanese quail (*Coturnix japonica*; Absil et al. 2001), and pigeon (Kiss and Péczely 1987; Durstewitz et al. 1998). DA neurons are distributed extensively through the hypothalamic nucleus, as well as in the infundibular nuclear complex (INF) where VIP neurons occur and in the preoptic area (POA) where GnRH neurons are localized.

DA activity in the hypothalamus is higher in incubating bantam hens than in laying or photorefractory hens (Macnamee and Sharp 1989). In ring dove, DA activity also increases in the periventricular region during the incubation period (Lea et al. 2001). To mediate the DA stimuli, VIP neurons express both D1 and D2 DA receptors, with the relative expression of each being modulated by physiological changes. Expression of D1 DA receptor is increased during incubation stage in the hypothalamus in turkey hens. In contrast, whereas D1 DA receptor and D2 DA receptor expresses in the anterior pituitary gland, D1 DA receptor is not altered, but D2 DA receptor is much reduced during incubation in turkey hens. There is also evidence that expression of the D2 DA receptor is increased in the pituitary gland of photorefractory hens, which characteristically exhibit hypoprolactinemia (Chaiseha et al. 2003).

10.2.1.2 Regulation of Gonadal Function During Broodiness

The plasma concentrations of LH and steroid hormones decrease during broodiness, resulting in regression of the ovary. This reduction in ovarian function may be associated with an increase in PRL (Opel and Proudman 1980; Zadworny et al. 1989). The PRL level increases at the end of the incubation period in all birds, and subsequently declines once the young have hatched. The hatching or introduction of chicks reduces the levels of circulating PRL in the parents, while the concentrations of LH and ovarian steroid hormones gradually increase (Opel and Proudman 1989; Lea et al. 1996; Richard-Yris 1998). The presence of altricial chicks stimulates PRL release in the parent (Silverin and Goldsmith 1984; Ketterson et al. 1990; Lea and Sharp 1991), while the removal of chicks immediately induces the release of LH and steroid hormones in the rearing chickens (Kuwayama et al. 1992). There is also evidence that non-rearing Thai native chicken hens recover laying faster than rearing hens. (Fig. 10.2)

Contact stimulation from chicks delays the decline in plasma PRL levels in the parent, and inhibits the release of gonadotropin and gonadal steroid hormones (Chaiyachet et al. 2013b). A rise in plasma gonadotropin is correlated with an increase in GnRH-1 mRNA in the hypothalamus in chicken (Dunn et al. 1996; Dunn and Sharp 1999). The decline in gonadal function during broodiness is presumably an inhibitory effect caused by PRL. Systemic administration of PRL reduces GnRH-I and GnRH-II mRNA expression in the hypothalamus, as well as plasma LH in turkey hens (Rozenboim et al. 1993). There is also evidence that PRL directly suppresses LHβ subunit mRNA expression in the anterior pituitary gland of turkey in culture (You et al. 1995b).

10.3 Molecular Biology of PRL

In the late 1980s, three independent groups cloned PRL cDNA from chicken (Hanks et al. 1989a; Watahiki et al. 1989), which marked the start of molecular-based analyses of the PRL system in birds. Avian PRL cDNA encodes a mature hormone of 199 amino acids and a 30 amino acid signal peptide, and the overall structure of avian PRL is highly conserved among vertebrate classes and species (Hanks et al. 1989a; Yasuda et al. 1991; Noso et al. 1992). cDNA cloning made it possible to produce recombinant protein in other organisms, and so recombinant-derived chicken PRL was expressed in *Escherichia coli*, allowing the immunological and biological activity of PRL to be characterized (Hanks et al. 1989b; Ohkubo et al. 1993). The estimated molecular mass of recombinant chicken PRL expressed in *E. coli* is 23 kDa, which is in close agreement with the predicted molecular mass based on chicken PRL cDNA (Watahiki et al. 1989) and turkey PRL cDNA (Karatzas et al. 1990). Recombinant chicken PRL specifically cross-reacted with anti-chicken PRL antibody (Hanks et al. 1989b) and anti-turkey PRL antibody (Ohkubo et al. 1993), and also stimulated proliferation of the pigeon crop sac mucosa with the same potency as authentic turkey PRL

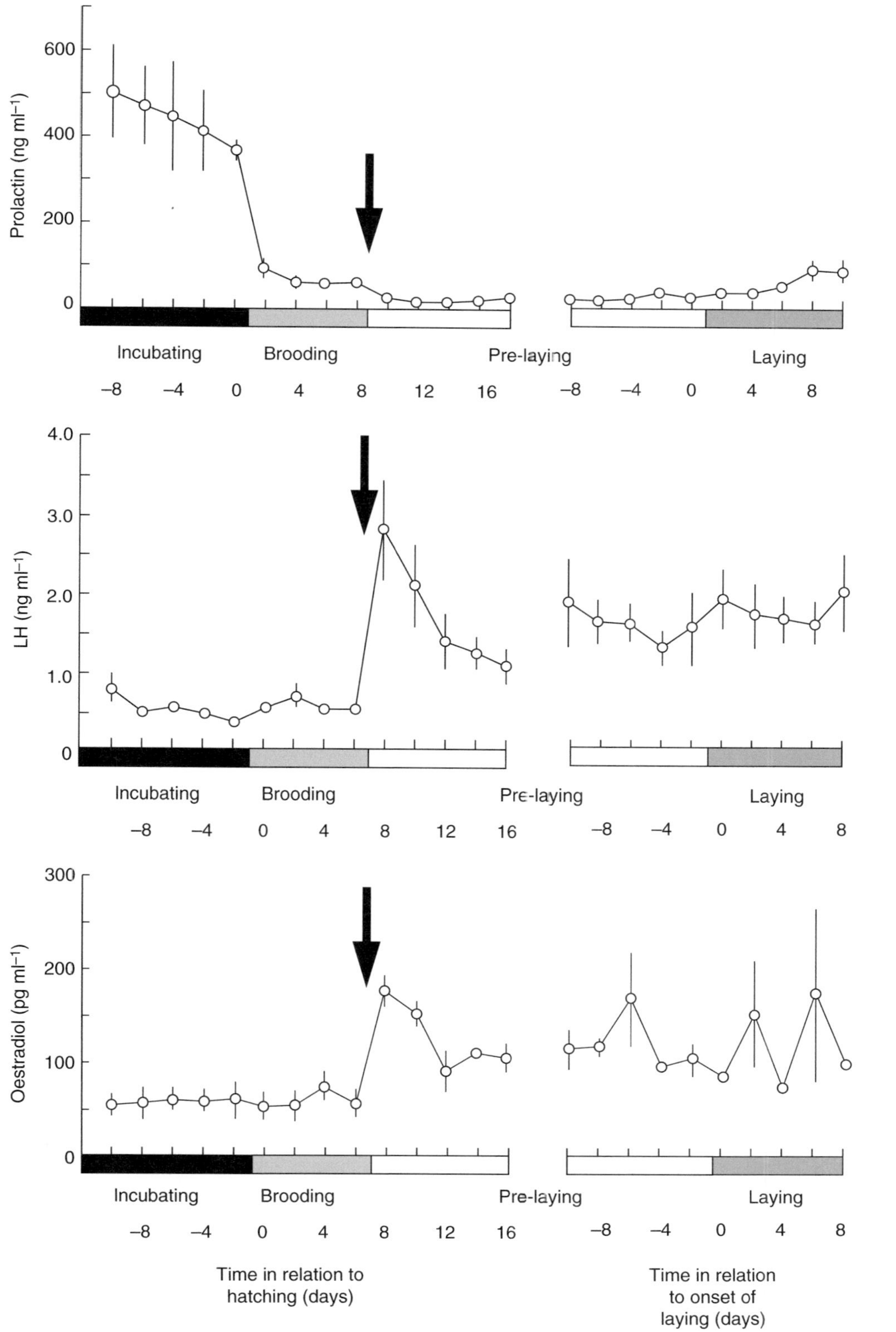

Fig. 10.2 In response to removal of chicks there was an immediate and rapid increase in concentration of LH and estradiol in the plasma of hens (adapted from Kuwayama et al. 1992)

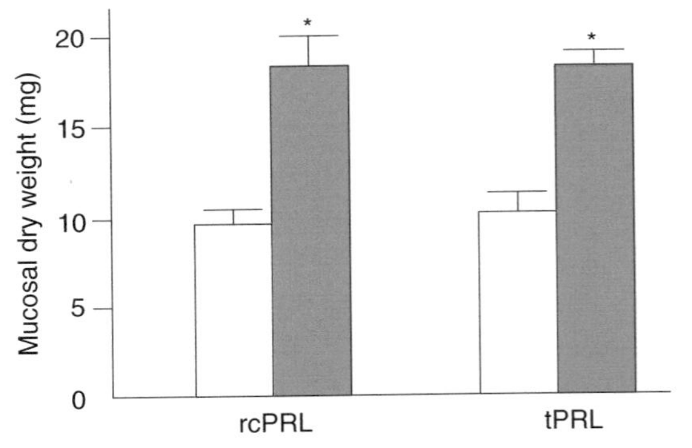

Fig. 10.3 Recombinant-derived chicken (White Leghorn) PRL increased the weight of pigeon crop sac mucosa to a degree comparable to that induced by turkey PRL purified from pituitary gland (adapted from Ohkubo et al. 1993)

purified from the anterior pituitary gland (Ohkubo et al. 1993) (Fig. 10.3).

The occurrence of multiple isoforms of PRL in the anterior pituitary gland has been established in turkey (Proudman and Corcoran 1981; Corcoran and Proudman 1991; Arámburo et al. 1992; Karatzas et al. 1993) and chicken (Arámburo et al. 1992), and is known to result from posttranslational modification. Some of these isoforms are known to be glycosylated (Proudman and Corcoran 1981; Corcoran and Proudman 1991; Karatzas et al. 1993). Glycosylated PRL has also been observed in the pituitary gland of mammals (Lewis et al. 1989; Young et al. 1990; Bollengier et al. 1993) and reptiles (Noso et al. 1992). In mammals, glycosylated PRL exhibits less biological activity in stimulating mitogenesis in the pigeon crop sac (Lewis et al. 1989) and the Nb2 cells (Markoff et al. 1988; Sinha et al. 1991) than non-glycosylated PRL, but exhibits greater lactogenetic activity (Young et al. 1990). The proportion of glycosylated PRL in the pituitary gland of turkey embryos increases alongside increases in the PRL content (Bédécarrats et al. 1999a). It should also be noted that, in turkey, there is a larger amount of glycosylated PRL in the pituitary gland during incubation than in any other physiological condition (Bédécarrats et al. 1999b). Thus, the occurrence of glycosylated PRL may modify physiological properties, such as incubation behavior.

The PRL gene has been cloned and characterized in a number of birds, including turkey (Kurima et al. 1995), chicken (Ohkubo et al. 2000), and duck (Kansaku et al. 2005). The avian PRL gene possesses common exon/intron arrangements and exhibits a length of approximately 6 kb from the first exon to the last exon in these species. The chicken PRL gene has been mapped to chromosome 2 in the chicken genome (Miao et al. 1999), while the turkey PRL gene is localized on chromosome 3.

10.3.1 Transcriptional Regulation of the PRL Gene

Lactotrophs and somatotrophs are mainly localized in the cephalic and caudal lobes, respectively, of the anterior pituitary gland in birds (Hansen and Hansen 1977; Józsa et al. 1979; Ramesh et al. 1996). PRL mRNA expression is markedly higher in incubating hens than in laying hens (Talbot et al. 1991). During incubation, the lactotrophs increase in size in the pituitary of turkey hens. In addition, localization of the lactotrophs and somatotrophs is altered, with PRL-producing cells becoming spread across half of the ventral region of the caudal lobe of the anterior pituitary gland in turkey (Ramesh et al. 1996). It is known that mammosomatotrophs that can differentiate into both lactotrophs and somatotrophs exist in the adenohypophysis in mammals (Porter et al. 1990), and the same characteristic cells are also observed in the pituitary gland of turkey, where they act as PRL-producing cells during incubation (Ramesh et al. 1998). The termination of incubation behavior causes a decline in the blood PRL level, the induction of apoptosis of the lactotrophs, and the disappearance of the mammosomatotrophs in turkey hens (Ramesh et al. 2001).

The regulation of PRL gene expression in association with broodiness has also been thoroughly investigated. During the incubation period, PRL mRNA in the pituitary gland is upregulated in chicken (Talbot et al. 1991; Shimada et al. 1991) and turkey (Wong et al. 1991; Ramesh et al. 1996). VIP is a major PRL-releasing factor in birds, which contributes to the maintenance of incubation behavior in chicken (Sharp et al. 1989). VIP also plays an important role in controlling PRL gene expression in the anterior pituitary gland in chicken (Sharp et al. 1989; Talbot et al. 1991, 1995; Kansaku et al. 1995) and turkey (Youngren et al. 1994; El Halawani et al. 1995). It has been shown that thyrotrophin-releasing hormone (TRH) stimulates PRL gene expression in the chicken pituitary gland (Shimada et al. 1991), as previously observed in mammals (Murdoch et al. 1985; Laverrière et al. 1988).

The transactivation of genes is complexly regulated by transcription factors and other regulatory molecules that are directly or indirectly associated with the target gene. Transcription of the chicken PRL gene has been extensively investigated (Ohkubo et al. 2000). Pituitary-specific transcription factor (Pit-1) is involved in regulating transcription of the PRL gene, as well as the growth hormone and thyroid-stimulating hormone-β (TSHβ) genes (Steinfelder et al. 1992). The PRL gene in mammals possesses several putative Pit-1 binding sites in the *PRL* promoter. The proximal promoter region of the PRL gene is highly conserved between chicken and mammals, with two putative Pit-1 binding sites corresponding to the P2 and P3 regions of the rat *PRL* promoter (Mangalam et al. 1989) being conserved in the chicken and turkey *PRL* promoter (Ohkubo et al. 2000). Pit-1 dependent gene expression is enhanced by cyclic adenosine monophosphate (cAMP) (Peers et al. 1991; Liang et al. 1992; Gaiddon et al. 1995) and other factors (Howard and Maurer 1995) interacting with the proximal promoter region. In an analysis that used the gel mobility shift assay and luciferase assay, it was shown that a putative Pit-1 binding element between −76 and −133 in chicken *PRL* binds with Pit-1 to activate its promoter, which is stimulated by an increase in intracellular cAMP (Ohkubo et al. 2000) (Fig. 10.4).

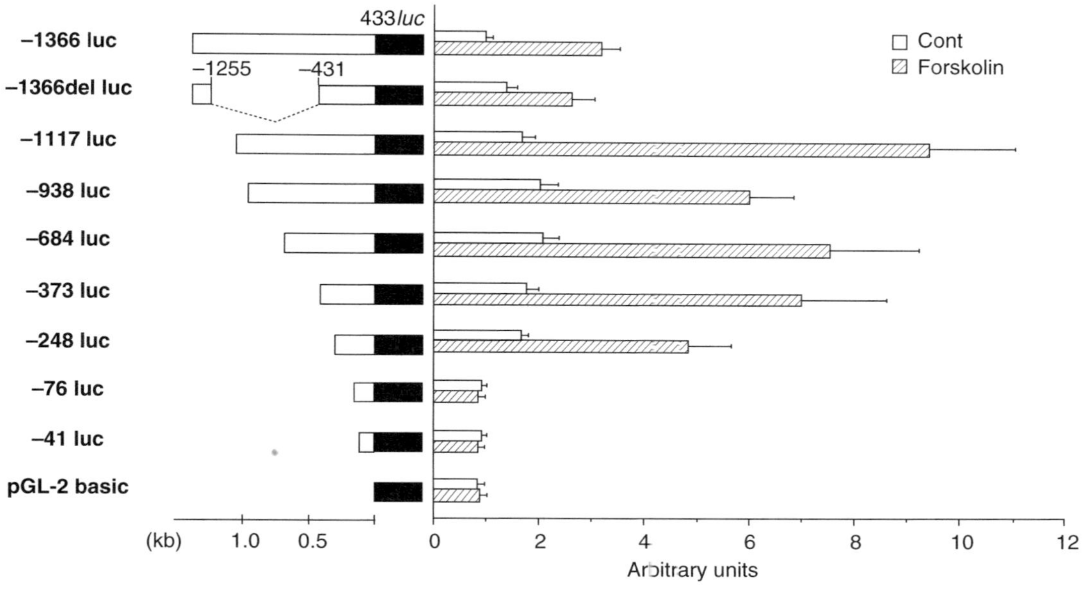

Fig. 10.4 Forskolin activated the *PRL* gene promoter in GH3 cells, which indicates that cAMP is involved in transactivating PRL gene (adapted from Ohkubo et al. 2000)

There is evidence that cyclic AMP responsive element binding protein (CBP) (Chrivia et al. 1993) and p/CAF (Yang et al. 1996) form a coactivator complex, which interacts with nuclear coactivator to mediate the nuclear receptor function (Torchia et al. 1997). This mechanism is involved in the Pit-1 dependent activation of the PRL gene. CBP stimulated by cAMP directly binds to the POU-specific domain of Pit-1 and induces *PRL* promoter activity (Tolón et al. 1998; Xu et al. 1998). It has been reported that VIP, a PRL-releasing factor in birds, increases the cellular cAMP concentration in the perfused chicken anterior pituitary gland (Kansaku et al. 1998), which induces PRL gene expression in the chicken anterior pituitary gland (Kansaku et al. 1995). Thus, the cAMP/protein kinase A (PKA) pathway may involve VIP-dependent PRL gene expression in the chicken anterior pituitary. In addition, the VIP receptor expresses predominantly in the cephalic lobe of the anterior pituitary gland, where PRL-producing cells are localized in chicken (Kansaku et al. 1998). Therefore, PRL gene expression in chicken may be cooperatively promoted by Pit-1 and cAMP-induced CBP, which probably form a complex, by binding to the proximal Pit-1 binding site.

10.3.2 PRL Receptor (PRLR)

The cDNA of avian PRLR was first isolated from chicken in 1992 (Tanaka et al. 1992), followed by pigeon (Chen and Horseman 1994), turkey (Zhou et al. 1996), and various other birds such as duck (*Anas platyrhynchos)* and great tit (*Parus major*). The PRLR gene has been mapped to the Z chromosome in chicken (Dunn et al. 1998). Avian PRLR possesses two repeated ligand-binding domains, as is found in reptilian PRLR (Kato et al. 2005). This double extracellular domain is the result of the duplication of exons 3–7 of mammalian PRLR, and so 5′-rapid amplification of cDNA ends (5′-RACE) suggested that at least ten non-coding exons are located upstream of the initiation codon in chicken PRLR (Bu et al. 2013), and tissue-specific transcription of PRLR is regulated by alternative promoters on the PRLR gene. In addition, truncated PRLRs that lack extracellular and transmembrane domains have been cloned in the testis of sexually mature chickens (Mao et al. 1999; Tanaka et al. 2000), and an alternative promoter is also involved in the transcription of these testis-specific isoforms (Tanaka et al. 2000).

It has been suggested that the variety of actions of PRL is accompanied by complex mechanisms involved in the transcription of PRLR in the target tissues of chicken. PRLR is categorized as a class I cytokine receptor, which has several features in common with intracellular signal transduction, such as the WSXWS motif and Box 1 (Bazan 1989, 1990; Itoh et al. 1990; Lebrun et al. 1995). The overall structure of avian PRLR is similar to that of mammals, except the extracellular domain has two homologous repeated units. Each repeated region has all cysteine pairs and a WSXWS motif that is characteristic of the conserved ligand-binding domain in mammalian PRLRs (Kelly et al. 1991). An in-vitro binding assay showed that the membrane-proximal PRL-binding domain specifically binds with PRL in chicken and pigeon PRLR (Chen and Horseman 1994; Bu et al. 2013).

PRLR mRNA is expressed widely in the central and peripheral tissues of chicken (Tanaka et al. 1992; Ohkubo et al. 1998a, b) and turkey (Zhou et al. 1996), which reflects the expression of PRL with its many physiological functions. It has been shown that the hypothalamus, which is expected to be a regulatory center for broodiness, has a high level of PRLR mRNA expression in chicken (Ohkubo et al. 1998a, b) and turkey (Zhou et al. 1996), as does the anterior pituitary gland (Ohkubo et al. 1998a, b; Zhou et al. 1996). PRL suppresses mRNA expression of the LHβ subunit in chicken (Kansaku et al. 1994) and turkey (You et al. 1995b), and the level of LHβ mRNA expression in the cephalic and caudal lobes of the anterior pituitary gland are inversely correlated with the expression levels of PRL mRNA in bantam hens (Kansaku et al. 1994).

Therefore, PRL may directly regulate pituitary function. In the peripheral tissues, a high level of PRL mRNA expression is found in the kidney and skin in chicken (Ohkubo et al. 1998a), and in the kidney and adrenal gland in turkey (Pitts et al. 2000). Localization of PRLR in these tissues is consistent with PRL playing a role in osmoregulation (Harvey et al. 1984; Murphy et al. 1986), brood-patch formation (Hutchison et al. 1967), and molting (Juhn and Harris 1958). In mammals, there are multiple isoforms of PRLR as a result of alternative splicing, which differ in the size of the intracellular domain and modulate physiological actions by altering intra-cellular signaling (Kelly et al. 1991). However, different isoforms have not been identified in birds.

10.3.2.1 Signal Transduction of PRL

In mammals, it has been established that PRL coupled with PRLR phosphorylates Janus kinase 2 (JAK2), which, in turn, phosphorylates signal transducer and activator of transcription 5 (STAT5) resulting in signal transduction (Gouilleux et al. 1994; Miyoshi et al. 2001). Avian PRLR also activates the JAK2-STAT5 pathway via PRL in vitro. Despite avian PRLR having two repeated ligand-binding domains, PRL specifically binds with the second domain that is close to the transmembrane region to transduce its signal (Bu et al. 2013).

In ring dove, phosphorylated STAT5 (pSTAT5) becomes localized in the brain as a result of various physiological states. The pSTAT5-immunoreactive (pSTAT5-ir) cells are distributed in the POA, suprachiasmatic nucleus (SCN), paraventricular nucleus (PVN), ventromedial nucleus (VMN), lateral hypothalamic region (LHy), and tuberal hypothalamic region (TU) in the hypothalamus, and in the lateral septum (LS) and lateral bed nucleus of the stria terminalis (BSTL), which are also known to contain localized PRLR. The density of pSTAT5-ir cells is higher in all of these regions except TU during the late incubation stage, when the PRL level peaks, than during the pre-laying stages, such as courtship and nest building, when plasma PRL concentrations are low (Buntin and Buntin 2014). Pigeon-administered PRL results in a similar distribution of pSTAT5-positive cells in the brain to that which occurs in pigeons during the breeding season, and also significantly increases the number of pSTAT5-ir cells in the POA, PVN, and TU. PRL binding sites are distributed in these areas, and c-fos immunoreactive cells are induced when the parent is introduced to its hatchling (Buntin et al. 2006). VIP-expressing neurons are localized in the TU (den Boer-Visser and Dubbeldam 2002) and mesotocin is expressed in the PVN (Goossens et al. 1977; Barth et al. 1997; Bons 1980; Thayananuphat et al. 2011). Mesotocin is also associated with parental behavior, with mesotocin neurons increasing during late incubation in turkey (Thayananuphat et al. 2011). Thus, multiple pathways are suggested in PRL signaling, with the JAK2-STAT5 pathway at least partly being involved in regulating parental behavior in birds.

10.4 The Genetics of Incubation Behavior

Broodiness is essential for the breeding success of birds. However, during domestication, several strains that never get broody have been established, such as White Leghorn. The red junglefowl (*Gallus gallus*), which is believed to be the ancestor of domestic chicken (Akishinonomiya et al. 1994, 1996), and other strains that have not undergone large amounts of genetic selection still exhibit incubation behavior, though. Incubation behavior in chicken is influenced by environmental factors, such as high temperatures, photoperiod, and the existence of a nest and egg. However, different individuals from a particular strain within the same environment exhibit variation in the occurrence and frequency of incubation behavior, indicating that there is a genetic basis to incubation behavior in chicken.

Spontaneously occurring mutations are accumulated in specific populations during the domestication process, allowing strains that do not exhibit incubation behavior to be established. Research of relationship between incubation behavior and heredity has been demonstrated for a long time. Punnett and Bailey (1920) pointed out that more than one independent autosomal gene is involved in the regulation of incubation behavior. Goodale et al. (1920) hypothesized that non-broody hens lack an allele that is involved in incubation behavior or possess an inhibitory gene(s) for the behavior—and this hypothesis is supported by the finding that the offspring that result from mating two different strains exhibit a higher incidence of broodiness (Hutt 1949). In addition, several researchers have reported that genes on the sex chromosome may be involved in broodiness (Warren 1930; Roberts and Card 1933; Saeki 1957; Saeki and Inoue 1979). However, Hays (1940) found no evidence for Z-linked genes being associated with broodiness in a study on a population of Rhode Island Red chickens. A recent analysis using an F1 generation cross of males from a non-broody breed (White Leghorn) and females from a broody breed (bantam) also showed that the major genes that affect incubation behavior may not be localized on the Z chromosome (Romanov et al. 1999)—and this finding is consistent with the reanalysis by Mead and Curnow (1983) of the data obtained by Saeki (1957) and Saeki and Inoue (1979).

PRL plays a fundamental role in the initiation and maintenance of incubation behavior in domestic chicken (El Halawani et al. 1986; Youngren et al. 1991; March et al. 1994). Hormonal action occurs via a specific receptor that is expressed in the target tissues, which also involve a specific signal transduction pathway. With regard to PRL signaling, the major pathway involves PRLR activating the JAK2-STAT5 cascade, and molecular analyses have revealed that the PRLR and JAK2 genes are localized on the Z chromosome (Dunn et al. 1998; Wong et al. 2004. Therefore, it is likely that genes on the Z chromosome are involved in incubation behavior. However, it is unlikely that it is a loss of function of PRLR and JAK2 that accounts for the difference in the manifestation of incubation behavior between non-broody and broody breeds. Northern blot analyses have shown that the PRLR transcripts expressed in the tissues of White Leghorn and bantam hens are the same size, and that the localization and amount of PRLR mRNA also does not differ between the two breeds (Ohkubo et al. 1998a). Furthermore, White Leghorn PRLR expressed in chicken hepatoma cells phosphorylates STAT5 and activates the STAT5 inducible gene after PRL stimulation (Fig. 10.5); and as with other genes related to the PRL system, both recombinant-derived White Leghorn and bantam PRL proliferate pigeon crop sac mucosa (Hanks et al. 1989b; Ohkubo et al. 1993), indicating that PRL is functional regardless of the behavioral difference. There is also evidence that VIP is able to induce PRL release from the anterior pituitary gland in White Leghorn (Ohkubo et al. 1998), strongly suggesting that the absence of incubation behavior in this breed cannot be explained by an inability of VIP to stimulate PRL release. Therefore, together these findings show that PRL and PRLR are not causative genes for the lack of broodiness in White Leghorn (Fig. 10.6).

A quantitative trait loci (QTL) analysis based on an F2 generation established from White Leghorn and Silkie, a broody breed, indicated that particular QTL that are associated with incubation behavior are not located on the Z chromosome or chromosome 2, where the PRLR and PRL genes respectively are localized (Basheer et al. 2015). By contrast, significant QTL that affect incubation status and the early onset of incubation behavior have been found on chromosomes 1, 5, 8, 13, 18, 19, and 26, and linkage group E22C19W28. The peak position of these loci on chromosome 5 matches the largest selective sweep between domesticated chicken breeds and red junglefowl (Rubin et al. 2010), and there are several genes associated with photoperiodism and reproduction, such as

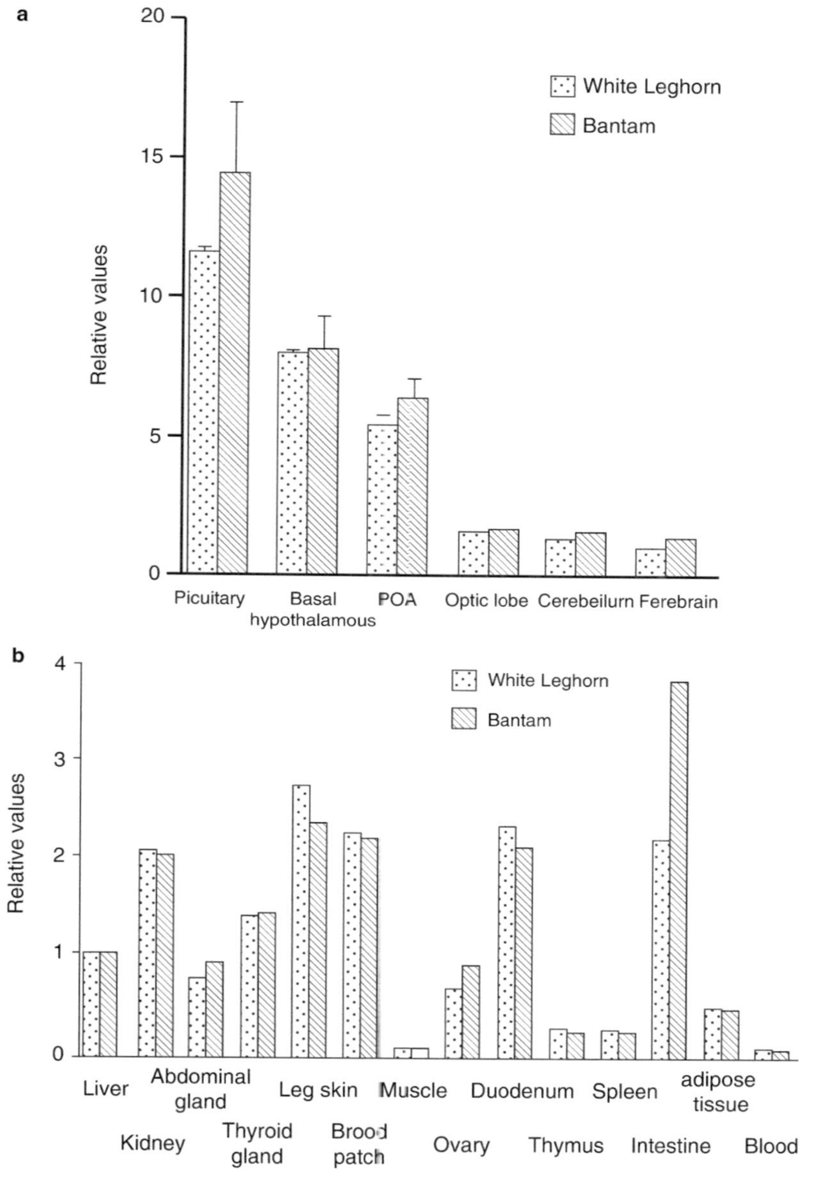

Fig. 10.5 PRLR mRNA is expressed widely in central (**a**) and peripheral (**b**) tissues. The largest amount of expression is found in the pituitary gland and hypothalamus in brain, which did not differ significantly between a non-broody breed, White Leghorns, and a broody breed, bantams, at laying period (adapted from Ohkubo et al. 1998a)

thyroid-stimulating hormone receptor (TSHR) and thyroid hormone deiodinase 2 (DIO2), in chicken (Nakao et al. 2008). A single nucleotide polymorphism (SNP) analysis showed that the peak position of the QTL is close to TSHR (Basheer et al. 2015), which is a characteristic locus for the selective sweep of the Silkie breed (Fan et al. 2013). As for DIO2, TSHR is a key trigger for the onset of seasonal breeding and increases the synthesis of triiodothyronine. Therefore, these findings suggest that the thyroid hormone axis is involved not only in seasonal reproduction, but also in the expression of incubation behavior in birds.

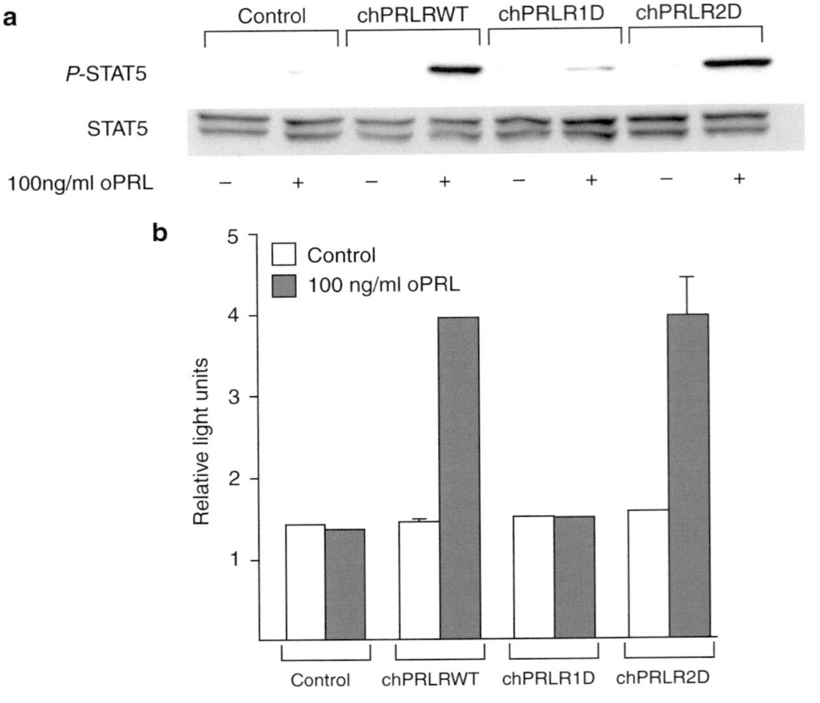

Fig. 10.6 PRL phosphorylates STAT5 (**a**) and activates STAT inducible reporter gene in Leghorm male hepatoma (LMH) cells. (**b**) PRL may specifically bind with the 2nd ligand domain (2D), but not with 1st domain (1D) to phosphorylate STAT5, and initiate following signal transduction

References

Absil P, Foidart A, Hemmings HC Jr, Steinbusch HW, Ball GF, Balthazart J. Distribution of DARPP-32 immunoreactive structures in the quail brain: anatomical relationship with dopamine and aromatase. J Chem Neuroanat. 2001;21:23–39.

Akishinonomiya F, Miyake T, Sumi S, Takada M, Ohno S, Kondo N. One subspecies of the red junglefowl (*Gallus gallus gallus*) suffices as the matriarchic ancestor of all domestic breeds. Proc Natl Acad Sci U S A. 1994;91:12505–9.

Akishinonomiya F, Miyake T, Takada M, Shingu R, Endo T, Gojobori T, Kondo N, Ohno S. Monophyletic origin and unique dispersal patterns of domestic fowls. Proc Natl Acad Sci U S A. 1996;93:6792–5.

Al-Zailaie K, El Halawani M. Neuroanatomical relationship between immunoreactive dopamine and vasoactive intestinal peptide neurons in the turkey hypothalamus. Poult Sci. 2000;79(Suppl 1):50.

Angelier F, Wingfield JC, Tartu S, Chastel O. Does prolactin mediate parental and life-history decisions in response to environmental conditions in birds? A review. Horm Behav. 2016;77:18–29.

Arámburo C, Montiel JL, Proudman JA, Berghman LR, Scanes CG. Phosphorylation of prolactin and growth hormone. J Mol Endocrinol. 1992;8:183–91.

Barth SW, Bathgate RA, Mess A, Parry LJ, Ivell R, Grossmann R. Mesotocin gene expression in the diencephalon of domestic fowl: cloning and sequencing of the MT cDNA and distribution of MT gene expressing neurons in the chicken hypothalamus. J Neuroendocrinol. 1997;9:777–87.

Basheer A, Haley CS, Law A, Windsor D, Morrice D, Talbot R, Wilson PW, Sharp PJ, Dunn IC. Genetic loci inherited from hens lacking maternal behaviour both inhibit and paradoxically promote this behaviour. Genet Sel Evol. 2015;47:100. doi:10.1186/s12711-015-0180-y.

Bazan JF. A novel family of growth factor receptors: a common binding domain in the growth hormone, prolactin, erythropoietin and IL-6 receptors, and the p75 IL-2 receptor beta-chain. Biochem Biophys Res Commun. 1989;164:788–95.

Bazan JF. Structural design and molecular evolution of a cytokine receptor superfamily. Proc Natl Acad Sci U S A. 1990;87:6934–8.

Bédécarrats G, Guémené D, Morvan C, Kühnlein U, Zadworny D. Quantification of prolactin messenger ribonucleic acid, pituitary content and plasma levels of prolactin, and detection of immunoreactive isoforms of prolactin in pituitaries from turkey embryos during ontogeny. Biol Reprod. 1999a;61:757–63.

Bédécarrats G, Guémené D, Morvan C, Crisóstomo-Pinto S, Kühnlein U, Zadworny D. In vitro release of isoforms of prolactin from pituitary glands of turkey hens at different physiological stages. Gen Comp Endocrinol. 1999b;113:105–11.

Ben-Jonathan N, Hnasko R. Dopamine as a prolactin (PRL) inhibitor. Endocr Rev. 2001;22:724–63.

Bollengier F, Geerts A, Matton A, Mahler A, Velkeniers B, Hooghe-Peters E, Vanhaeist L. Identification and localization of 23,000 and glycosylated rat prolactin in subcellular fractions of rat anterior pituitary and purified secretory granules. J Neuroendocrinol. 1993;5:669–76.

Bons N. The topography of mesotocin and vasotocin systems in the brain of the domestic mallard and Japanese quail: immunocytochemical identification. Cell Tissue Res. 1980;213:37–51.

Bu G, Ying Wang C, Cai G, Leung FC, Xu M, Wang H, Huang G, Li J, Wang Y. Molecular characterization of prolactin receptor (cPRLR) gene in chickens: gene structure, tissue expression, promoter analysis, and its interaction with chicken prolactin (cPRL) and prolactin-like protein (cPRL-L). Mol Cell Endocrinol. 2013;370:149–62.

Buntin JD. Neural and hormonal control of parental behavior in birds. Adv Stud Behav. 1996;25:161–213.

Buntin JD, Buntin L. Increased STAT5 signaling in the ring dove brain in response to prolactin administration and spontaneous elevations in prolactin during the breeding cycle. Gen Comp Endocrinol. 2014;200:1–9.

Buntin JD, Ruzycki E. Characteristics of prolactin binding sites in the brain of the ring dove (*Streptopelia risoria*). Gen Comp Endocrinol. 1987;65:243–2553.

Buntin JD, Lea RW, Figge GR. Reductions in plasma LH concentration and testicular weight in ring doves following intracranial injection of prolactin and growth hormone. J Endocrinol. 1988;118:33–40.

Buntin JD, Becker GM, Ruzycki E. Facilitation of parental behavior in ring doves by systemic or intracranial injections of prolactin. Horm Behav. 1991;25:424–44.

Buntin JD, Ruzycki E, Witebsky J. Prolactin receptors in dove brain: autoradiographic analysis of binding characteristic in discrete brain regions and accessibility to blood-borne prolactin. Neuroendocrinology. 1993;57:738–50.

Buntin L, Berghman LR, Buntin JD. Patterns of *fos*-like immunoreactivity in the brains of parent ring doves (*Streptopelia risoria*) given tactile and nontactile exposure to their young. Behav Neurosci. 2006;120:651–64.

Chaiseha Y, Youngren O, Al-Zailaie K, El Halawani M. Expression of D1 and D2 dopamine receptors in the hypothalamus and pituitary during the turkey reproductive cycle: colocalization with vasoactive intestinal peptide. Neuroendocrinology. 2003;77:105–18.

Chaiyachet OA, Chokchaloemwong D, Prakobsaeng N, Sartsoongnoen N, Kosonsiriluk S, Rozenboim I, El Halawani ME, Porter TE, Chaiseha Y. Neuroendocrine regulation of rearing behavior in the native Thai hen. Acta Histochem. 2013b;115:209–18.

Chen XJ, Horseman ND. Cloning, expression and mutational analysis of the pigeon prolactin receptor. Endocrinology. 1994;135:269–76.

Chrivia JC, Kwok RP, Lamb N, Hagiwara M, Montminy MR, Goodmam RH. Phosphorylated CREB binds specifically to the nuclear protein CBP. Nature. 1993;365:855–9.

Cloues R, Ramos C, Silver R. Vasoactive intestinal polypeptide-like immunoreactivity during reproduction in doves: influence of experience and number of offspring. Horm Behav. 1990;24:215–31.

Collias NE. The vocal repertoire of the red junglefowl—a spectrographic classification and the code of communication. Condor. 1987;89:510–24.

Collias N, Joos M. The spectrographic analysis of sound signals of the domestic fowl. Behaviour. 1953; 5:175–88.

Corcoran DH, Proudman JA. Isoforms of turkey prolactin: evidence for differences in glycosylation and in trypic peptide mapping. Comp Biochem Physiol. 1991;99B:563–70.

den Boer-Visser AM, Dubbeldam JL. The distribution of dopamine, substance P, vasoactive intestinal polypeptide and neuropeptide Y immunoreactivity in the brain of the collared dove, *Streptopelia decaocto*. J Chem Neuroanat. 2002;23:1–27.

Duckworth RA, Badyaev AV, Parlow AF. Elaborately ornamented males avoid costly parental care in the house finch (*Carpodacus mexicanus*): a proximate perspective. Behav Ecol Sociobiol. 2003;55:176–83.

Dufty AM Jr, Goldssmith AR, Wingfield JC. Prolactin secretion in a brood parasite, the brown-headed cowbird, *Molothrus ater*. J Zool. 1987;212:669–75.

Dunn IC, Sharp PJ. Photo-induction of hypothalamic gonadotrophin-releasing hormone-I mRNA in the domestic chicken: a role for oestrogen? J Neuroendocrinol. 1999;11:371–5.

Dunn IC, Beattie KK, Maney D, Sang HM, Talbot RT, Wilson PW, Sharp PJ. Regulation of chicken gonadotropin-releasing hormone-I mRNA in incubating, nest-deprived and laying bantam hens. Neuroendocrinology. 1996;63:504–13.

Dunn IC, McEwan G, Okhubo T, Sharp PJ, Paton IR, Burt DW. Genetic mapping of the chicken prolactin receptor gene: a candidate gene for the control of broodiness. Br Poult Sci. 1998;39(Suppl):S23–4.

Durstewitz D, Kröner S, Hemmings HC Jr, Güntürkün O. The dopaminergic innervation of the pigeon telencephalon: distribution of DARPP-32 and cooccurrence with glutamate decarboxylase and tyrosine hydroxylase. Neuroscience. 1998;83:763–79.

El Halawani ME, Burke WH, Dennison PT. Effect of nest-deprivation on serum prolactin level in nesting female turkeys. Biol Reprod. 1980;23:118–23.

El Halawani ME, Burke WH, Millam JR, Fehrer SC, Hargis BM. Regulation of prolactin and its role in gallinaceous bird reproduction. J Exp Zool. 1984;232:521–9.

El Halawani ME, Silsby JL, Behnke EJ, Fehrer SC. Hormonal induction of incubation in ovariectomized female turkey (*Meleagris gallopavo*). Biol Reprod. 1986;35:59–67.

El Halawani ME, Silsby JL, Mauro LJ. Vasoactive intestinal peptide is a hypothalamic prolactin-releasing neuropeptide in the turkey (*Meleagris gallopavo*). Gen Comp Endocrinol. 1990;78:66–73.

el Halawani ME, Youngren OM, Rozenboim I, Pitts GR, Silsby JL, Phillips RE. Serotonergic stimulation of prolactin secretion is inhibited by vasoactive intestinal

peptide immunoneutralization in the turkey. Gen Comp Endocrinol. 1995;99:69–74.
el Halawani ME, Pitts GR, Sun S, Silsby JL, Sivanandan V. Active immunization against vasoactive intestinal peptide prevents photo-induced prolactin secretion in turkeys. Gen Comp Endocrinol. 1996;104:76–83.
el Halawani ME, Youngren OM, Pitts GR. Vasoactive intestinal peptide as the avian prolactin releasing factor. In: Harvey S, Etches R, editors. Prospectives in Avian Endocrinology, Bristol: The Society of Endocrinology. 1997. pp 403–416.
Fan WL, Ng CS, Chen CF, Lu MY, Chen YH, Liu CJ, Wu SM, Chen CK, Chen JJ, Mao CT, Lai YT, Lo WS, Chang WH, Li WH. Genome-wide patterns of genetic variation in two domestic chickens. Genome Biol Evol. 2013;5:1376–92.
Field SE, Rickard NS, Toukhsati SR, Gibbs ME. Maternal hen calls modulate memory formation in the day-old chick: the role of noradrenaline. Neurobiol Learn Mem. 2007;88:321–30.
Gaiddon C, Mercken L, Bancroft C, Loeffler JP. Transcriptional effects in GH3 cells of Gs alpha mutants associated with human pituitary tumors: stimulation of adenosine 3′,5′-monophosphate response element-binding protein-mediated transcription and of prolactin and growth hormone promoter activity via protein kinase A. Endocrinology. 1995;136:4331–8.
Georgiou GG, Sharp PJ, Rea RW. [^{14}C]-2 deoxyglucose uptake in the brain of ring dove (*Streptopelia risoria*). II. Differential uptake at the onset of incubation. Brain Res. 1995;700:137–41.
Goldsmith AR. Prolactin in avian reproductive cycles. In: Balthazart J, Prove JE, Gilles R, editors. Hormones and behaviour in higher vertebrates. Berlin: Springer-Verlag; 1983. p. 375–87.
Goodale HD, Sanborn R, White D. Broodiness in domestic fowl: data concerning its inheritance in the Rhode Island Red breed. Massachusetts Agr Exp Sta Bull. 1920;199:94–116.
Goossens N, Blähser S, Oksche A, Vandesande F, Dierickx K. Immunocytochemical investigation of the hypothalamo-neurohypophysial system in birds. Cell Tissue Res. 1977;184:1–13.
Gouilleux F, Wakao H, Mundt M, Groner B. Prolactin induces phosphorylation of Tyr694 of Stat5 (MGF), a prerequisite for DNA binding and induction of transcription. EMBO J. 1994;13:4361–9.
Guémené D, Williams JB. Comparison of the basal and luteinising hormone-releasing hormone induced luteinising hormone release by perifused hypophyses from turkey hens (*Meleagris gallopavo*) at different physiological stages. Br Poult Sci. 1992;33:153–63.
Haller RW, Cherms FL Jr. A comparison of several treatments on terminating broodiness in broad breasted bronze turkeys. Poult Sci. 1961;40:155–63.
Hanks MC, Alonzi JA, Sharp PJ, Sang HM. Molecular cloning and sequence analysis of putative chicken prolactin cDNA. J Mol Endocrinol. 1989a;2:21–30.
Hanks MC, Talbot RT, Sang HM. Expression of biologically active recombinant-derived chicken prolactin in Escherichia coli. J Mol Endocrinol. 1989b;3(1):15–21.
Hansen GN, Hansen BL. Light and electron microscopic identification of pituitary cells containing growth hormone and prolactin in the pigeon (*Columba livia*), using the immunoglobulin-enzyme bridge technique. Gen Comp Endocrinol. 1977;32:99–107.
Harvey S, Hall TR, Chadwick A. Growth hormone and prolactin secretion in water-deprived chickens. Gen Comp Endocrinol. 1984;54:46–50.
Hays FA. Inheritance of broodiness in Rhode Island Reds. Massachusetts Agr Exp Sta Bull. 1940;377:1–11.
Hnasko RM, Buntin JD. Functional mapping of neural sites mediating prolactin-induced hyperphage in doves. Brain Res. 1993;623:257–66.
Howard PW, Maurer RA. A composite Ets/Pit-1 binding site in the prolactin gene can mediate transcriptional responses to multiple signal transduction pathway. J Biol Chem. 1995;270:20930–6.
Hutchison RE. Effect of ovarian steroids and prolactin on the sequential development of nesting behavior in female budgerigars. J Endocrinol. 1975;67:29.
Hutchison RE, Hinde RA, Steel E. The effect of oestrogen, progesterone and prolactin on brood patch formation in ovariectomized canaries. J Endocrinol. 1967;39:379–85.
Hutt FB. Genetics of the fowl. New York, NY: McGraw-Hill Book Company Inc.; 1949.
Itoh N, Yonehara S, Schreurs J, Gorman DM, Maruyama K, Ishii A, Yahara I, Arai K, Miyajima A. Cloning of an interleukin-3 receptor gene: a member of a distinct receptor gene family. Science. 1990;247:324–7.
Jones RE. The incubation patch of birds. Biol Rev. 1971;46:315–39.
Józsa R, Scanes CG, Vigh S, Mess B. Functional differentiation of the embryonic chicken pituitary gland studied by immunohistological approach. Gen Comp Endocrinol. 1979; 39(2):158–163.
Juhn M, Harris PC. Molt of capon feathering with prolactin. Proc Soc Exp Biol Med. 1958;98:669–72.
Kansaku N, Shimada K, Terada O, Saito N. Prolactin, growth hormone, and luteinizing hormone-beta subunit gene expression in the cephalic and caudal lobes of the anterior pituitary gland during embryogenesis and different reproductive stages in the chicken. Gen Comp Endocrinol. 1994;96:197–205.
Kansaku N, Shimada K, Saito N. Regionalized gene expression of prolactin and growth hormone in the chicken anterior pituitary gland. Gen Comp Endocrinol. 1995;99:60–8.
Kansaku N, Shimada K, Saito N, Hidaka H. Effects of protein kinase A inhibitor (H-89) on VIP- and GRF-induced release and mRNA expression of prolactin and growth hormone in the chicken pituitary gland. Comp Biochem Physiol C Pharmacol Toxicol Endocrinol. 1998;119:89–95.
Kansaku N, Ohkubo T, Okabayashi H, Guémené D, Kuhnlein U, Zadworny D, Shimada K. Cloning of

duck PRL cDNA and genomic DNA. Gen Comp Endocrinol. 2005;141:39–47.

Karatzas CN, Zadworny D, Kuhnlein U. Nucleotides sequence of turkey prolactin. Nucleic Acids Res. 1990;18:3071.

Karatzas CN, Guémené D, Zadworny D, Kühnlein U. Production and characterization of recombinant turkey prolactin. Comparative Biochemistry and Physiology Part B: Comparative Biochemistry 1993;106(2):273–280.

Kato K, Ikemoto T, Park MK. Identification of the reptilian prolactin and its receptor cDNAs in the leopard gecko, *Eublepharis macularius*. Gene. 2005;346:267–76.

Kelly PA, Djiane J, Postel-Vinay MC, Edery M. The prolactin/growth hormone receptor family. Endocr Rev. 1991;12:235–51.

Ketterson ED, Nolan V Jr, Wolf L, Goldsmith AR. Effect of sex, stage of reproduction, season, and mate removal on prolactin in dark-eyed juncos. Condor. 1990;92:922–30.

Khan MZ, McNabb FM, Walters JR, Sharp PJ. Patterns of testosterone and prolactin concentrations and reproductive behavior of helpers and breeders in the cooperatively breeding red-cockaded woodpecker (*Picoides borealis*). Horm Behav. 2001;40:1–13.

Kiss JZ, Péczely P. Distribution of tyrosine-hydroxylase (TH)-immunoreactive neurons in the diencephalon of the pigeon (*Columba livia domestica*). J Comp Neurol. 1987;257:333–46.

Knigge KM, Piekut DT. Distribution of CRF- and tyrosine hydroxylase-immunoreactive neurons in the brainstem of the domestic fowl (*Gallus domesticus*). Peptides. 1985;6:97–101.

Kurima K, Proudman JA, El Halawani ME, Wong EA. The turkey prolactin-encoding gene and its regulatory region. Gene. 1995;156:309–10.

Kuwayama T, Shimada K, Saito N, Ohkubo T, Sato K, Wada M, Ichinoe K. Effects of removal of chicks from hens on concentrations of prolactin, luteinizing hormone and oestradiol in plasma of brooding Gifujidori hens. J Reprod Fertil. 1992;95:617–22.

Laverrière JN, Tixier-Vidal A, Buisson N, Morin A, Martial JA, Gourdji D. Preferential role of calcium in the regulation of prolactin gene transcription by thyrotropin-releasing hormone in GH3 pituitary cells. Endocrinology. 1988;122:333–40.

Lea RW, Sharp PJ. Plasma prolactin concentrations in broody turkeys: lack of agreement between homologous chicken and turkey prolactin radioimmunoassay. Br Poult Sci. 1982;23:451–9.

Lea RW, Sharp PJ. Effects of presence of squabs upon plasma concentrations of prolactin and LH and length of time of incubation in ringdoves on "extended" incubatory patterns. Horm Behav. 1991;25:275–82.

Lea RW, Dods AS, Sharp PJ, Chadwick A. The possible role of prolactin in the regulation of nesting behaviour and the secretion of luteinizing hormone in broody bantams. J Endocrinol. 1981;91:89–97.

Lea RW, Richard-Yris MA, Sharp PJ. The effect of ovariectomy on concentrations of plasma prolactin and LH and parental behavior in the domestic fowl. Gen Comp Endocrinol. 1996;101:115–21.

Lea RW, Clark JA, Tsutsui K. Changes in central steroid receptor expression, steroid synthesis, and dopaminergic activity related to the reproductive cycle of the ring dove. Microsc Res Tech. 2001;55:12–26.

Lebrun JJ, Ali S, Goffin V, Ullrich A, Kelly PA. A single phosphotyrosine residue of the prolactin receptor is responsible for activation of gene transcription. Proc Natl Acad Sci U S A. 1995;92:4031–5.

Lewis UJ, Singh RN, Lewis LJ. Two forms of glycosylated human prolactin have different pigeon crop sac-stimulating activities. Endocrinology. 1989;124:1558–63.

Liang J, Kim KE, Schoderbek WE, Maurer RA. Characterization of a nontissue-specific, 3′, 5′-cyclic adenosine monophosphate-responsive element in the proximal region of the rat prolactin gene. Mol Endocrinol. 1992;6:885–92.

Lormée H, Jouventin P, Chastel O, Mauget R. Endocrine correlates of parental care in an Antarctic winter breeding seabird, the emperor penguin, *Aptenodytes forsteri*. Horm Behav. 1999;35:9–17.

Macnamee MC, Sharp PJ. The functional activity of hypothalamic dopamine in broody bantam hens. J Endocrinol. 1989;121:67–74.

Macnamee MC, Sharp PJ, Lea RW, Sterling RJ, Harvey S. Evidence that vasoactive intestinal polypeptide is a physiological prolactin-releasing factor in the bantam hen. Gen Comp Endocrinol. 1986;62:470–8.

Mangalam HJ, Albert VR, Ingraham HA, Kapiloff M, Wilson L, Nelson C, Elsholtz H, Rosenfeld MG. A pituitary POU domain protein, Pit-1, activates both growth hormone and prolactin promoters transcriptionally. Genes Dev. 1989;3:946–58.

Mao JN, Burnside J, Li L, Tang J, Davolos C, Cogburn LA. Characterization of unique truncated prolactin receptor transcripts, corresponding to the intracellular domain, in the testis of the sexually mature chicken. Endocrinology. 1999;140:1165–74.

March JB, Sharp PJ, Wilson PW, Sang HM. Effect of active immunization against recombinant-derived chicken prolactin fusion protein on the onset of broodiness and photoinduced egg laying in bantam hens. J Reprod Fertil. 1994;101:227–33.

Markoff E, Sigel MB, Lacour N, Seavey BK, Friesen HG, Lewis UJ. Glycosylation selectively alters the biological activity of prolactin. Endocrinology. 1988;123:1303–6.

Massaro M, Setiawan AN, Davis LS. Effects of artificial eggs on prolactin secretion, steroid levels, brood patch development, incubation onset and clutch size in the yellow-eyed penguin (*Megadyptes antipodes*). Gen Comp Endocrinol. 2007;151:220–9.

Mauro LJ, Youngren OM, Proudman JA, Phillips RE, el Halawani ME. Effects of reproductive status, ovariectomy, and photoperiod on vasoactive intestinal peptide in the female turkey hypothalamus. Gen Comp Endocrinol. 1992;87:481–93.

Mead R, Curnow RN. Statistical methods in agriculture and experimental biology. London, UK: Chapman and Hall; 1983.

Miao Y-W, Burt DW, Paton IR, Sharp PJ, Dunn IC. Mapping of the chicken prolactin gene to chromosome 2. Anim Genet. 1999;30:462–78.

Mills AD, Crawford LL, Domjan M, Faure JM. The behavior of the Japanese or domestic quail *Coturnix japonica*. Neurosci Biobehav Rev. 1997;21:261–81.

Miyoshi K, Shillingford JM, Smith GH, Grimm SL, Wagner KU, Oka T, Rosen JM, Robinson GW, Hennighausen L. Signal transducer and activator of transcription (Stat) 5 controls the proliferation and differentiation of mammary alveolar epithelium. J Cell Biol. 2001;155:531–42.

Moons L, D'Hondt E, Pijcke K, Vandesande F. Noradrenergic system in the chicken brain: immunocytochemical study with antibodies to noradrenaline and dopamine-beta-hydroxylase. J Comp Neurol. 1995;360:331–48.

Murdoch GH, Waterman M, Evans RM, Rosenfeld MG. Molecular mechanisms of phorbol ester, thyrotropin-releasing hormone and growth factor stimulation of prolactin gene expression. J Biol Chem. 1985;260:11852–8.

Murphy MJ, Brown PS, Brown SC. Osmoregulatory effects of prolactin and growth hormone in embryonic chicks. Gen Comp Endocrinol. 1986;62:485–92.

Nakao N, Ono H, Yamamura T, Anraku T, Takagi T, Higashi K, Yasuo S, Katou Y, Kageyama S, Uno Y, Kasukawa T, Iigo M, Sharp PJ, Iwasawa A, Suzuki Y, Sugano S, Niimi T, Mizutani M, Namikawa T, Ebihara S, Ueda HR, Yoshimura T. Thyrotrophin in the pars tuberalis triggers photoperiodic response. Nature. 2008;452:317–22.

Nelson RJ. An introduction to behavioral endocrinology. Sunderland, MA, USA: Sinauer Associates; 1995.

Nilsson Ann. Structure of the vasoactive intestinal octacosapeptide from chicken intestine. The amino acid sequence. FEBS Letters. 1975;60(2):322–326.

Noso T, Swanson P, Lance VA, Kawauchi H. Isolation and characterization of glycosylated and non-glycosylated prolactins from alligator and crocodile. Int J Pept Protein Res. 1992;39:250–7.

Numan M, Woodside B. Maternity: neural mechanisms, motivational processes, and physiological adaptations. Behav Neurosci. 2010;124:715–41.

Ohkubo T, Tanaka M, Nakashima K, Shimada K, Saito N, Sato K. High-level expression of biologically active chicken prolactin in *E. coli*. Comp Biochem Physiol Comp Physiol. 1993;105:123–8.

Ohkubo T, Tanaka M, Nakashima K, Talbot RT, Sharp PJ. Prolactin receptor gene expression in the brain and peripheral tissues in broody and nonbroody breeds of domestic hen. Gen Comp Endocrinol. 1998a;109:60–8.

Ohkubo T, Tanaka M, Nakashima K, Sharp PJ. Relationship between prolactin receptor mRNA in the anterior pituitary gland and hypothalamus and reproductive state in male and female bantams (*Gallus domesticus*). Gen Comp Endocrinol. 1998b;111:167–76.

Ohkubo T, Tanaka M, Nakashima K. Molecular cloning of the chicken prolactin gene and activation by Pit-1 and cAMP-induced factor in GH3 cells. Gen Comp Endocrinol. 2000;119:208–16.

Opel H. Induction of incubation behavior in the hen by brain implants of prolactin. Poult Sci. 1971;50:1613.

Opel H, Proudman JA. Failure of mammalian prolactin to induce incubation behavior in chickens and turkeys. Poult Sci. 1980;59:2550–8.

Opel H, Proudman JA. Stimulation of prolactin release in turkeys by vasoactive intestinal peptide. Proc Soc Exp Biol Med. 1988;187:455–60.

Opel H, Proudman JA. Plasma prolactin levels in incubating turkey hens during pipping of the eggs and after introduction of poults into the nest. Biol Reprod. 1989;40:981–7.

Peers B, Monget P, Nalda MA, Voz ML, Berwaer M, Belayew A, Martial JA. Transcriptional induction of the human prolactin gene by cAMP requires two cis-acting elements and at least the pituitary-specific factor Pit-1. J Biol Chem. 1991;266:18127–34.

Pitts GR, You S, Foster DN, El-Halawani ME. Evidence for multiple prolactin receptor transcripts in the turkey. Poult Sci. 2000;79:355–62.

Porter TE, Hill JB, Wiles CD, Frawley LS. Is the mammosomatotrope a transitional cell for the functional interconversion of growth hormone- and prolactin-secreting cells? Suggestive evidence from virgin, gestating, and lactating rats. Endocrinology. 1990;127:2789–94.

Proudman JA, Corcoran DH. Turkey prolactin: purification by isotachophoresis and partial characterization. Biol Reprod. 1981;25:375–84.

Punnett RC, Bailey PG. Genetic studies in poultry: II. Inheritance of egg-colour and broodiness. J Genet. 1920;10:277–92.

Ramesh R, Proudman JA, Kuenzel WJ. Changes in pituitary somatotroph and lactotroph distribution in laying and incubating turkey hens. Gen Comp Endocrinol. 1996;104:67–75.

Ramesh R, Solow R, Proudman JA, Kuenzel WJ. Identification of mammosomatotrophs in the turkey hen pituitary: increased abundance during hyperprolactinemia. Endocrinology. 1998;139:781–6.

Ramesh R, Kuenzel WJ, Proudman JA. Increased proliferative activity and programmed cellular death in the turkey hen pituitary gland following interruption of incubation behavior. Biol Reprod. 2001;64:611–8.

Richard-Yris MA, Leboucher G. Responses to successive test of induction of maternal behaviour in hens. Behav Processes. 1987;15:17–26.

Richard-Yris MA, Sharp PJ, Wauters AM, Guémené D, Richard JP, Forasté M. Influence of stimuli from chicks on behavior and concentrations of plasma prolactin and luteinizing hormone in incubating hens. Horm Behav. 1998;33:139–48.

Riddle O, Bates RW, Dykshorn SW. The preparation, identification and assay of prolactin—a hormone of the anterior pituitary. Am J Physiol. 1933;105:191–216.

Riddle O, Bates RW, Lahr EL. Prolactin induces broodiness in fowl. Am J Physiol. 1935;111:352–60.

Roberts E, Card LE. Inheritance of broodiness in the domestic fowl. Proceedings of the 5th World's Poultry Congress; Vol 2; 1933. p. 353–8.

Romanov MN, Talbot RT, Wilson PW, Sharp PJ. Inheritance of broodiness in the domestic fowl. Br Poult Sci. 1999;40(Suppl):S20–1.

Romanov MN, Talbot RT, Wilson PW, Sharp PJ. Genetic control of incubation behavior in the domestic chicken. Poult Sci. 2002;81:928–31.

Rosenblatt JS, Mayer AD, Siegel HI. Maternal behavior among nonprimate mammals. In: Adler NT, Pfaff D, Goy RW, editors. Handbook of behavioral neurobiology. New York: Plenum; 1985. p. 229–98.

Rozenboim I, Silsby JL, Tabibzadeh C, Pitts GR, Youngren OM, el Halawani ME. Hypothalamic and posterior pituitary content of vasoactive intestinal peptide and gonadotropin-releasing hormones I and II in the turkey hen. Biol Reprod. 1993;49:622–6.

Rubin CJ, Zody MC, Eriksson J, Meadows JR, Sherwood E, Webster MT, Jiang L, Ingman M, Sharpe T, Ka S, Hallböök F, Besnier F, Carlborg O, Bed'hom B, Tixier-Boichard M, Jensen P, Siegel P, Lindblad-Toh K, Andersson L. Whole-genome resequencing reveals loci under selection during chicken domestication. Nature. 2010;464:587–91.

Saeki Y. Inheritance of broodiness in Japanese Nagoya fowl, with special reference to sex-linkage and notice in breeding practice. Poult Sci. 1957;36:378–83.

Saeki Y, Inoue Y. Body growth, egg production, broodiness, age at first age and egg size in red jungle fowls, and attempt at their genetic analyses by reciprocal crossing with White Leghorns. Jpn Poult Sci. 1979;16:121–5.

Saeki Y, Tanabe Y. Changes in prolactin content of fowl pituitary during broody periods and some experiments on the induction of broodiness. Poult Sci. 1955;34:909–19.

Saldanha CJ, Silver R. Intraventricular prolactin inhibits hypothalamic vasoactive intestinal polypeptide expression in doves. J Neuroendocrinol. 1995;11:881–7.

Sharp PJ. Female reproduction. In: Epple A, Stetson MH, editors. Avian endocrinology. London and New York: Academic Press; 1980. p. 435–54.

Sharp PJ, Scanes CG, Williams JB, Harvey S, Chadwick A. Variations in concentrations of prolactin, luteinizing hormone, growth hormone and progesterone in the plasma of broody bantams (*Gallus domesticus*). J Endocrinol. 1979;80:51–7.

Sharp PJ, Dunn IC, Talbot RT. Sex differences in the LH responses to chicken LHRH-I and -II in the domestic fowl. J Endocrinol. 1987;115:323–31.

Sharp PJ, Macnamee MC, Sterling RJ, Lea RW, Pedersen HC. Relationships between prolactin, LH and broody behaviour in bantam hens. J Endocrinol. 1988;118:279–86.

Sharp PJ, Sterling RJ, Talbot RT, Huskisson NS. The role of hypothalamic vasoactive intestinal polypeptide in the maintenance of prolactin secretion in incubating bantam hens: observations using passive immunization, radioimmunoassay and immunohistochemistry. J Endocrinol. 1989;122:5–13.

Sharp PJ, Talbot RT, Main GM, Dunn IC, Fraser HM, Huskisson NS. Physiological roles of chicken LHRH-I and -II in the control of gonadotrophin release in the domestic chicken. J Endocrinol. 1990;124:291–9.

Sharp PJ, Li Q, Georgiou G, Lea RW. Expression of *fos*-like immunoreactivity in the hypothalamus of the ring dove (*Streptopelia risoria*) at the onset of incubation. J Neuroendocrinol. 1996;8:291–8.

Sherry DF, Mrosovsky N, Hogan J. Weight loss and anorexia during incubation in birds. J Comp Physiol Psycobiol. 1980;94:89–98.

Shimada K, Ishida H, Sato K, Seo H, Matsui N. Expression of prolactin gene in incubating hens. J Reprod Fertil. 1991;91:147–54.

Silver R, Witkovsky P, Horvath P, Alones V, Barnstable CJ, Lehman MN. Coexpression of opsin- and VIP-like-immunoreactivity in CSF-contacting neurons of the avian brain. Cell Tissue Res. 1988;253:189–98.

Silverin B, Goldsmith A. The effects of modifying incubation on prolactin secretion in free-living pied flycatchers. Gen Comp Endocrinol. 1984;55:239–44.

Sinha YN, DePaolo LV, Haro LS, Singh RN, Jacobsen BP, Scott KE, Lewis UJ. Isolation and biochemical properties of four forms of glycosylated porcine prolactin. Mol Cell Endocrinol. 1991;80:203–13.

Slawski BA, Buntin JD. Preoptic area lesions disrupt prolactin-induced parental feeding behavior in ring doves. Horm Behav. 1995;29:248–66.

Smiley KO, Adkins-Regan E. Prolactin is related to individual differences in parental behavior and reproductive success in a biparental passerine, the zebra finch (*Taeniopygia guttata*). Gen Comp Endocrinol doi. 2016; doi:10.1016/j.ygcen.2016.03.006. [Epub ahead of print].

Sockman KW, Sharp PJ, Schwabl H. Orchestration of avian reproductive effort: an integration of the ultimate and proximate bases for flexibility in clutch size, incubation behaviour, and yolk androgen deposition. Biol Rev Camb Philos Soc. 2006;81:629–66.

Steinfelder HJ, Radovick S, Mroczynski MA, Hauser P, McClaskey JH, Weintraub BD, Wondisford FE. Role of a pituitary-specific transcription factor (Pit-1/GHF-1) or a closely related protein in cAMP regulation of human thyrotropin-beta subunit gene expression. J Clin Invest. 1992;89:409–19.

Talbot RT, Hanks MC, Sterling RJ, Sang HM, Sharp PJ. Pituitary prolactin messenger ribonucleic acid levels in incubating and laying hens: effects of manipulating plasma levels of vasoactive intestinal polypeptide. Endocrinology. 1991;129:496–502.

Talbot RT, Dunn IC, Wilson PW, Sang HM, Sharp PJ. Evidence for alternative splicing of the chicken vasoactive intestinal polypeptide gene transcript. J Endocrinol. 1995;15:81–91.

Tanaka M, Maeda K, Okubo T, Nakashima K. Double antenna structure of chicken prolactin receptor deduced from the cDNA sequence. Biochem Biophys Res Commun. 1992;188:490–6.

Tanaka M, Yamamoto I, Hayashida Y, Nakao N, Ohkubo T, Wakita M, Nakashima K. Two novel first exons in the prolactin receptor gene are transcribed in a tissue-specific and sexual maturation-dependent manner to encode multiple 5′-truncated transcripts in the testis of the chicken. Biochim Biophys Acta. 2000;1491:279–84.

Thayananuphat A, Youngren OM, Kang SW, Bakken T, Kosonsiriluk S, Chaiseha Y, El Halawani ME. Dopamine and mesotocin neurotransmission during the transition from incubation to brooding in the turkey. Horm Behav. 2011;60:327–35.

Tolón RM, Castillo AI, Aranda A. Activation of the prolactin gene by peroxisome proliferator-activated receptor-alpha appears to be DNA binding-independent. J Biol Chem. 1998;273:26652–61.

Torchia J, Rose DW, Inostroza J, Kamel Y, Westin S, Glass CK, Rosenfeld MG. The transcriptional co-activator p/CIP binds CBP and mediates nuclear-receptor function. Nature. 1997;387:677–84.

Vleck CM, Mays NA, Dawson JW, Goldsmith AR. Hormonal correlates of parental and helping behavior in cooperatively breeding Harris' hawks (*Parabuteo unicinctus*). Auk. 1991;108:638–48.

Wang Q, Buntin JD. The roles of stimuli from young, previous breeding experience, and prolactin in regulating parental behavior in ring doves (*Streptopelia risoria*). Horm Behav. 1999;35:241–53.

Warren DC. Cross bred poultry, Kansas Agr Exp Sta Bull. 1930;252:5–54.

Watahiki M, Tanaka M, Masuda N, Sugisaki K, Yamamoto M, Yamakawa M, Nagai J, Nakashima K. Primary structure of chicken pituitary prolactin deduced from the cDNA sequence. Conserved and specific amino acid residues in the domains of the prolactins. J Biol Chem. 1989;264:5535–9.

Wauters AM, Richard-Yris MA. Mutual influence of the maternal hen's food calling and feeding behavior on the behavior of her chicks. Dev Psychobiol. 2002;41:25–36.

Wong EA, Ferrin NH, Silsby JL, el Halawani ME. Cloning of a turkey prolactin cDNA: expression of prolactin mRNA throughout the reproductive cycle of the domestic turkey (*Meleagris gallopavo*). Gen Comp Endocrinol. 1991;83:18–26.

Wong GK, Liu B, Wang J, Zhang Y, Yang X, Zhang Z, Meng Q, Zhou J, Li D, Zhang J, Ni P, Li S, Ran L, Li H, Zhang J, Li R, Li S, Zheng H, Lin W, Li G, Wang X, Zhao W, Li J, Ye C, Dai M, Ruan J, Zhou Y, Li Y, He X, Zhang Y, Wang J, Huang X, Tong W, Chen J, Ye J, Chen C, Wei N, Li G, Dong L, Lan F, Sun Y, Zhang Z, Yang Z, Yu Y, Huang Y, He D, Xi Y, Wei D, Qi Q, Li W, Shi J, Wang M, Xie F, Wang J, Zhang X, Wang P, Zhao Y, Li N, Yang N, Dong W, Hu S, Zeng C, Zheng W, Hao B, Hillier LW, Yang SP, Warren WC, Wilson RK, Brandström M, Ellegren H, Crooijmans RP, van der Poel JJ, Bovenhuis H, Groenen MA, Ovcharenko I, Gordon L, Stubbs L, Lucas S, Glavina T, Aerts A, Kaiser P, Rothwell L, Young JR, Rogers S, Walker BA, van Hateren A, Kaufman J, Bumstead N, Lamont SJ, Zhou H, Hocking PM, Morrice D, de Koning DJ, Law A, Bartley N, Burt DW, Hunt H, Cheng HH, Gunnarsson U, Wahlberg P, Andersson L, Kindlund E, Tammi MT, Andersson B, Webber C, Ponting CP, Overton IM, Boardman PE, Tang H, Hubbard SJ, Wilson SA, Yu J, Wang J, Yang H, International Chicken Polymorphism Map Consortium. A genetic variation map for chicken with 2.8 million single-nucleotide polymorphisms. Nature. 2004;432:717–22.

Xu L, Lavinsky RM, Dasen JS, Flynn SE, Mclnerney EM, Mullen T-M, Heinzel T, Szeto D, Korzus E, Kurokawa R, Aggarwal AK, Rose DW, Glass CK, Rosenfeld MG. Signal-specific co-activator domain requirements for Pit-1 activation. Nature. 1998;395:301–6.

Yamashina Y. Notes on experimental brooding induced by prolactin injections in the domestic cock. Annot Zool Japon. 1952;25:135–42.

Yang X-Y, Ogryzko VV, Nishikawa J, Howard BH, Nakatani Y. p300/CBP-associated factor that competes with the adenoviral oncoprotein E1A. Nature. 1996;382:319–24.

Yasuda A, Yamaguchi K, Kobayashi T, Yamamoto K, Kikuyama S, Kawauchi H. The complete amino acid sequence of prolactin from bullfrog (*Rana catesbeiana*). Gen Comp Endocrinol. 1991;83:218–26.

You S, Silsby JL, Farris J, Foster DN, el Halawani ME. Tissue-specific alternative splicing of turkey preprovasoactive intestinal peptide messenger ribonucleic acid, its regulation, and correlation with prolactin secretion. Endocrinology. 1995a;136:2602–10.

You S, Foster LK, Silsby JL, el Halawani ME, Foster DN. Sequence analysis of the turkey LH beta subunit and its regulation by gonadotrophin-releasing hormone and prolactin in cultured pituitary cells. J Mol Endocrinol. 1995b;14:117–29.

Young KH, Buhi WC, Horseman N, Davis J, Kraeling R, Linzer D, Bazer FW. Biological activities of glycosylated and nonglycosylated porcine prolactin. Mol Cell Endocrinol. 1990;71:155–62.

Youngren OM, El Halawani ME, Phillips RE, Silsby JL. Effects of preoptic and hypothalamic lesions in female turkeys during a photoinduced reproductive cycle. Biol Reprod. 1989;41:610–7.

Youngren OM, El Halawani ME, Silsby JL, Phillips RE. Intracranial prolactin perfusion induces incubation behavior in turkey hens. Biol Reprod. 1991;44:425–43.

Youngren OM, Silsby JL, Rozenboim I, Phillips RE, el Halawani ME. Active immunization with vasoactive intestinal peptide prevents the secretion of prolactin induced by electrical stimulation of the turkey hypothalamus. Gen Comp Endocrinol. 1994;95:330–6.

Youngren OM, Pitts GR, Phillips RE, el Halawani ME. The stimulatory and inhibitory effects of dopamine on prolactin secretion in the turkey. Gen Comp Endocrinol. 1995;98:111–7.

Youngren OM, Pitts GR, Phillips RE, el Halawani ME. Dopaminergic control of prolactin secretion in the turkey. Gen Comp Endocrinol. 1996;104:225–30.

Zadworny D, Walton JS, Etches RJ. The relationship between plasma concentrations of prolactin and consumption of feed and water during the reproductive cycle of the domestic turkey. Poult Sci. 1985;64:401–10.

Zadworny D, Shimada K, Ishida H, Sumi C, Sato K. Changes in plasma levels of prolactin and estradiol, nutrient intake, and time spent nesting during the incubation phase of broodiness in the Chabo hen (Japanese bantam). Gen Comp Endocrinol. 1988;71:406–12.

Zadworny D, Shimada K, Ishida H, Sato K. Gonadotropin-stimulated estradiol production in small ovarian follicles of the hen is suppressed by physiological concentrations of prolactin in vitro. Gen Comp Endocrinol. 1989;74:468–73.

Zhou JF, Zadoworny D, Guémené D, Kuhnlein U. Molecular cloning, tissue distribution, and expression of the prolactin receptor during various reproductive states in *Meleagris gallopavo*. Biol Reprod. 1996;55:1081–90.

11 Sperm Storage in the Female Reproductive Tract: A Conserved Reproductive Strategy for Better Fertilization Success

Mei Matsuzaki and Tomohiro Sasanami

Abstract

In internal fertilizers including mammals, fertilization success depends on the timely arrival of sperm and egg at the site of fertilization. Males should transfer their spermatozoa to the female reproductive tract by copulation during or prior to ovulation in order to achieve this aim. However, such a collaborative mating behavior is often disconnected from the efficiency of the sperm–egg encounter, i.e., ovulation by females occurs independently from insemination by males in many species. To compensate for this time lag, females are capable of storing spermatozoa in their reproductive tracts until the eggs are ready to be fertilized. In avian species, simple tubular invaginations referred to as sperm storage tubules (SSTs) are located between the vagina and uterus as sperm storage sites. Spermatozoa, once ejaculated, migrate to and are thereafter stored in the lumen of the SSTs without loss of fertilizing capacity for up to 15 weeks at a body temperature of 41 °C. This is astonishing, because terminally differentiated cells that lack new protein synthesis are still capable of being functional for a long period at a high temperature; however, the actual mechanism has been an enigma for more than half a century. In this chapter, we will first

M. Matsuzaki
Department of Applied Biological Chemistry, Faculty of Agriculture, Shizuoka University, 836 Ohya, Shizuoka 422-8529, Japan

United Graduate School of Agricultural Science, Gifu University, 1-1 Yanagido, Gifu 501-1193, Japan

T. Sasanami (✉)
Department of Applied Biological Chemistry, Faculty of Agriculture, Shizuoka University, 836 Ohya, Shizuoka 422-8529, Japan
e-mail: atsasan@shizuoka.ac.jp

T. Sasanami (ed.), *Avian Reproduction*, Advances in Experimental Medicine and Biology 1001,
DOI 10.1007/978-981-10-3975-1_11

describe the physiological importance and adoptive benefits of sperm storage in the female genital tract for successful fertilization in animals, and next, we will describe our recent findings in birds with regard to the specific mechanism of sperm uptake into the SST, sperm maintenance within it, and controlled release from it.

Keywords
Oviduct • Sperm storage • Sperm storage tubules • Utero-vaginal junction • Sperm • Sperm motility • L-lactic acid • Progesterone • Prostaglandin $F_{2\alpha}$

11.1 Introduction

Fertilization success results from a spatiotemporally regulated encounter between sperm and egg, therefore, a diverse reproductive strategy that ensures the sperm–egg encounter has evolved in many species. In external fertilizers such as sea urchins and ascidians, the spawning of gametes from both sexes usually occurs simultaneously, and the spermatozoa exhibit chemotaxis toward eggs by detecting a chemical gradient of chemoattractant released from the spawned eggs (Yoshida et al. 2008).

Such a synchrony in the release of gametes, together with sperm chemotaxis, leads to better fertilization outcomes in external fertilizers such as marine invertebrates. In the case of internal fertilizers, including mammals, one of the most important mechanisms in achieving better fertilization success is that the gametes from both sexes present together at the site of fertilization at the appropriate time. It is not likely to occur coincidently in vivo because sperm must travel a long distance within the female genital tract and must reach the site of fertilization when the oocytes are ready to be fertilized. In addition, the timing of ovulation in the female does not always coincide with that of insemination by the male, except for a few exceptions such as rabbits and domestic cats where ovulation is initiated by copulatory stimulation (Dal Bosco et al. 2011; Brown 2011). Indeed, it is reported in poultry, such as chicken and quail, that the ovulated oocytes quickly lose their fertilizability because the chalaza-layer, also referred to as the outer layer of the vitelline membrane secreted from the infundibulum part of the oviduct, overlays the surface of the egg immediately after ovulation (Wishart 1997). The sperm no longer interact with these oocytes thereafter and are trapped in the chalaza-layer or egg albumen (Wishart 1997). In order to increase the chance of the encounter of the gametes, the female stores the sperm in the reproductive tract. By storing the sperm, the female can regulate the release of the sperm from the storage site according to the timing of her ovulation. This ability to store the sperm in the female reproductive tract is frequently found in a variety of animals including insects, fish, amphibians, reptiles, birds, and mammals (Birkhead and Møller 1993; Holt and Lloyd 2010; Holt and Fazeli 2016; Orr and Zuk 2012). In these animals, females possess specialized structures in their genital tracts, such as sperm reservoirs in mammalian species (Suarez 2008), sperm storage tubules in birds and reptiles (Sasanami et al. 2013; Sever and Hamlett 2002), spermathecae in amphibians (Sever 2002) or spermathecae and seminal receptacles in insects (Heifetz and Rivlin 2010; Wolfner 2011), and these structures hold the sperm until the time of ovulation or when the oocytes are ready to be fertilized. Although the mechanism of sperm storage in such sperm storage organs is not fully understood, similar features in species of different lineages have evolved convergently. Sperm storage is an astonishing phenomenon, because some species of insect or reptile have evolved the ability to store fertilizable sperm in the genital tract for years and even up to a decade (Mangusson 1979; Birkhead and Møller 1993; Heifetz and Rivlin 2010; Wolfner 2011). Despite recent advances in molecular biology and improvements in microscope resolution, there is no reasonable explanation of how spermatozoa maintain their fertility for such a long period of time at ambient or body temperature in any species.

Until recently, such sperm storage was thought to solely occur in animals; however, recent findings suggested that some plants also use a similar mechanism for successful fertilization. Indeed, in flowering plants, when the first pollen tube failed to fertilize, a second pollen tube that was quiescent in the pistil started to move toward the ovule and rescue the fertilization failure (Kasahara et al. 2012). In the case of Fagaceae, fertilization is delayed from 4 days up to more than a year after pollination and until such time as the ovule becomes fully developed (Sogo and Tobe 2006). These phenomena appear to be very similar to sperm storage in the female reproductive tracts of animals in the aspect of decoupling the timing of pollination and fertilization. From these viewpoints, we would like to emphasize that sperm storage is a pivotal event in achieving better fertilization success and is evolutionally conserved in many species, including plants.

In this chapter, we will first describe the physiological importance and adoptive benefits of sperm storage in the female genital tract for successful fertilization in animals, and next, we will introduce our recent findings on the mechanism of sperm storage in the oviduct of avian species.

11.2 Sperm Storage Is an Evolutionary Conserved Reproductive Strategy to Ensure Fertilization Success

Sperm storage typically occurs in female animals whose ovulatory cycles do not coincide with the timing of insemination by males. In the case of mammalian species such as sheep, pigs, and cows, viable spermatozoa are trapped in the isthmus part of the oviduct referred to as the "sperm reservoir" until the time of ovulation (Suarez 2008). It is known that this trapping is mediated by the specific interaction of spermatozoa and the epithelial surface of the caudal isthmus. A series of studies using boars and cows demonstrated that the binding is mediated via cell adhesion molecules on the sperm head and carbohydrate moieties on the epithelial surface of the sperm reservoir (Ignotz et al. 2001). The sperm release from the storage site is facilitated by the stimulation of an ovarian steroid before ovulation, in which the increasing concentration of the circulating progesterone reactivates sperm motility sufficient for breaking the tethered sperm away from the sperm reservoir (Hunter 2008). These observations suggest that the sperm storage is acting to synchronize the timing of sperm arrival at the site of fertilization with that of ovulation. The duration of the sperm storage in mammalian species is relatively short (i.e., up to several days); however, some bats store sperm in the oviduct for 6 months (Racey 1979). Most of these bats reside in temperate zones where a period of hibernation interrupts the reproductive cycle. Typically, they copulate in autumn; however, fertilization does not take place immediately after copulation. The female bats store the sperm in their genital tract for several months and fertilize in the spring of the following year. By controlling the timing of sperm release from the storage site, female bats are able to optimize the timing of the copulation, as well as adjusting the time of birth to early summer, a suitable season for raising their offspring. A recent study in the Greater yellow bat (*Scotophilus heathi*) indicated that the sperm storage in this species is dependent upon the maintenance of a high level of circulating androgen concentration produced by the ovaries together with the androgen receptor in the utero-tubal junction where sperm storage has occurred (Roy and Krishna 2010). Although details are still not clear, this phenomenon leads to the expression of B-cell lymphoma factor 2 (Bcl-2), a key regulator of apoptosis in the utero-tubal junction (Roy and Krishna 2011). From these results, it could be hypothesized that androgen-dependent expression of Bcl-2 might act as an anti-apoptotic factor in the utero-tubal junction and may play some role in long-term sperm survival in the storage site.

Sperm storage in reptiles such as turtles, snakes, and alligators has been reported to last up to 4 or even 7 years (Birkhead and Møller 1993;

Holt 2011). Such long-term sperm storage beyond one breeding season (i.e., more than a year) allows the female animals to produce offspring without additional mating in the next breeding season, which could function as an insurance against not finding a mating partner in subsequent breeding seasons. In addition to these terrestrial animals, sperm storage in various elasmobranch species is known to last for between days and years, which allows females to decouple mating from fertilization (Fitzpatrick et al. 2012). It should be noted that parthenogenesis may be an alternative explanation for such observations of embryos produced in isolation from the male; therefore, we should carefully consider long-term sperm storage in these animals as to whether the embryo derived from the isolated female resulted from parthenogenesis or sexual reproduction.

As described above, the sperm-storage period of higher vertebrates is usually much shorter than the lifespan of a female; however, the queens of eusocial Hymenoptera (ants, bees, and wasps) often store sperm for more than a decade, i.e., throughout their lives (den Boer et al. 2009a, b). Sperm is stored in the storage organs (spermathecae) early in adult life following copulation, but mating does not occur again during the remainder of the lifespan. Thus, the queens produce fertilized eggs throughout their lifespan with the stored sperm obtained from a single mating flight, and the resident sperm must remain fertilizable throughout the years of storage. The spermathecae are often accompanied by glands, and it has been hypothesized that their secretions may play a role in the survival of the stored sperm, though the molecular composition of these secretions as well as their functions have not been elucidated. In these species, queens initially store several hundred million sperm, which leads to a reduction in the risks to the queen caused by repeating unnecessary mating (such as infection or predation during the mating flight). Although a direct relationship with the benefit from sperm storage has not been elucidated, it was reported in *Drosophila melanogaster* that females undergoing repeated mating survived a shorter time than females in a non-mating control group, with a reduced propensity to re-mate and increased egg-laying rate due to the exposure to seminal fluid products from the male accessory gland (Chapman et al. 2003; Liu and Kubli 2003). This finding indicates that unnecessary mating appears to bear a heavy cost for females. Thus, sperm storage may also lead to avoidance of such negative influences from the male.

In addition to their function for better fertilization success, sperm storage organs are also thought to work for the sperm selection site in the female reproductive tract. Recent findings indicated that the female reproductive tract actively works to eliminate the spermatozoa. Indeed, mice lacking the seminal vesicle secretion 2 (SVS2) protein are infertile and their sperm are killed by uterus-derived cytotoxic factors; it is believed that the SVS2 protein coats the sperm surface and protects the sperm from the attack by the uterus (Kawano et al. 2014). Thus the sperm, which protect themselves from female attack, are successfully stored in the storage sites and win the race for fertilization. This sperm/male selection by the female may maximize the genetic quality of a descendant and the number of offspring by selecting the sperm in the oviduct deposited by different males at different periods (i.e., cryptic female choice, Eberhard 2009; Holt and Fazeli 2010). From the above, the female reproductive tract acts as not only a path for sperm migration to the site of fertilization, but also provides an natural machinery ensuring that eggs are fertilized with favored sperm.

11.3 Sperm Storage in Birds

11.3.1 Sperm Uptake into the SST

In external fertilizers such as marine invertebrates, the released spermatozoa swim up toward the oocytes by chemotactic stimulation. However, in most internal fertilizers, including avian species, sperm do not directly migrate to the site of fertilization where the oocytes exist, but they initially enter/attach into the sperm storage site. In avian species, sperm storage tubules (SSTs), simple tubular invaginations located between the

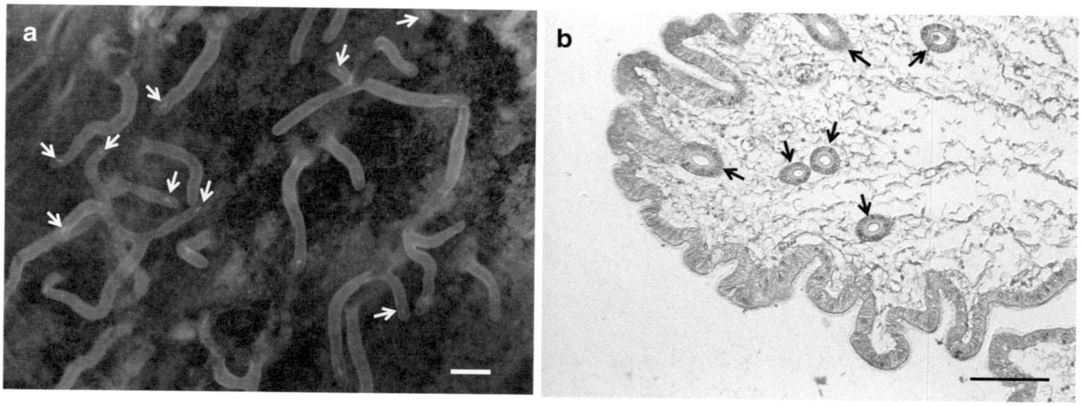

Fig. 11.1 Appearance of the SST in the Japanese quail. (**a**) Whole-mount specimen of the UVJ isolated from the bird. The SSTs appear as a simple tubular structure. *Arrows* indicate the resident spermatozoa in the SST. (**b**) Histochemical analysis of the SST in the UVJ sections. The section was lightly stained with H&E. *Arrows* indicate the SST. *Bar* = 50 μm

vagina and uterus (utero-vaginal junction: UVJ), serve as sperm storage sites (Bobr et al. 1964; Sasanami et al. 2013; Fig. 11.1). Then, the spermatozoa are transported to the infundibulum, which is the site of fertilization. It is reported that additional sperm storage tubules exist and serve as a secondary sperm storage site in the infundibulum (Bakst 1981). In this chapter, we will describe the mechanism of sperm storage in the UVJ because there is very little knowledge of sperm storage in the infundibulum. Once ejaculated, spermatozoa migrate to and are stored in the lumen of SSTs, where they can remain without loss of fertilizability for long periods of time (up to 15 weeks) at normal body temperatures (41 °C). This phenomenon is astonishing, because the terminally differentiated cells (spermatozoa) are able to maintain their biological function for long periods of time at high temperatures without new protein synthesis. Sperm storage in SSTs is thought to be an essential step for fertilization in birds, as SSTs protect sperm from noxious conditions, such as anti-sperm immune responses (Das et al. 2006, 2008), vaginal sperm ejection (Bakst 2011), and mechanical flushes caused by egg extrusion; therefore, it is believed that only spermatozoa successfully stored in SSTs are capable of fertilizing ova in domestic birds. The mechanism of how the ejaculated spermatozoa migrate to reside in the SST is not understood due to the difficulty of observation given the small inner diameter of the tubules (ca. 10–20 μm) and their location (i.e., embedded in the UVJ mucosa). As a result, only indirect methods, such as counting of the sperm on the sections of the SSTs after insemination, or determining the sperm-filling rate of the SST, can be used to elucidate the mechanism of sperm migration in the oviduct (Burke and Ogasawara 1969; Compton et al. 1978; Brillard 1993). After natural mating, the ejaculated sperm are deposited into the vagina. However, in the turkey, it has been reported that more than 80% of the sperm are rejected from the vagina soon after mating (Bakst 2011). In addition, less than 1% of the sperm inseminated into the vagina enter the SST (Bakst 1994). Das et al. observed the rate of SST filling after a single insemination event, and the sperm artificially introduced into the vagina of a chicken reached the SST within an hour (2006). With regard to the mechanism of sperm uptake into the SST, the intrinsic motility of sperm may be one of the important factors (Brillard 1993; Froman 2003). However, a specific molecule required for sperm transport to the SST has not been discovered. Recently, in Japanese quail, we reported that cloacal gland secretion (CGS), a major seminal plasma in male quail, contains prostaglandin $F_{2\alpha}$ ($PGF_{2\alpha}$) that contributes to successful fertilization (Sasanami et al. 2015). $PGF_{2\alpha}$ in the CGS

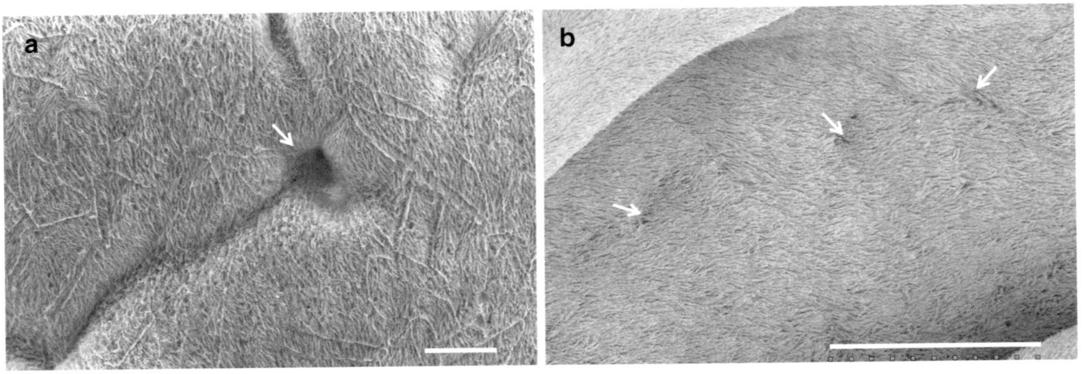

Fig. 11.2 $PGF_{2\alpha}$ induced the opening of the entrance of the SST. (**a**) Observation of the entrance of the SST isolated from the $PGF_{2\alpha}$-treated bird. The entrance of the SST becomes hollow due to the $PGF_{2\alpha}$ treatment. (**b**) The entrance of the SST of a bird without treatment. No morphological change shown in panel (**a**) was observed. *Bar* = 50 μm

induced vaginal contractions and caused the opening of the entrance of the SST (Fig. 11.2). When we measured the sperm-filling rate in the SST of quail which had received an intra-vaginal injection of CGS or $PGF_{2\alpha}$ before artificial insemination (AI) of the washed sperm, both CGS and $PGF_{2\alpha}$ treatment increased SST sperm filling. These results indicated that the male bird transports $PGF_{2\alpha}$ into the vagina at the time of mating, and this substance supports the sperm uptake into the SST in quail. Indeed, the removal of CGS from the male before mating reduced fertility because sperm are not allowed to enter into the SST. Because the supplementation of CGS or $PGF_{2\alpha}$ rescued the subfertility of the male, efficient sperm uptake into the SST with the aid of $PGF_{2\alpha}$ activity appears to be a key event for the achievement of fertilization in Japanese quail. Furthermore, we detected a constitutive expression of the receptor for $PGF_{2\alpha}$ during the ovulatory cycle. This expression pattern is reasonable because females can copulate with a male whenever they wish. In agreement with this assumption, the sperm-filling rate of the SST was constant irrespective of the timing of mating during the ovulatory cycle. Although such knowledge is limited in other birds, the concentrations of $PGF_{2\alpha}$ found in the seminal plasma of the turkey and rooster are reported to be 100–140 pg/ml and less than 100 pg/ml respectively (Kennedy et al. 2003; Hertelendy 1974). Although those values are approximately 7 times lower than that of quail CG, the $PGF_{2\alpha}$ levels in both turkey and chicken seminal plasma were sufficient to induce vaginal contractions in the quail (Sasanami et al. 2015). The question of whether $PGF_{2\alpha}$ is functional during sperm storage in other birds remains to be confirmed in future studies.

The above-mentioned observations indicate that male factors such as sperm motility and the constituents of the seminal plasma appear to contribute to the mechanism of sperm entrance to the SST. However, the female factor required for sperm migration to the SST is still not known. The surface epithelial cells located on the UVJ are covered with well-developed cilia (Fig. 11.3), and a vigorous movement of the cilia was observed on the dissected UVJ in vitro. Although direct evidence is lacking, these ciliary movements of the surface epithelium of the UVJ probably facilitate the sperm transport to the SST. We currently do not know whether a chemotactic guidance of the sperm by the SST is involved in the process of the sperm uptake. We have reported that the filling rate of the SSTs after a single insemination in the quail was approximately 30–50% and that the sperm-filled SSTs were not uniformly distributed in the UVJ (Ito et al. 2011), indicating a difference in each SST in terms of the ability for sperm uptake. What is the difference between the sperm-filled SST and the empty

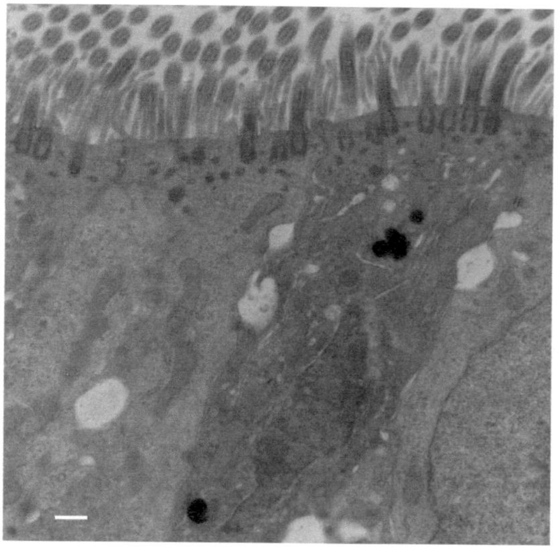

Fig. 11.3 Electron microscopical analysis of the surface epithelium of the UVJ in Japanese quail. On the apical surface of the cells, well-developed cilia are observed. *Bar* = 500 nm

SST? How does the SST select the population of the spermatozoa? The mechanism of the sperm uptake into the SST is still to be uncovered in future studies.

11.3.2 Sperm Maintenance Within the SST

The phenomenon of sperm storage in the SST of poultry was discovered in the 1960s, and has been extensively studied because the mechanism of sperm maintenance within the SST directly links to the development of a new strategy for long-term sperm preservation in vitro. Due to the presence of the SST, female birds are able to produce fertilized eggs for a long period of time following a single insemination event. The period of fertility in birds is correlated with the population of sperm-filled SSTs in the UVJ, and the fertile duration differs in different species. It has been reported that the differences in the duration of fertility between domestic fowl (2–3 weeks) and turkeys (10–15 weeks) are, in part, related to their respective numbers of SSTs (the mean numbers of SSTs for chickens and turkeys are 4893 and 30,566 respectively) (Birkhead and Møller 1992). We also observed that the fertile period of the female Japanese quail was 5–10 days after a single insemination, and the numbers of SSTs in the UVJ are approximately 1000 per bird. The underlying mechanisms of how the resident spermatozoa maintain their fertilizability at a high temperature for an extended period are a long-standing enigma. Several proteins including carbonic anhydrase (Holm et al. 1996), avidin (Foye-Jackson et al. 2011), aquaporins (Zaniboni and Bakst 2004), and alkaline phosphatase (Bakst and Akuffo 2007), as well as fatty acids (Huang et al. 2016), have been suggested to have potential roles in sperm maintenance in the SSTs, although no direct implication in sperm storage has been demonstrated.

In the lumen of the SSTs, sperm appear as packed, parallel bundles with their heads directed toward the blind tubular end (Bobr et al. 1964; Tingari and Lake 1973). In addition, it is reported that the sperm stored in the SSTs are typically immotile (Bakst 1987; Holm et al. 2000) and thus thought to be metabolically quiescent as a result of lowered ATP production/consumption. This is a reasonable theory because it leads to reduction of reactive oxygen species generated by sperm respiration, and it may reduce damage to the resident sperm in the SSTs. Thus, the SST function that allows the stored sperm to be in a quiescent state in motility may be one of the most important factors in the maintenance of the sperm.

Recently, we have also investigated the mechanism of sperm maintenance in the SST of Japanese quail. It is reported that sperm at the utero-tubal junction (UTJ) in the bovine oviduct bind to the surface of epithelial cells, and that this binding ensures the tethering of the sperm at the UTJ until the time of ovulation (Hunter 2008; Suarez 2010). In contrast to the situation in mammalian species, the resident sperm seem to be detached from SST epithelial cells in the quail oviduct (Fig. 11.4). This finding led us to hypothesize that unknown materials in the lumen of the SSTs may affect sperm mobility. We prepared UVJ extracts, and ejaculated sperm were incubated in the presence or absence of the UVJ extracts. The flagellar movement of the sperm was recorded using a high-speed camera. When

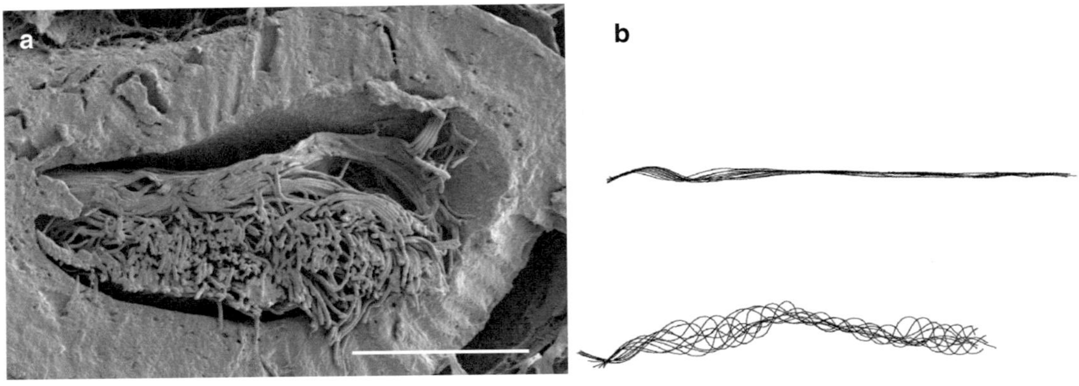

Fig. 11.4 (**a**) Ultrastructural observation of the resident sperm in the SST. Thick sections of the UVJ were prepared, and the surface of the cross section was observed by scanning electron microscopy. *Bar* = 20 μm. (**b**) Effects of UVJ extracts on the sperm motility. The ejaculated sperm were incubated in the presence (*upper*) or absence (*lower*) of the UVJ extracts (300 μg/ml) at 39 °C for 10 min. The flagellar movements of the sperm were recorded by high-speed camera, and the flagella of randomly selected sperm were traced; five images taken every 1/20th of a second were overlaid

the sperm were incubated in the absence of the UVJ extracts, a vigorous flagellar movement was observed (Fig. 11.4, *lower* panel). However, in the presence of the UVJ extracts, we found that the flagellar movements were relatively quiescent, and that the amplitude of the flagellar movement, as well as the linear velocity of the sperm, decreased (Fig. 11.4, *upper* panel). From these facts, we hypothesized that a bioactive substance responsible for sperm quiescence is present in the UVJ extracts. The bioactive substance was successfully purified by liquid chromatography using size-exclusion and reversed-phase columns followed by preparative thin-layer chromatography (TLC). Analytical TLC of the sample in the final step of purification detected only one major spot, which was thereafter assigned as lactic acid by the spectrum analysis. In addition, the extracts of the SSTs were assayed for L-lactic acid, which was found to be five times greater in SSTs than in non-SST epithelial cells. These results indicated that L-lactic acid is a sperm motility inactivation factor in the SST of Japanese quail.

How do the SSTs produce and release such large amounts of lactic acid? To answer this question, we performed a RNA-seq analysis of the SSTs. When the isolated SST was incubated in the presence of glucose, SST released lactic acid over time. Because this release was completely blocked in the presence of 2-deoxy glucose, an inhibitor of glycolysis, we believed that the enhanced glycolysis was the main reason for the robust synthesis of lactic acid. However, our RNA-seq analysis revealed that the expression levels of key glycolytic enzymes (such as hexokinase, phosphofructokinase, and pyruvate kinase) as well as several glucose transporters were not significantly higher in SSTs than those in non-SST cells. Because hypoxia can be an alternative cause for the promotion of lactic acid production in the cells (Pelicano et al. 2006), we investigated whether the SST cells are subjected to hypoxia in vivo. To test this, we employed Hypoxiprobe™ that allows us to detect the hypoxic cells and tissues by immunohistochemistry. The results demonstrated SST cells labeled by Hypoxiprobe™, and the signals were much stronger than those of the non-SST epithelial cells (Fig. 11.5). In addition, our RNA-seq analysis revealed the up-regulation of several oncogenes such as *c-Fos*, *c-Jun*, and *v-Myc* in the SST, which was in good agreement with the response observed in other hypoxic cells (Müller et al. 1997). In addition, a marked decrease in the mitochondrial activity of the SST cells also supports the idea that oxygen supply is limited in the SST (Fig. 11.5). From these facts, we concluded that hypoxia is the main cause for the large quantity of lactic acid production in the SSTs.

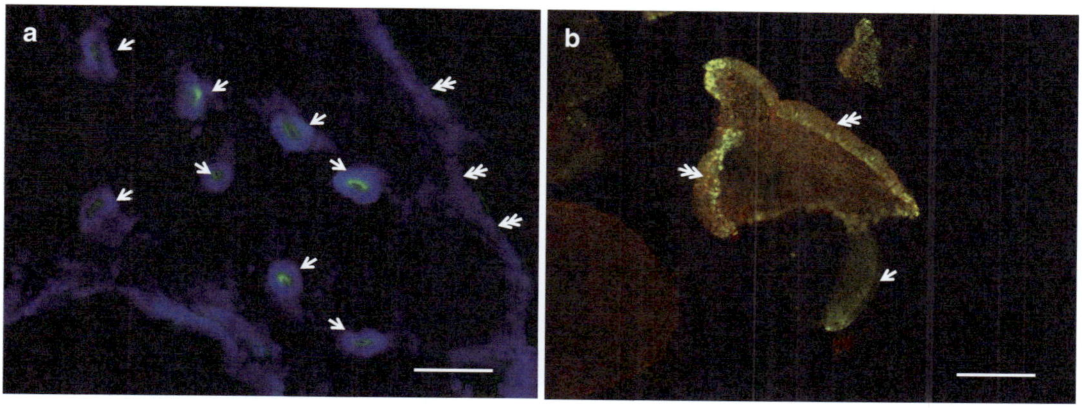

Fig. 11.5 The SSTs are subjected to hypoxic condition in vivo. (**a**) Immunohistochemical detection of hypoxic cells. UVJ tissues isolated from birds injected with Hypoxyprobe™ were stained with anti-pimonidazole antibody (*green*). The nuclei were counterstained with 4′,6-diamidino-2-phenylindole (DAPI; *blue*). *Arrows* and *double arrows* indicate SSTs and surface epithelium, respectively. *Scale bar* = 50 μm. (**b**) Mitochondrial activities of SSTs and surface epithelium assessed by JC-1 fluorescence. Surface epithelium displayed high mitochondrial activities (*double arrows*), while low mitochondrial activities were exhibited by SSTs (*arrow*), indicating that the oxygen supply is limited in the SSTs. *Scale bar* = 50 μm

Another important factor supporting sperm storage in the SSTs is protection of the resident sperm from anti-sperm immune responses in the oviduct, because the immune system is well developed to protect against infection by various microorganisms. This may also affect survivability of the sperm in the SSTs, because the sperm are recognized as foreign bodies in the oviduct. Immunocompetent cells for acquired immunity (i.e., macrophages, antigen-presenting cells expressing MHC class II, CD41, CD81 T cells and premature B and plasma cells) are reported to exist in the oviduct mucosal tissue in sexually matured chickens to protect the tissues from infection (Zheng et al. 1998; Zheng and Yoshimura 1999). Significant increases in the number of lymphocytes, as well as antigen-presenting cells expressing MHC class II in the stroma of UVJ in low-fertility hens, were observed (Yoshimura et al. 1997). Das et al. (2006) demonstrated that resident sperm in the SSTs are protected from immune responses by SST structures and transforming growth factor β (TGFβ), the expression of which increased when the SSTs were filled with sperm. Since TGFβ and the receptors for TGFβ are also expressed in sperm, the enhanced expression of TGFβ and its receptors may protect sperm in the SSTs by suppressing anti-sperm immunoreactions, presumably by suppressing the proliferation of T- and B-lymphocytes (Das et al. 2008). Thus, the elimination of anti-sperm immune responses by the TGFβ system appears to be one of the factors responsible for sperm maintenance in the SSTs.

11.3.3 Sperm Release from the SST

To achieve better fertilization success, the resident sperm in the SST must be released at an appropriate timing to reach the fertilization site in accordance with the ovulation timing. However, there are several reports indicating that sperm release from the SST is not regulated, but occurs in response to the mechanical pressures of a passing ovum, because no contractile elements associated with the SST were found (Van Krey et al. 1967; Mero and Ogasawara 1970). In addition, Burke and Ogasawara (1969), who recovered the sperm from an inseminated hen oviduct, concluded that the sperm release from the SST is a slow and continuous event that occurs constitutively during the ovulatory cycle. On the other hand, Bobr et al. (1964), who investigated the distribution of spermatozoa in the hen oviduct after insemination, reported that the resident spermatozoa were discharged from the SST close to the times of ovulations and/or ovipositions. In addition, Mero and Ogasawara (1970) reported that the tubular enlargement of the SST is associated with sperm release in the chicken. Although we do not know the reason for these discrepancies in early studies, later observations indicated that the sperm release from the SST may be regulated to coordinate with the ovulations and/or ovipositions.

We also previously observed in Japanese quail that the release of the sperm from the SST occurred only within a 20 h period after oviposition (Ito et al. 2011). In addition, the sperm-filling rate of the SST significantly decreased between 14 and 25 h after egg-laying. Based on these results, we hypothesized that sperm release from the SST in the Japanese quail is a controlled event during the ovulatory cycle. To test whether the hormonal stimulation relates to the sperm release from the SST, we injected the birds with various steroid hormones and calculated the SST filling rate. The percentage of SST with sperm only significantly decreased when the animals were treated with greater than 0.8 μg/ml progesterone compared to the control (vehicle alone). Interestingly, the morphology of the progesterone-treated SST had been changed, that is, the treated SST showed contraction-like morphological changes. Scanning electron microscopical observation revealed that the bundle of the sperm tail released from the SST was observed when the birds were administered progesterone (Fig. 11.6). The expression of membrane progestin receptor α, both at mRNA and protein levels, was also confirmed in the SST. These results strongly indicated that progesterone is a sperm-releasing factor from the SST in Japanese quail. If sperm are released from the SST without any regulation, most sperm ascending the oviduct may be trapped by the descending egg. Bobr et al. (1964) found that the sperm after insemination had usually disappeared from the oviduct lumen within 24 h of being deposited in a laying hen, in spite of their potency for producing an additional fertilized egg. That observation might support our suggestion that the resident sperm in the SST are protected from egg extrusion by being released from the SST for a restricted time under the control of progesterone. Brillard and Bakst (1990) reported that the SST of turkey hens inseminated within 1 week before the expected onset of egg production possessed more spermatozoa than those in the SST of turkey hens inseminated after the onset of laying. Similarly, a significantly lower number of empty SSTs in non-laying adult hens as compared to the laying hens was observed (McIntyre and Christensen 1983). These obser-

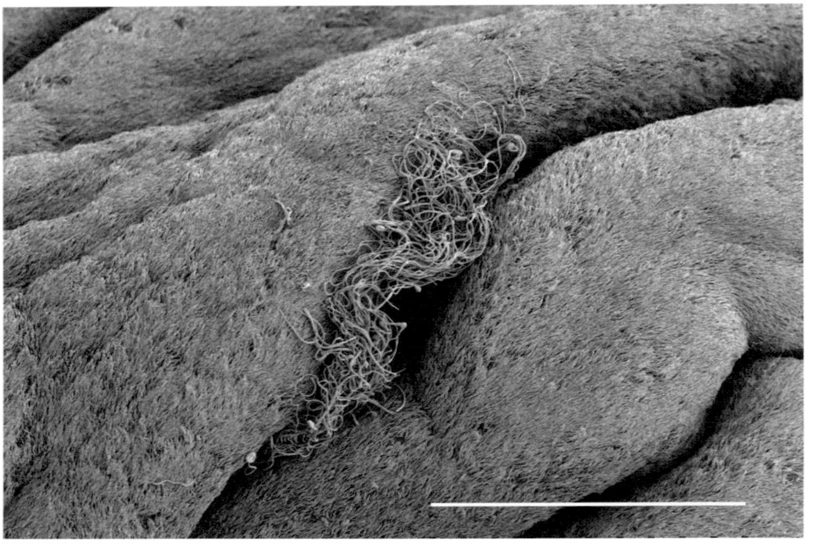

Fig. 11.6 Ultrastructural observation of the UVJ surface treated with progesterone. After mating, the animals were intravenously injected with 0.8 μg/ml progesterone. The SST entrance area was observed by scanning electron microscopy. *Scale bar* = 100 μm

vations indicate that the increased numbers of resident spermatozoa in the SST of hens without egg production may be due to the lack of a functional level of circulating progesterone in relation to the ovulatory cycle.

In our immunohistochemical observations using anti-cuticle materials, the outermost covering of the eggshell, which protect the egg from microorganisms, the secretory activity of the UVJ ciliated cells for the cuticle materials occurred in concurrence with the sperm release. Moreover, the progesterone treatment stimulated the release of cuticle materials from the cells. Rahman et al. (2007), who investigated the role of cuticle in egg formation by electron microscopy, demonstrated that the surface of the luminal epithelia in the uterus as well as the outside of the egg shell at 7 h after oviposition showed a coarse appearance, whereas both became much smoother after cuticle secretion at 20 h. From these observations, it was suggested that the cuticle might function as a lubricant that facilitates egg rotation in the uterus. Considering these observations along with our results, we suggest that the cuticle materials secreted from the ciliated cells of the UVJ may support the sperm movement, in ascending the oviduct, by facilitating the rotation of the egg, as well as function as a fluid material for the sperm swimming/transport after release from the SST by stimulation with progesterone.

In birds, it is reasonable to suppose that two events, the sperm release from the SST and the cuticle secretion, are stimulated simultaneously by progesterone for efficient fertilization, because there is at least a 5-h grace period before the next ovulation and the released sperm can reach the infundibulum, the site of fertilization, without hindrance by the descending egg. Via this process, birds acquired an excellent strategy to achieve better fertilization success in the case of sperm release from the SSTs.

In mammalian species, the sperm captured on the surface of the UTJ are reported to break this tethering by activating motility from the stimulation of progesterone. Because stored sperm in the avian SST are quiescent in motility, re-activation after release is also probably necessary. However, in our test, progesterone did not stimulate sperm motility in vitro, indicating that an alternative mechanism may exist. Our preliminary proteomic analyses on the soluble extracts of the UVJ and vagina mucosa have shown that several proteins were abundant in the UVJ mucosa. We produced antisera against these proteins, and the ability to disturb fertilization in quail was tested after intra-vaginal injection of the antibody. When the birds were treated with nonspecific rabbit IgG, fertility was >70%; however, fertility decreased to <10% when birds were injected with an anti-heat shock protein 70 (HSP70) antibody (Hiyama et al. 2013). By additional experiments, we showed that the bacterially expressed HSP70 activates sperm motility in vitro. Analysis of HSP70 expression revealed that *HSP70* mRNA and the HSP70 protein were expressed in the UVJ, where the protein was primarily localized in the surface epithelium. Moreover, HSP70-binding was detected on the entire surface of the sperm. Since the timing of sperm release from the SSTs temporally coincides with a peak in HSP70 expression, it is possible that HSP70 derived from surface epithelium of the UVJ reactivates sperm, facilitating sperm migration toward the infundibulum (Hiyama et al. 2013). Further experiments will be required if this in-vitro result reflects the in-vivo situation occurring in avian oviducts.

Conclusion

In sexual reproduction, sperm are subject to selection by the female and sperm competition, and these factors drive the development of various reproductive tactics through evolution. In this chapter, we introduced sperm storage in birds in terms of an excellent reproductive tactic in internally fertilizing animals. The mechanism of sperm storage and release in the avian oviduct has been obscure since its discovery in the 1960s. During the last decade, several key molecules such as $PGF_{2\alpha}$, lactic acid and progesterone that contribute to sperm uptake into, maintenance within, and release from the SST respectively were discovered in Japanese quail (Fig. 11.7). However, we are unable to answer many important questions on

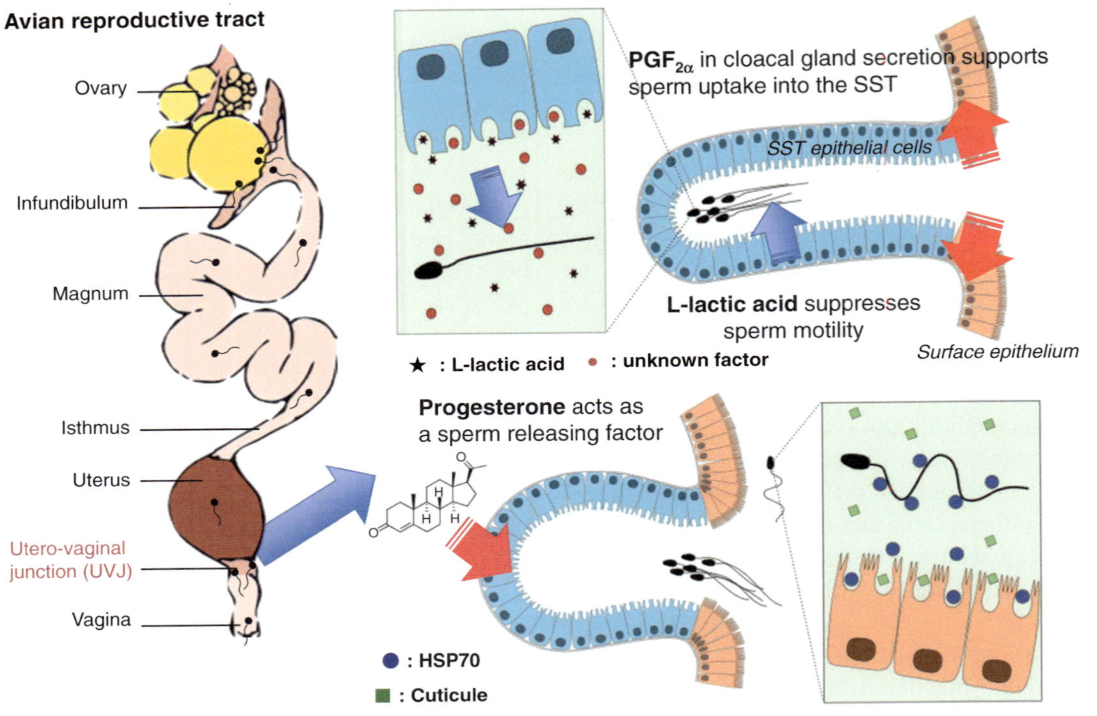

Fig. 11.7 Suggested sperm storage mechanism in the SST of Japanese quail. After copulation, $PGF_{2\alpha}$ in the cloacal gland secretion induces the contraction of the female vagina and it also enlarges the entrance of the SST, which supports sperm uptake. In the lumen of the SST, a high concentration of lactic acid exists because of the mass production of lactic acid by SST cells due to hypoxic conditions, and it makes any resident sperm quiescent. We expect the presence of an unknown factor that plays a major role in the maintenance of long-term sperm storage in the SSTs. When ovulation becomes imminent, the blood concentration of progesterone increases. This induces morphological changes in the SSTs, which squeeze out the resident sperm into the lumen of the oviduct. Simultaneously, cuticle materials are also released under progesterone control and the cuticle may function as a lubricant that facilitates egg rotation in the uterus. Thus, progesterone acts as a sperm-releasing factor in birds. From the surface epithelium of the UVJ, HSP70 protein is secreted, bound to the surface of the sperm and reactivates sperm motility, which may help sperm migration to the fertilization site

the mechanism of sperm storage. How do SSTs choose the sperm from the ejaculates that consist of a population of several hundred million? In the case of promiscuous copulation, which male wins the race for the residency in the SSTs? The elucidation of this process may directly lead to our understanding on the molecular mechanism of cryptic female choice; a mechanism that is difficult to uncover due to the lack of a suitable experimental model. Avian sperm storage is undoubtedly a suitable model for this elucidation. In regard to the mechanism of sperm maintenance, novel methods that will allow us to maintain sperm longevity for extremely long periods of time at ambient or body temperature can be engineered by mimicking the circumstances in the lumen of the SSTs. This technique will be of importance in the field of reproductive therapy in humans and conservation of endangered animals. As such, investigations into the mechanism of sperm storage may open up new opportunities with enormous implications for industry, agriculture, and conservation, as well as in the field of basic life science. We hope that the mystery of the avian sperm storage mechanism will be further elucidated in the near future.

References

Bakst MR. Sperm recovery from oviducts of turkeys at known intervals after insemination and oviposition. J Reprod Fertil. 1981;62(1):159–64.

Bakst MR. Anatomical basis of sperm-storage in the avian oviduct. Scanning Microsc. 1987;1(3):1257–66.

Bakst MR. Fate of fluorescent stained sperm following insemination: new light on oviducal sperm transport and storage in the turkey. Biol Reprod. 1994;50(5):987–92.

Bakst MR. Physiology and endocrinology symposium: role of the oviduct in maintaining sustained fertility in hens. J Anim Sci. 2011;89(5):1323–9.

Bakst MR, Akuffo V. Alkaline phosphatase reactivity in the vagina and uterovaginal junction sperm-storage tubules of turkeys in egg production: implications for sperm storage. Br Poult Sci. 2007;48(4):515–8.

Birkhead TR, Møller AP. Numbers and size of sperm storage tubules and the duration of sperm storage in birds: a comparative study. Biol J Linnean Soc. 1992;45:367–72.

Birkhead TR, Møller AP. Sexual selection and the temporal separation of reproductive events: sperm storage data from reptiles, birds and mammals. Biol J Linnean Soc. 1993;50:295–311.

Bobr LW, Lorenz FW, Ogasawara FX. Distribution of spermatozoa in the oviduct and fertility in domestic birds. J Reprod Fertil. 1964;8:39–47.

Brillard JP. Sperm storage and transport following natural mating and artificial insemination. Poult Sci. 1993;72(5):923–8.

Brillard JP, Bakst MR. Quantification of spermatozoa in the sperm-storage tubules of turkey hens and the relation to sperm numbers in the perivitelline layer of eggs. Biol Reprod. 1990;43(2):271–5.

Brown JL. Female reproductive cycles of wild female felids. Anim Reprod Sci. 2011;124(3–4):155–62.

Burke WH, Ogasawara FX. Presence of spermatozoa in uterovaginal fluids of the hen at various stages of the ovulatory cycle. Poult Sci. 1969;48(2):408–13.

Chapman T, Bangham J, Vinti G, Seifried B, Lung O, Wolfner MF, Smith HK, Partridge L. The sex peptide of *Drosophila melanogaster*: female post-mating responses analyzed by using RNA interference. Proc Natl Acad Sci USA. 2003;100(17):9923–8.

Compton MM, Van Krey HP, Siegel PB. The filling and emptying of the uterovaginal sperm-host glands in the domestic hen. Poult Sci. 1978;57(6):1696–700.

Dal Bosco A, Rebollar PG, Boiti C, Zerani M, Castellini C. Ovulation induction in rabbit does: current knowledge and perspectives. Anim Reprod Sci. 2011;129(3–4):106–17.

Das SC, Isobe N, Nishibori M, Yoshimura Y. Expression of TGFβ-isoforms and their receptors in utero-vaginal junction of hen oviduct in presence or absence of resident sperm with reference to sperm storage. Reproduction. 2006;132(5):781–90.

Das SC, Isobe N, Yoshimura Y. Mechanism of prolonged sperm storage and sperm survivability in hen oviduct: a review. Am J Reprod Immunol. 2008;60(6):477–81.

den Boer SP, Baer B, Dreier S, Aron S, Nash DR, Boomsma JJ. Prudent sperm use by leaf-cutter ant queens. Proc Biol Sci. 2009a;276(1675):3945–53.

den Boer SP, Boomsma JJ, Baer B. Honey bee males and queens use glandular secretions to enhance sperm viability before and after storage. J Insect Physiol. 2009b;55(6):538–43.

Eberhard WG. Postcopulatory sexual selection: Darwin's omission and its consequences. Proc Natl Acad Sci USA. 2009;106(Suppl 1):10025–32.

Fitzpatrick JL, Kempster RM, Daly-Engel TS, Collin SP, Evans JP. Assessing the potential for post-copulatory sexual selection in elasmobranchs. J Fish Biol. 2012;80(5):1141–58.

Foye-Jackson OT, Long JA, Bakst MR, Blomberg LA, Akuffo VG, Silva MV, Guthrie HD, McMurtry JP. Oviductal expression of avidin, avidin-related protein-2, and progesterone receptor in turkey hens in relation to sperm storage: effects of oviduct tissue type, sperm presence, and turkey line. Poult Sci. 2011;90(7):1539–47.

Froman D. Deduction of a model for sperm storage in the oviduct of the domestic fowl (*Gallus domesticus*). Biol Reprod. 2003;69(1):248–53.

Heifetz Y, Rivlin PK. Beyond the mouse model: using *Drosophila* as a model for sperm interaction with the female reproductive tract. Theriogenology. 2010;73(6):723–39.

Hertelendy F. Effects of prostaglandins, cyclic AMP, seminal plasma, indomethacin, and other factors on oviposition in the Japanese quail, *Coturnix japonica*. J Reprod Fertil. 1974;40:87–93.

Hiyama G, Matsuzaki M, Mizushima S, Dohra H, Ikegami K, Yoshimura T, Shiba K, Inaba K, Sasanami T. Sperm activation by heat shock protein 70 supports migration of sperm released from sperm storage tubules in Japanese quail (*Coturnix japonica*). Reproduction. 2013;147(2):167–78.

Holm L, Ekwall H, Wishart GJ, Ridderstråle Y. Localization of calcium and zinc in the sperm storage tubules of chicken, quail and turkey using X-ray microanalysis. J Reprod Fertil. 2000;118(2):331–6.

Holm L, Ridderstråle Y, Knutsson PG. Localisation of carbonic anhydrase in the sperm storing regions of the domestic hen oviduct. Acta Anat (Basel). 1996;156(4):253–60.

Holt WV. Mechanisms of sperm storage in the female reproductive tract: an interspecies comparison. Reprod Domest Anim. 2011;46(Suppl 2):68–74.

Holt WV, Fazeli A. The oviduct as a complex mediator of mammalian sperm function and selection. Mol Reprod Dev. 2010;77(11):934–43.

Holt WV, Fazeli A. Sperm storage in the female reproductive tract. Annu Rev Anim Biosci. 2016;4(11):291–310.

Holt WV, Lloyd RE. Sperm storage in the vertebrate female reproductive tract: how does it work so well? Theriogenology. 2010;73(6):713–22.

Huang A, Isobe N, Yoshimura Y. Expression of lipases and lipid receptors in sperm storage tubules and possible role of fatty acids in sperm survival in the hen oviduct. Theriogenology. 2016;85(7):1334–42.

Hunter RH. Sperm release from oviduct epithelial binding is controlled hormonally by peri-ovulatory graafian follicles. Mol Reprod Dev. 2008;75(1):167–74.

Ignotz GG, Lo MC, Perez CL, Gwathmey TM, Suarez SS. Characterization of a fucose-binding protein from bull sperm and seminal plasma that may be responsible for formation of the oviductal sperm reservoir. Biol Reprod. 2001;64(6):1806–11.

Ito T, Yoshizaki N, Tokumoto T, Ono H, Yoshimura T, Tsukada A, Kansaku N, Sasanami T. Progesterone is a sperm-releasing factor from the sperm-storage tubules in birds. Endocrinology. 2011;152(10):3952–62.

Kasahara RD, Maruyama D, Hamamura Y, Sakakibara T, Twell D, Higashiyama T. Fertilization recovery after defective sperm cell release in *Arabidopsis*. Curr Biol. 2012;22(12):1084–9.

Kawano N, Araki N, Yoshida K, Hibino T, Ohnami N, Makino M, Kanai S, Hasuwa H, Yoshida M, Miyado K, Umezawa A. Seminal vesicle protein SVS2 is required for sperm survival in the uterus. Proc Natl Acad Sci USA. 2014;111(11):4145–50.

Kennedy JH, Korn N, Thurston RJ. Prostaglandin levels in seminal plasma and sperm extracts of the domestic turkey, and the effects of cyclooxygenase inhibitors on sperm motility. Reprod Biol Endocrinol. 2003;1:74.

Liu H, Kubli E. Sex-peptide is the molecular basis of the sperm effect in *Drosophila melanogaster*. Proc Natl Acad Sci USA. 2003;100(17):9929–33.

Mangusson WE. Production of an embryo by *Acrochordas vavanicus* isolated for seven years. Copeia. 1979;1979:744–5.

McIntyre DR, Christensen VL. Filling rates of the uterovaginal sperm storage gland in the turkey. Poult Sci. 1983;62:1652–6.

Mero KN, Ogasawara FX. Dimensions of uterovaginal sperm-storage tubules of the chicken and their possible significance in sperm release. Poult Sci. 1970;49(5):1304–8.

Müller JM, Krauss B, Kaltschmidt C, Baeuerle PA, Rupec RA. Hypoxia induces c-fos transcription via a mitogen-activated protein kinase-dependent pathway. J Biol Chem. 1997;272(37):23435–9.

Orr TJ, Zuk M. Sperm storage. Curr Biol. 2012;22(1):R8–R10.

Pelicano H, Martin DS, RH X, Huang P. Glycolysis inhibition for anticancer treatment. Oncogene. 2006;25(34):4633–46.

Racey PA. The prolonged storage and survival of spermatozoa in *Chiroptera*. J Reprod Fertil. 1979;56(1):391–402.

Rahman MA, Baoyindeligeer IA, Yoshizaki N. Mechanism of chalaza formation in quail eggs. Cell Tissue Res. 2007;330:535–43.

Roy VK, Krishna A. Evidence of androgen-dependent sperm storage in female reproductive tract of *Scotophilus heathii*. Gen Comp Endocrinol. 2010;165(1):120–6.

Roy VK, Krishna A. Sperm storage in the female reproductive tract of *Scotophilus heathii*: role of androgen. Mol Reprod Dev. 2011;78(7):477–87.

Sasanami T, Izumi S, Sakurai N, Hirata T, Mizushima S, Matsuzaki M, Hiyama G, Yorinaga E, Yoshimura T, Ukena K, Tsutsui K. A unique mechanism of successful fertilization in a domestic bird. Sci Rep. 2015;5:7700.

Sasanami T, Matsuzaki M, Mizushima S, Hiyama G. Sperm storage in the female reproductive tract in birds. J Reprod Dev. 2013;59(4):334–8.

Sever DM. Female sperm storage in amphibians. J Exp Zool. 2002;292(2):165–79.

Sever DM, Hamlett WC. Female sperm storage in reptiles. J Exp Zool. 2002;292(2):187–99.

Sogo A, Tobe H. Delayed fertilization and pollen-tube growth in pistils of *Fagus japonica* (*Fagaceae*). Am J Bot. 2006;93(12):1748–56.

Suarez SS. Regulation of sperm storage and movement in the mammalian oviduct. Int J Dev Biol. 2008;52(5–6):455–62.

Suarez SS. How do sperm get to the egg? Bioengineering expertise needed! Exp Mech. 2010;50:1267–74.

Tingari MD, Lake PE. Ultrastructural studies on the uterovaginal sperm-host glands of the domestic hen, *Gallus domesticus*. J Reprod Fertil. 1973;34(3): 423–31.

Van Krey HP, Ogasawara FX, Pangborn J. Light and electron microscope studies of possible sperm gland emptying mechanisms. Poult Sci. 1967;46:69–78.

Wishart GJ. Quantitative aspects of sperm: egg interaction in chickens and turkeys. Anim Reprod Sci. 1997;48(1):81–92.

Wolfner MF. Precious essences: female secretions promote sperm storage in Drosophila. PLoS Biol. 2011;9:e1001191.

Yoshida M, Kawano N, Yoshida K. Control of sperm motility and fertility: diverse factors and common mechanisms. Cell Mol Life Sci. 2008;65(21):3446–57.

Yoshimura Y, Okamoto T, Tamura T. Localisation of MHC class II, lymphocytes and immunoglobulins in the oviduct of laying and moulting hens. Br Poult Sci. 1997;38(5):590–6.

Zaniboni L, Bakst MR. Localization of aquaporins in the sperm storage tubules in the turkey oviduct. Poult Sci. 2004;83(7):1209–12.

Zheng WM, Yoshimura Y. Localization of macrophages in the chicken oviduct: effects of age and gonadal steroids. Poult Sci. 1999;78(7):1014–8.

Zheng WM, Yoshimura Y, Tamura T. Effects of age and gonadal steroids on the localization of antigenpresenting cells, T and B cells in the chicken oviduct. J Reprod Fertil. 1998;114(1):45–54.

Avian Biotechnology

12

Yoshiaki Nakamura

Abstract

Primordial germ cells (PGCs) generate new individuals through differentiation, maturation and fertilization. This means that the manipulation of PGCs is directly linked to the manipulation of individuals, making PGCs attractive target cells in the animal biotechnology field. A unique biological property of avian PGCs is that they circulate temporarily in the vasculature during early development, and this allows us to access and manipulate avian germ lines. Following the development of a technique for transplantation, PGCs have become central to avian biotechnology, in contrast to the use of embryo manipulation and subsequent transfer to foster mothers, as in mammalian biotechnology. Today, avian PGC transplantation combined with recent advanced manipulation techniques, including cell purification, cryopreservation, depletion, and long-term culture in vitro, have enabled the establishment of genetically modified poultry lines and ex-situ conservation of poultry genetic resources. This chapter introduces the principles, history, and procedures of producing avian germline chimeras by transplantation of PGCs, and the current status of avian germline modification as well as germplasm cryopreservation. Other fundamental avian reproductive technologies are described, including artificial insemination and embryo culture, and perspectives of industrial applications in agriculture and pharmacy are considered, including poultry productivity improvement, egg modification, disease resistance impairment and poultry gene "pharming" as well as gene banking.

Keywords

Artificial insemination • Chickens • Embryo culture • Cryopreservation • Genetic modification • Gemline chimeras • Poultry • Primordial germ cells • Transplantation

Y. Nakamura
Division of Germ Cell Biology, National Institute for Basic Biology, Okazaki, Japan
e-mail: ynsu@nibb.ac.jp

T. Sasanami (ed.), *Avian Reproduction*, Advances in Experimental Medicine and Biology 1001,
DOI 10.1007/978-981-10-3975-1_12

12.1 Introduction

The germ cell is the only cell type capable of transmitting genetic information to the next generation. Therefore, they have been often the objects of manipulation in various organisms in the field of biotechnology. For example, many mammalian biotechnologies are based on the manipulation of embryos followed by embryo transfer into the uterus of a pseudopregnant female recipient, with the intention of establishing a viable pregnancy. However, the development of avian biotechnologies has lagged far behind that of mammalian species because of the difficulty in accessing and manipulating the large yolky zygotes. As an alternative, a unique biological property of avian primordial germ cells (PGCs), progenitor cells for gametes (spermatozoa and ova), which circulate temporarily in the vasculature during early development, allows us to access and manipulate the avian germline. The technical development of PGC transplantation to produce germline chimeras was a major breakthrough in the field of avian biotechnology. To date, several techniques for PGC manipulation have been developed, such as purification, cryopreservation, sterilization, and long-term culture, and for domestic chickens in particular. PGC transplantation combined with recent advanced PGC manipulation techniques has enabled the establishment of genetically modified poultry lines and ex-situ conservation of poultry genetic resources. PGC-mediated avian biotechnologies provide insight into industrial applications in agriculture and pharmacy as well as to the promotion of studies in basic biology. This chapter introduces the principles, history, and procedures of PGC-mediated avian germline chimera production and its application for germline modification as well as genetic resource conservation. This chapter also details other fundamental avian reproductive technologies, including artificial insemination and embryo culture, and presents a perspective of industrial applications in agriculture and pharmacy, including poultry productivity improvement, egg modification, disease resistance impairment and gene "pharming" as well as poultry gene banking.

12.2 Artificial Insemination

Artificial insemination is the technique of manually introducing ejaculated spermatozoa into the reproductive tract, such as the uterus or cervix of a female recipient, for the purpose of achieving fertilization in vivo. Successful insemination requires the deposition of sufficient number of spermatozoa with adequate motility into the female reproductive tract at an appropriate time. It has provided many benefits to the livestock industry, including the widespread use of genetically superior sires, faster and increased genetic improvement, increased safety for animals and farmers, reduced disease transmission, reduced costs, and accelerated progeny testing. Since Spallanzani reported the first artificial insemination in mammals (specifically in dog, *Canis lupus familiaris*) in 1780, this technique has been the most commonly used reproductive technology in the livestock industry, particularly in cattle (*Bos taurus*) production. There are two reasons for this: (1) bull semen can be cryopreserved in a relatively easy manner, and (2) a large amount of semen can be collected from a single ejaculate (5–8 ml), which can be diluted and extended to inseminate hundreds of cows. For example, the prevailing rates of artificial insemination in dairy and beef cattle production in Japan are approximately 99 and 96%, respectively. Today, the use of artificial insemination has extended to human infertility clinics.

12.2.1 Artificial Insemination in Poultry

Artificial insemination has been used in poultry, in particular for the turkey (*Meleagris gallopavo*) and chicken (*Gallus domesticus*), since the 1830s due to the simplification of the insemination process through the establishment of methods for semen collection and vaginal insemination. The average semen volumes collected from birds such as the chicken, turkey, and duck are considerably less than from farm animals, at approximately 100–300 μl. In contrast, the average sperm concentration of poultry semen such as chicken (3–5 × 10^9 spermatozoa/ml) is much higher than from farm animals such as bulls (0.3–2 × 10^9 spermatozoa/ml) and boars (0.3–3.5 × 10^8 spermatozoa/ml). In chicken, insemination of a hen with approximately 100 million spermatozoa can result in a ~90% fertilization rate. Chicken semen begins to lose fertility when stored for more than 60 min. Extenders enable more precise control of insemination and short-term storage, with maintenance of acceptable fertility for up to 24 h at 4 °C. Several commercial semen extenders as well as a phosphate-buffered semen diluent known as Beltsville Poultry Semen Extender are available and used routinely, particularly for turkeys.

Artificial insemination is widely used for commercial turkey breeding management. This is because the larger and heavily muscled birds (about 13 kg body weight for males and 9 kg for females) are physically unable to complete the natural mating process. Apart from turkeys, artificial insemination is also extensively used for the production of guinea fowls (*Numida meleagris*) and "mule" ducks, the sterile hybrid progeny from a cross between female common ducks (*Anas platyrynchos*) and male Muscovy ducks (*Cairina moschata*). In contrast, artificial insemination is not commonly used for chicken breeding because natural mating results in adequate fertility levels, but it is routinely used for special breeding and research purposes. However, the use of artificial insemination is becoming more common in breeding broilers, in accordance with the recent advancement of genetic selection for their body confirmation as well as faster growth rates, which hinder natural mating. Artificial insemination has been applied to test the possibility of interspecific hybridization. Interspecific crossing is possible between chickens and several species of Phasianidae, including the Japanese quail (*Coturnix japonica*), the common pheasant (*Phasianus colchicus*), the green pheasant (*Phasianus versicolor*), turkeys and guinea fowls. Interspecific hybridization allows us to study the homologies between gene loci, with visible effects between the two species. For example, the artificial insemination of Japanese quail with chicken semen has been used for testing homologies of loci for plumage colors, and homologies of the *sex-linked imperfect albinism* (*AL*) as well as the *Lavender* (*LAV*) loci in the chicken and the Japanese quail have been demonstrated (Silversides and Mérat 1991; Minvielle et al. 2002).

12.2.2 Poultry Semen Cryopreservation

The greatest progress in commercial semen cryopreservation has been achieved by the dairy and beef cattle industries. Development of methods for semen cryopreservation has made it possible to distribute semen obtained from genetically superior bulls widely, both geographically and temporally; in this way, genetic selection has been accelerated remarkably. In cattle, pregnancy rates achieved by artificial insemination using cryopreserved semen are similar to those achieved after natural mating. In contrast, the fertility rates of frozen poultry semen are far lower than any of the domesticated mammalian species. It has been estimated that post-thawing chicken semen retains only 1.6% of the fertilizing capacity of fresh semen (Whishart 1985). The overall fertility rates with post-thawed poultry semen are highly variable and therefore not reliable enough for use in commercial production or genetic resource conservation. Further complicating the situation the post-thawing viability and fertility of chicken semen varies among breeds and lines (Tajima et al. 1990; Alexander et al. 1993). Up to now, several empirical studies involving different cryoprotectant types and packaging methods, as well as cooling and warming rates, have been carried out for poultry semen cryopreservation. It appears from these studies that of the currently available methods, semen frozen slowly in straws with glycerol or semen frozen rapidly in pellets with dimethylacetamide (DMA) are the most efficient protocols for chicken semen cryopreservation (Tselutin et al. 1999; Chalah et al. 1999). Glycerol has been the most widely used cryoprotectant for chicken semen cryopreservation because of the relatively high post-thawing fertility rates (Lake et al. 1981). However, it must be removed from semen at thawing to avoid its contraceptive effects. Furthermore, a decarease in fertility was found with weekly insemination of post-thawed chicken semen in these protocols. Recently reported protocols showed that chicken semen frozen rapidly as straws with methyl acetamide (MA) could result in improved fertility duration while maintaining acceptable fertility rates (Hanzawa et al. 2010). For example, this freezing protocol has been used in the ex-situ management of foundation stock chickens in the National Livestock Breeding Center, Japan. Turkey semen has also been tested for cryogenic storage employing different cryoprotectants such as glycerol, dimethylsulfoxide (DMSO), ethylene glycol (EG), and DMA, freezing the specimens either as pellets or in straws. However, the fertility levels of frozen turkey semen were lower than those of frozen chicken semen, indicating that turkey spermatozoa are much more sensitive to damage caused by freeze–thawing (Blanco et al. 2000). For duck species, it seems that the spermatozoa of Muscovy ducks are more resistant to freeze–thawing than those of common ducks. Guinea fowl spermatozoa seem very sensitive to cryogenic damage (Váradi et al. 2013). Cryopreservation of quail spermatozoa caused sperm immobility but still maintained fertility following intracytoplasmic sperm injection (ICSI) (Kang et al. 2015). These emphasize the need for optimization of cryopreservation protocols suitable for each poultry species.

The poor fertility of post-thawed poultry sperm can be attributed to several factors such as higher sensitivity to freeze–thawing, deleterious effects of the cryoprotectant on survival, and lower survival during storage in the sperm storage tubules of the oviduct. Carbohydrates on the plasma membrane of avian spermatozoa are significantly altered following cryopreservation, and the degrees of such alterations were affected by the type of cryoprotectant and cooling–warming rates (Peláez et al. 2011). The ability of chicken spermatozoa to undergo the acrosome reaction was also altered by the type of cryoprotectant and freeze–thawing (Mocé et al. 2010). The fluidity of the sperm plasma membrane is one of the predictors for the success rate of chicken semen cryopreservation. Membrane fluidity is affected with a significant decrease in cholesterol/phospholipid ratio following freeze–thawing (Blesbois et al. 2005). Retention of mitochondrial integrity and function are also critical for successful cryopreservation of avian spermatozoa because mitochondria provide energy for the motility and survival of spermatozoa in the oviduct. Further studies are required to develop a method for poultry semen cryopreservation that maintains the integrity of spermatozoa upon freeze–thawing and their longevity once inside the sperm storage tubules of the oviduct.

12.3 Culture of Avian Embryos

In most mammals, fertilization and subsequent embryo development occur within the mother. In avian species, fertilization takes place in the infundibulum of the oviduct, after which a yolk-laden zygote undergoes blastoderm formation while egg formation is completed by the deposition of albumin, uterine fluid, chalaza, shell membrane, and shell. After oviposition, an avian embryo develops in a hard-shelled egg until hatching. The embryonic lifespan of the chicken is 22 days. Development in the first day takes place in the oviduct, and in the remaining 21 days in the egg. The developmental stage at the timing of oviposition differs between avian species. Zebra Finch (Taeniopygia guttata) eggs are laid at an earlier developmental stage than in the chicken (Mak et al. 2015). The hatching time also varies between avian species. For example, the hatching time-periods for Japanese quail, Mallard (Anas boschas), Adelie penguin (Pygoscelis adeliae) and Emu (Dromaius novaehollandiae) are 16, 28, 35 and 50–56 days respectively. The embryonic development of these species is relatively well conserved morphologically, but the speed varies widely.

Currently, most of the methodologies used in avian biotechnology involve the accession to and manipulation of developing embryos. There are two methods of accessing developing avian embryos: the shell windowing method and the ex-vivo culture method.

12.3.1 Shell Windowing

A freshly oviposited avian embryo can be accessed through a hole or "window" in its own egg-shell. Historically, a window approximately 5 mm in diameter was ground in the shell at the top of the horizontal of a freshly oviposited egg and sealed with the shell membrane after manipulation. However, this classical windowing method severely compromises avian embryo survival, with hatching rates as low as a few percent, probably resulting from air bubbles trapped in the albumin during windowing. Entrapment at the blastoderm stage has been shown to cause deaths and embryonic defects. Improvements in hatchability (around 30–45%) of manipulated freshly oviposited embryos have been achieved by avoiding the entrance of air bubbles during opening and sealing of a window (Speksnijder and Ivarie 2000). The shell windowing method is also applicable to access avian embryos at the somitogenesis stage. Briefly, a window with a diameter of ~20 mm is opened in the narrow end after pricking the shell at the blunt end. After manipulation of an embryo, a window is sealed with cling-film wrap using thin egg albumin as a paste. This conventional windowing method at the somitogenesis stage yields a hatchability of chicken embryos at as high as >70% (Nakamura et al. 2010a). Therefore, this is the best method for producing germline chimeras through intra-vascular injection of PGCs (see Sect. 12.4.1).

12.3.2 Ex-Vivo Culture of Avian Embryos

An ex-vivo culture system of an embryo from the single-cell stage—normally in the oviduct—through to hatching was first established for the chicken by Perry (1988). Here, an updated ex-vivo culture protocol for chicken embryos from the zygote stage to hatching is introduced. Chicken embryonic development is divided into three periods for the purpose of ex-vivo culture: fertilization to blastoderm formation (I) lasts for 1 day, embryogenesis (II) lasts for 3 days, and embryonic growth (III) lasts 18 days. The culture system is divided into three steps, corresponding to these three periods. The first culture system (I), covering the oviductal phase, was designed to provide an environment similar to that in vivo. The culture medium consists of thin albumin obtained from freshly laid eggs, glucose, and a salt solution. The composition of salts in the culture medium (50 mM $KHCO_3$, 30 mM $NaHCO_3$, 10 mM KCl, 2.5 mM $MgCl_2 \cdot 6H_2O$, 0.7 mM $CaCl_2 \cdot 2H_2O$) is formulated, based on the differences in ionic composition of albumin in uterine eggs and in laid eggs. The concentration of glucose in the culture medium (11 mM) is maintained at a level compatible with that in uterine fluid. This culture medium can be replaced with freshly collected thin albumin. The fertilized chicken egg surrounded by thick albumin (8–16 ml) obtained from the magnum is cultured for 24 h at 41–42 °C in a tightly sealed 60-ml glass jar or plastic cup, with the culture medium added to the equatorial level of the ovum. To extend the course of development into embryogenesis, transfer of a cultured embryo to a host eggshell is required (system II). After transferring the contents of a glass jar in system I to a surrogate eggshell prepared from a freshly laid egg by cutting off the sharp end (diameter of hole, 33 mm), the shell is filled with thin albumin, and sealed with cling-film secured by plastic rings and rubber bands. The hatching rate is increased by removing the thick albumin from cultured eggs at the blastoderm stage (Naito et al. 1990). The reconstituted eggs are incubated for around 3 days at 38 °C and 50–60% relative humidity, with 90° tilting in hourly cycles. For further development, an embryo cultured in system II is transferred again to a larger host eggshell containing an air space to expose the vascular systems of the extra-embryonic membranes to the atmosphere (system III). After transferring the contents of a reconstituted egg from system II to a larger surrogate shell prepared by cutting off the blunt end (diameter of hole, 40–45 mm), a portion of culture medium or thin albumin is discarded to adjust to the normal volume of the egg. The open surface of the shell is sealed tightly and the embryos are cultured for 15 days at 38 °C and 50–60% relative humidity, with 30° tilting in hourly cycles, then maintained at 37.5 °C in a stationary position for the remaining period. The embryos cultured ex vivo rarely hatch out without human assistance. When an embryo has held the chorioallantoic membrane with its beak, about 10 holes are pricked in the cling-film with a fine needle. Immediately before hatching, the cling-film is removed and the open surface is covered with a plastic dish. Embryos will get out from the surrogate shell by themselves. Ex-vivo culture of zygotes obtained from the oviduct (systems I–III) yielded about 35 and 25% hatchability in chicken and Japanese quail respectively. Surrogate eggshell culture of freshly oviposited fertilized eggs (systems II and III) resulted in about 50 and 40% hatchability in chicken and Japanese quail respectively (Naito et al. 1990; Ono et al. 1994). This surrogate eggshell culture system from the blastoderm stage through to hatching has now extended to the blue-breasted quail (*Coturnix chinensis*) and bobwhite quail (*Colinus virginianus*), with hatching rates of about 25 and 40% respectively (Ono et al. 2005; Kato et al. 2013). Artificial vessels such as those made from gas-permeable membranes can be used as substitutes for system III surrogate eggshells (Kamihira et al. 1998; Tahara and Obara 2014). In the shell-less culture system, calcium supplementation throughout development, as well as oxygen supplementation for the last 4 days until hatching, are critically important for hatching. A recently modified shell-less culture method for chicken embryos using a food wrap (polymethylpentene film) that corresponds to Perry's system III yielded a hatchability of 57.1% (Tahara and Obara 2014).

Ex-vivo culture enables more easy access and precise manipulation of the developing avian embryos. Successful generation of transgenic chickens generated by DNA microinjection into the cytoplasm of the germinal disc of zygotes, followed by ex-vivo culture, has been reported (Love et al. 1994). The two-step system (systems II and III) has been applied widely to produce somatic as well as germline chimeras by transplantation of donor blastoderm cells to the subgerminal cavity of recipient embryos after oviposition (see Sect. 12.4.1).

Microinsemination is a technique for fertilizing ova by micromanipulation. ICSI is now a well-established technique in mammals, and has been used successfully to produce offspring in 16 mammalian species, including humans (*Homo sapiens*). Recently, it was reported that the first ICSI-generated birds were produced by microinjection of a single spermatozoon together with a sperm extract, followed by ex-vivo culture (Mizushima et al. 2014; see Chap. 7). In chickens, sex reversal from a genotypic female to a phenotypic male bird can be introduced by treating an early chicken embryo with a non-steroidal aromatase inhibitor. These sex-reversed chickens possess bilateral testes that are capable of complete spermatogenesis, but with a fully developed oviduct, not a vas deferens. Interestingly, sex-reversed chickens produce spermatids and spermatozoa carrying the female-specific W chromosome and their fertility has been demonstrated, but the fertilization efficiency remains unknown (Takagi et al. 2007). In future research, the fertilizing and developmental abilities of chicken W-bearing spermatozoa should be elucidated using ICSI coupled with ex-vivo culture.

12.4 Avian Germline Chimeras

Cell-based technologies have expanded the limits of what is possible, beyond what occurs in normal reproduction. In mammals, cloning by somatic cell nuclear transfer has been used to produce offspring in at least 20 species, and has been widely used to produce model animals for research, in particular for the pig (*Sus scrofa domesticus*). In mice (*Mus musculus*), stem cells represented by embryonic stem (ES) cells and induced pluripotent stem (iPS) cells, that have the ability to differentiate into any cell type exclusive of the placental lineage, have been used widely not only to generate several genetically modified animals particularly in mice, but also in regenerative medicine. Until now, these fundamental techniques widely used in some mammalian species have not been established for birds.

The term "chimera" is derived from the name of the Greek mythical monster that had the body of a lion, the head of a goat, and a tail ending in a snake's head. In biology, "chimera" is an informal term for a single organism composed of more than two cells from different zygotes. Chimera production technologies are now essential for avian biotechnologies.

12.4.1 Production of Avian Germline Chimeras

One of the most important aspects of cell-based technologies is that manipulated cells can contribute to the germline in particular gametes. Unlike in mice, where the germline is specified by inductive signals from surrounding somatic cells controlled by zygotic genes—the "epigenesis" model—the avian germline is thought to be specified by the incorporation of maternal determinants (germplasm): the "preformation" model (see Chap. 1). Therefore, it follows that the germ cell is the target cell type for avian biotechnologies. Following the development of PGC transplantation in chickens (Tajima et al. 1993), avian biotechnologies have developed based on the micromanipulation of PGCs. Avian PGCs first appear in the center of the blastoderm at the freshly oviposited stage and then traverse through several developing somatic tissues and vasculature on their migration to the emerging gonads (Nakamura et al. 2007, 2013a). Germline chimeras can be produced through the transplantation of PGCs to the appropriate sites of developing recipient embryos where endogenous PGCs will be located. There are generally two ways of producing germline chimeras: transplantation of donor cells into the subgerminal cavity at the blastoderm stage, or into the bloodstream at the somitogenesis stage.

12.4.1.1 Subgerminal Cavity Injection

The subgerminal cavity is a space between the blastoderm and yolk of avian eggs, which is created when the blastoderm cells absorb water from the albumin and secrete the fluid between themselves and the yolk (Fig. 12.1a, b). Somatic as well as germline chimeras can be produced by transfer of blastoderm cells into the subgerminal cavity of recipient embryos at stage X (Roman numerals from I to XIV are used to classify embryonic stages before formation of the primitive streak, according to Eyal-Giladi and Kochav 1976; Arabic numerals from 2 to 46 are used for later embryos to the hatching stage following Hamburger and Hamilton 1951). Because the area pellucida of stage X blastoderm forms most of the actual embryos, transplanted blastodermal cells can contribute to all somatic tissues as well as the germline in recipients. For the production of germline chimeras, donor cells should be taken from the central part of the area pellucida of stage X blastoderm where PGCs are localized (Petitte et al. 1990; Kagami et al. 1997). Ex-vivo culture (systems II and III above) allows the precise manipulation of recipient embryos, and thereby facilitates the production of both somatic and germline chimeras (Fig. 12.1c–e). Somatic chimerism is not useful in predicting germline chimerism because of a mismatch between their degrees. To ensure germline chimeras, a fairly large number of somatic chimeras should be produced. Although isolated PGCs can repopulate the gonads after transplantation into the subgerminal cavity (Mozdziak et al. 2006), this is an indirect and impractical method.

Injection into the subgerminal cavity is a useful method for assaying the contributions of transferred cells to three germ layers (ectoderm, mesoderm, and endoderm) and germline in vivo. Chicken blastoderm cells can be cultured and maintained in vitro in an undifferentiated state on a feeder of irradiated Sandoz inbred mouse-derived thioguanine-resistant and ouabain-resistant cells (STO; a mouse embryonic fibroblast cell line) in the presence of leukemia inhibitory factor (LIF), basic fibroblast growth factor (bFGF), insulin growth factor (IGF)-1, stem cell factor (SCF), interleukin (IL)-6 and IL-11. Long-term cultured chicken blastoderm cells can contribute to the three germ layers both in vitro and in vivo, but germline competency seems to be lost rapidly in culture systems (Pain et al. 1996). This could be because of the preformation system of avian germline specification in that their linage has already been determined as the

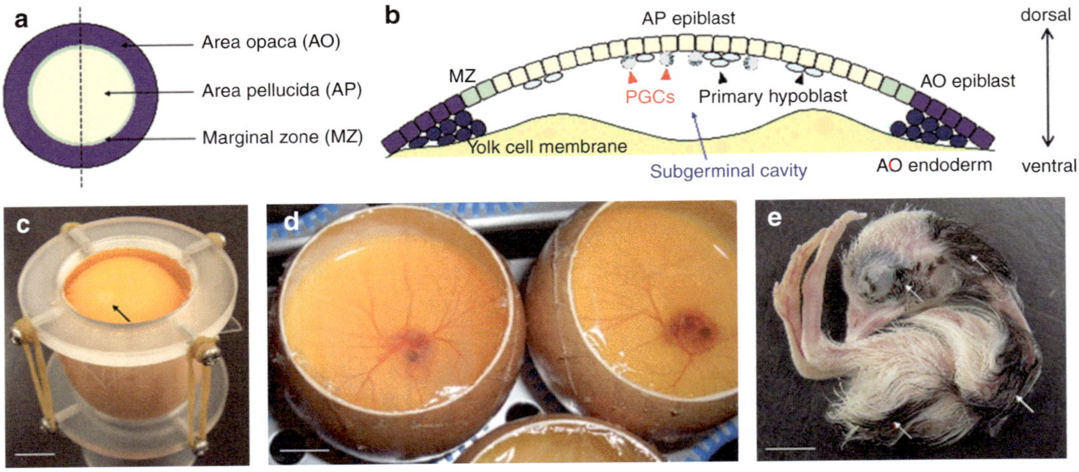

Fig. 12.1 Structure of the chicken blastoderm, and somatic chimera production by subgerminal cavity injection of donor blastoderm cells. (**a**) Dorsal view of chicken blastoderm stage X. (**b**) Sagittal view of the same. (**c**, **d**) Chicken embryos cultured ex vivo using systems II (**c**) and III (**d**) after subgerminal cavity injection of donor blastoderm cells (see Sect. 12.3.2). The *black arrow* indicates the blastoderm at stage X. (**e**) A somatic chimera embryo produced by transplantation of blastoderm cells from a Barred Plymouth chicken embryo (adults have black feathers) into the subgerminal cavity of a recipient White Leghorn chicken embryo (white feathers). *White arrows* indicate donor-derived black feathers. *Scale bars* = 1 cm; *PGCs* primary germ cells

germline by the incorporation of germplasm, and thereby separated from a somatic fate. In this culture system, chicken PGCs do not survive, while the remaining epiblast-derived cells, which are already committed to somatic lineages, propagate for the long-term and maintain their abilities to contribute to all three germ layers.

12.4.1.2 Intravascular Injection

Unlike other species, in avian and some reptile embryos, PGCs use the blood circulation for transport to the gonadal anlage. This unique biological property and accessibility of avian PGCs during early development provides an opportunity to manipulate them. Germline chimeras can be produced by intravascular injection of PGCs. Because the blood circulation of avian PGCs is transitory, PGC transplantation is feasible only during their circulation period. For example, the entrance of chicken PGCs from the germinal crescent region to the vascular network starts at stage 10 and is completed at stage 13. Conversely, the migration of circulating PGCs to the gonadal anlage starts at stage 15 and is complete at stage 17 (Nakamura et al. 2007). However, access to the developing vasculature is technically difficult until a chicken embryo reaches stage 13. In addition, intravascular transplantation of chicken PGCs to stage 17 and later embryos to be lowering or fail of germline transmission. Taking these into consideration, the appropriate time for transplanting chicken PGCs is in embryo stages 13–16.

Avian PGCs obtained from different sites such as the germinal crescent region and gonads can migrate to the gonads and then differentiate into functional gametes after intravascular transplantation (Vick et al. 1993; Tajima et al. 1998). However, the germinal crescent region is the least useful source for chimera production, given the low numbers of PGCs during this period. Both male and female PGCs collected from 5- to 7-day-old chicken embryos (stages 27–31) could be used as donor cells for the production of germline chimeric chickens (Tajima et al. 1998; Nakajima et al. 2014). Male germ cells obtained from adult chicken testes, possibly spermatogonial stem cells (SSCs), have the ability to migrate toward recipient embryonic gonads, and subsequently differentiate into functional spermatozoa following transplantation (Jung et al. 2010). Thus, avian germ cells have developmental plasticity. In contrast to spermatogenesis in chickens, oogenesis is not supported by a stem cell system because all of the germ cells in female gonads enter meiosis and stop at the diplotene stage (the fourth stage of meiotic prophase) after hatching. In germ cell transplantation, only some of the mitotic germ cells would contribute to repopulating the recipient gonads. Germ cell meiosis in female chicken embryos occurs well after the onset of gonadal somatic sex differentiation, which starts at 6 days of incubation (stage 29). Meiosis of female chicken germ cells begins at 15.5 days of incubation (stage 41), when ovarian differentiation is well advanced, and meiotic germ cells are abundant by 18.5 days of incubation (stage 44; Smith et al. 2008). Although a few female germ cells at 20 days of incubation still retain the ability to repopulate gonads after intravascular transplantation (Naito et al. 2007), PGCs obtained from earlier stages in development are much more useful for chimera production.

12.4.2 Improvement of Germline Transmission of Chimeras

The practicality of applied studies such as the generation of genetically modified birds and conservation of avian genetic resources is affected by the efficiency of producing functional donor-derived gametes in germline chimeras. In the gonads of such chimeras, there is competition between the two populations of donor and recipient germ cells. Three factors are involved in this competition: (1) the sex combinations of donor PGCs and the recipient embryo, (2) the combinations of donor and recipient breeds or lines, and (3) the proportion of donor and recipient PGCs.

In fish, the male germ cells (probably SSCs) and female germ cells (probably oogonial stem cells) that are transplanted to the opposite sex of newly hatched embryos can differentiate into functional eggs and spermatozoa (Okutsu et al. 2006; Yoshizaki et al. 2010). Although chicken PGCs can differentiate into functional gametes in the gonads of opposite-sex recipients, the efficiency is much lower than in same-sex recipients (Naito et al. 1999). Histological analysis suggested that female PGCs in male chicken gonads are capable of passing through the first and second meiotic divisions, but rarely complete spermatid elongation (Tagami et al. 2007) (see Chap. 1). For the purpose of efficient germline chimera production, the sex of the donor PGCs should be matched to that of the recipient.

Germline chimerism is affected by the combinations of donor and recipient breeds or lines. For example, intravascular transplantation of PGCs obtained from White Leghorn chicken embryos to recipient embryos of the Barred Plymouth Rock chicken breed resulted in a 3.5-fold higher germline transmission rate than the opposite combination of donor and recipient chicken breeds (Naito et al. 1994b). A significant difference in germline chimerism was also reported when PGCs obtained from an indigenous chicken breed were transplanted into recipient embryos of two different White Leghorn chicken lines (Nakamura et al. 2010b). The selection of recipient breeds or lines is important for the efficient production of donor-derived offspring from germline chimeras.

The proportion of donor-derived gametes is basically determined by the ratio between the numbers of donor and recipient germ cells. Two approaches are available to improve germline transmission rate: (1) increasing the numbers of donor PGCs transplanted, and (2) decreasing the numbers of endogenous PGCs in recipient embryos. The effective outcomes in the former and the latter cases are enrichment of PGCs and removal of PGCs prior to transplantation respectively. Recent advances in chicken PGC culture in vitro (see Chap. 1) enabling the transplantation of a huge number of donor germ cells also contribute to the former approach (van de Lavoir et al. 2006; Whyte et al. 2015).

12.4.2.1 Purification of Avian PGCs

PGCs comprise a very small cell population during early avian development (Fig. 12.2a). In the chicken, PGCs comprise less than 0.02% of blood cells at 52 h of incubation (stages 13–16)

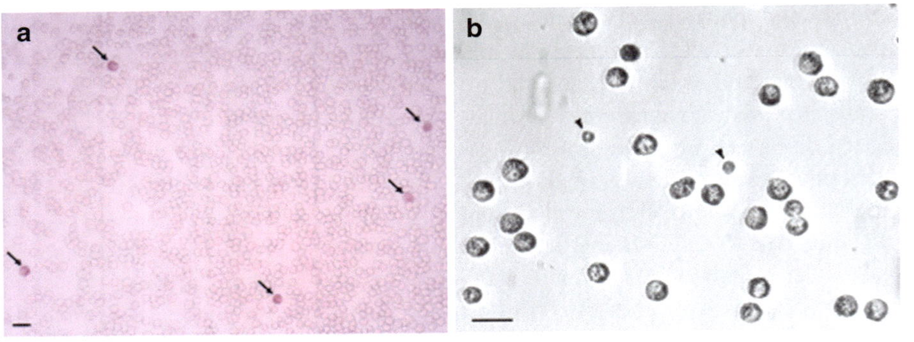

Fig. 12.2 Purification of chicken PGCs from early embryonic blood. (**a**) A small number of periodic acid-Schiff (PAS) positive PGCs (*arrows*) and numerous PAS-negative erythrocytes are found in the blood of 2-day-old chicken embryos. (**b**) Chicken PGCs can be purified from erythrocytes (*arrowheads*) by density gradient centrifugation. *Scale bars* = 20 μm

and 2% of gonadal cells at 5 days (stage 27). Because of this very low concentration of PGCs, transfer of intact blood or gonadal cells to recipient embryos results in a very low efficiency of producing germline chimeras. To date, there are roughly four types of method to concentrate avian PGCs.

The first is a purification method based on antigen–antibody reactions such as magnetic-activated cell sorting (MACS) or florescence-activated cell sorting (FACS). Recent investigations of cell surface markers antigens of chicken PGCs such as stage-specific embryonic antigen-1 (SSEA-1), embryonic mouse antigen-1 (EMA-1), and c-kit allow the enrichment of chicken PGCs for transplantation (Karagenç et al. 1996; Miyahara et al. 2016). The QCR1 antibody (Aoyama et al. 1992), which reacts with an epitope in PGCs of Japanese quail and the common pheasant, but does not react with chicken PGCs, enables PGC isolation in these species (Ono and Machida 1999; Kang et al. 2008). The primary advantage of this method is that PGCs can be isolated from various tissues that contain them, such as blastoderm, germinal crescent, blood, and gonads. In future, further investigations of cell surface markers of PGCs in several avian species other than the chicken, Japanese quail, and common pheasant are expected.

The second method is purification of avian PGCs by density gradient centrifugation. This method has been used to concentrate PGCs from early embryonic blood (Fig. 12.2b). Although Ficoll and Nycodenz have been used in density gradients, Nycodenz has advantageous chemical properties over Ficoll such as low toxicity, high water solubility, and stability when autoclaved. In addition, Nycodenz gradient centrifugation achieves both high recovery (over 70%) and purity (70–90%) rates for PGCs in the chicken and Japanese quail (Zhao and Kuwana 2003). Nycodenz gradient centrifugation is further applicable to PGC enrichment in various avian species such as ducks, the green pheasant, and the common pheasant (Nakamura 2016). In addition to this, Percoll density gradient centrifugation is useful for purifying cultured chicken PGCs from feeder cells (Oishi 2010).

The third method is to enrich avian PGCs by hemolysis. An ammonium chloride–potassium (ACK) lysis buffer was originally used in mammals to purify bone marrow cells from whole blood. An ACK lysis buffer consisting of 150 mM NH_4Cl, 1 mM $KHCO_3$, and 1 μM EDTA can be used to disrupt chicken embryonic erythrocytes and thereby enable the enrichment of PGCs (Yamamoto et al. 2007).

The fourth method is avian PGC enrichment from gonadal tissues. In this method, chicken PGCs are discharged from embryonic gonads following incubation in Dulbecco's phosphate-buffered saline without Ca^{2+} or Mg^{2+} (D-PBS(−)) (Nakajima et al. 2011). This is the simplest method currently available to enrich chicken PGCs and is potentially applicable to other avian species. In future, further improvement of the currently unpredictable PGC recovery rate is expected.

12.4.2.2 Removal of Endogenous PGCs

The ideal recipient for PGC transplantation is an infertile embryo. Indeed, several infertile animal models, such as germ-cell-deficient mutant mice and triploid fishes, have been used widely in germ cell transplantation studies to produce only donor-derived offspring (Brinster and Zimmermann 1994; Okutsu et al. 2007). As far as I know, genetic mutants either lacking germ cells or with deficient gametogenesis have not been reported in avian species. ZZZ triploid chickens develop and grow normally but are infertile (Lin et al. 1995). Therefore, such embryos have potential applications to use as a recipient for PGC transplantation, but are obtained only rarely. Therefore, partial if not complete elimination of endogenous PGCs prior to PGC transplantation is an effective approach to increase the efficiency of germline chimerism in recipients. Presently, three types of methods are capable of reducing the numbers of endogenous PGCs in recipient embryos.

In the first method, a piece of tissue containing PGCs is removed surgically from recipient embryos. Surgical removal of blastodermal cells from the center of the area pellucida at stage X is an effective method to increase both somatic and germline chimerism following the transplantation of donor blastoderm cells (Kagami et al. 1997). Drawing blood from embryos at stages 13–16 prior to transplantation of donor PGCs enhances germline chimerism (Naito et al. 1994b). Because collected tissue samples are available for sexing, these procedures enable PGC transplantation while knowing the sex combinations of donors and recipients.

The second method involves irradiation. Electromagnetic waves such as ultraviolet, X-, and gamma rays are known to induce DNA damage, cell cycle arrest, and apoptosis. Susceptibility to irradiation varies depending on the type of cells and tissues, and germ cells are much more sensitive than somatic cells. The effects of such sterilization on endogenous PGCs does not much affect the developmental stages of embryos, whereas the hatchability increases with the age of the embryos at the time of irradiation. Exposure of chicken embryos at stages 13–15 to 3 Gy of X-rays is one of the effective protocols to improve the germline transmission of donor PGCs with acceptable hatchability (Nakamura et al. 2012). However, irradiation in the mid-point of incubation of recipient embryos is only feasible if appropriate radiation apparatus is in the laboratory, because the transplantable time-period is very limited. A protocol using the irradiation of freshly oviposited, non-incubated eggs with 5–7 Gy of gamma rays has been used to produce high-grade germline chimeras (Carsience et al. 1993; Oishi et al. 2016). In this protocol, irradiated eggs can be stored at 14 °C for about 2 weeks, and are thereby available to shipping after irradiation.

The third method is to remove endogenous PGCs following application of a drug. Busulfan (1,4-butanediol dimethanesulfonate), a DNA alkylating agent, is the most commonly used to remove endogenous germ cells in mammals. For example, in mouse SSC transplantation, a single intraperitoneal injection of >40 mg/kg busulfan to an adult male mouse is the standard protocol for preparing a recipient lacking endogenous germ cells. Busulfan also sterilizes avian germ cells during development (Reynaud 1969; Aige-Gil and Simkiss 1991). However, variable degrees of germ cell depletion because of inefficient drug delivery have prevented the use of busulfan in avian species. A unique drug delivery method was developed for chicken embryos by utilizing the biological property of the chicken embryo, which remains on the top of the yolk even when the egg is rotated (Nakamura et al. 2008). The principle of this method is as follows: a sustainable emulsion containing busulfan rises readily when it is injected into the yolk, then makes contact with the chicken embryos, which are lighter than the yolk contents (Fig. 12.3a). This method allows the depletion of PGCs at a constant level in a dose-dependent manner (Nakamura et al. 2010a). Administration of 100 μg of busulfan in 50 μl of vehicle is a reliable protocol for preparing recipient chicken embryos lacking

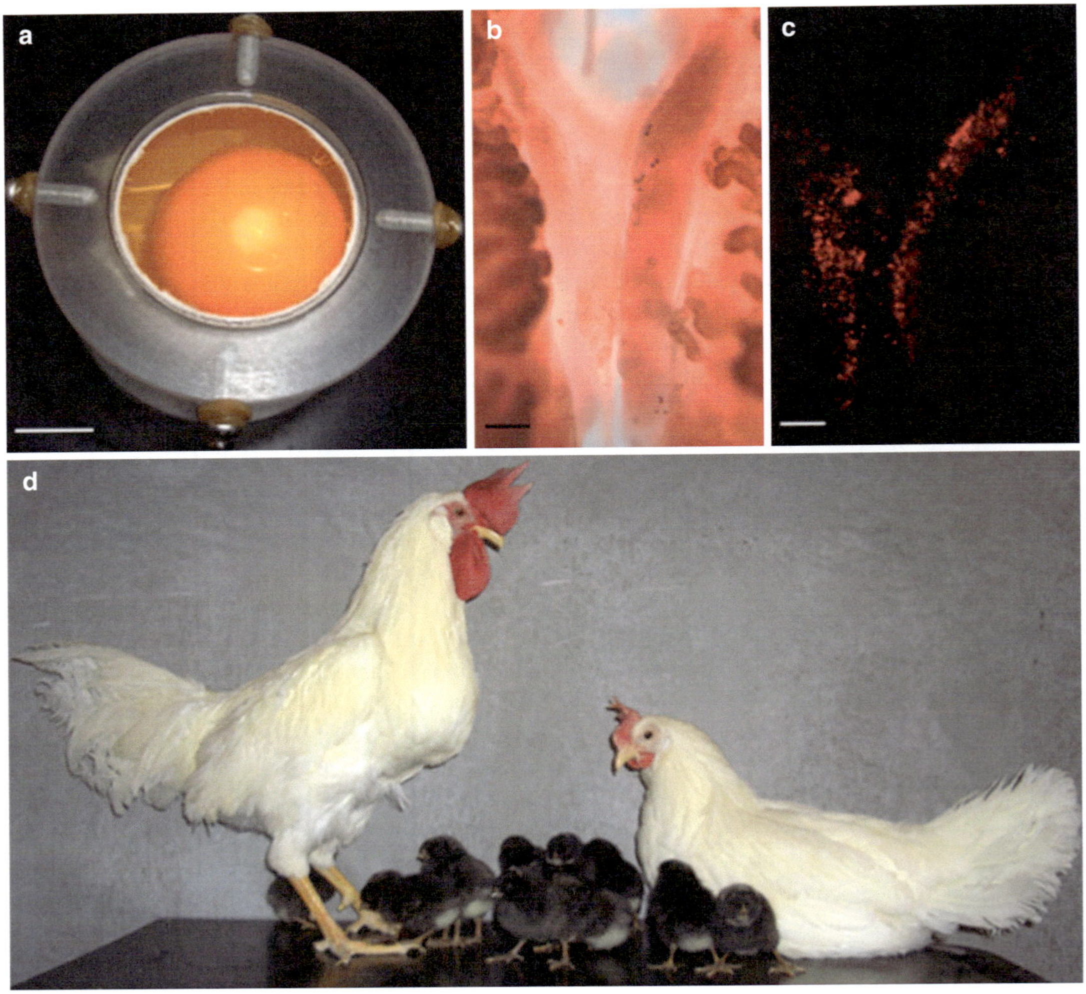

Fig. 12.3 Production of germline chimeric chickens that produce only donor-derived gametes. (**a**) An emulsion containing busulfan makes contact with the chicken embryos rapidly after injection into the yolk. (**b**) This method allows the continuing production of sterile chicken embryos. *Black spots* indicate PGCs stained with an antibody raised against Vasa, a pan-germ cell marker. (**c**) Donor PGCs can repopulate gonads of sterile recipient embryos following transplantation. *Red spots* indicate donor PGCs stained with a fluorescent dye PKH26. (**d**) Male (*left*, *white chicken*) and female (*right*, *white chicken*) germline chimeras that yielded only donor-derived offspring (*center*, *black chicks*). *Scale bars* = 1 cm (**a**) and 100 μm (**b** and **c**). (**b–d** from Nakamura Y et al. (2010b), courtesy of Tagami T)

endogenous PGCs, with acceptable hatchability (Fig. 12.3b). After transplantation, donor PGCs can migrate toward and repopulate the sterile gonads (Fig. 12.3c). Germline chimeric chickens that yield only donor-derived gametes can be produced following transplantation of 200 intact PGCs (Fig. 12.3d).

In the future, the generation of genetically modified chicken lines where the germ cells are depleted completely is expected for more efficient production of recipient chickens that produce only donor-derived offspring. For example, in combination with transgenesis of cultured PGCs in vitro (see Sect. 12.5.1), generation of germ-cell-specific knockouts mediated by a Cre recombinase/loxP system or avian PGC ablation via germline chimeras would be the best approaches.

12.4.3 Interspecific Germline Chimeras

In vertebrates, germline chimeras can be produced by germ cell transplantation to the same species (allogeneic) and in cross-species situations (xenogeneic). In fish, successful production of interspecific germline chimeras by transplantation of PGCs or SSCs has been reported in salmonid species between the rainbow trout (*Oncorhynchus mykiss*) and masu salmon (*Oncorhynchus masou*), and in cyprinid species between the pearl danio (*Danio albolineatus*) and zebrafish (*Danio rerio*) (Okutsu et al. 2007; Saito et al. 2008). These xenogeneic fish germline chimeras produced functional donor-derived gametes. Mammalian xenogeneic germline chimeras can be produced by microinjection into blastocysts of inner cell mass cells, or pluripotent stem cells such as ES or iPS cells, or by the transplantation of SSCs into seminiferous tubules (hitherto "intratubular" transfer). Mammalian xenogeneic spermatogenesis in mouse seminiferous tubules has been well studied using the latter method. Transplantation of SSCs obtained from rodents such as rats (*Rattus norvegicus*) and Syrian hamsters (*Mesocricetus auratus*) could lead to spermatogenesis in the testes of immunodeficient mice (Clouthier et al. 1996; Ogawa et al. 1999), whereas meiotic differentiation could not be achieved in case of SSC transplantation obtained from non-rodent animals such as rabbits (*Oryctolagus cuniculus*), cows, horses (*Equus caballus*), baboons (*Simia hamadryas*), and humans (Dobrinski et al. 1999, 2000; Nagano et al. 2002). Hamster spermatogenesis in mouse seminiferous tubules is less efficient, and more defective spermatozoa are observed compared with rat spermatogenesis in the mouse. Rats and mice diverged 10–11 million years ago, and the mouse–hamster lineage separation was about 16 million years ago (Catzeflis et al. 1993). It appears that increased phylogenetic distance between the donor and recipient species in mammals leads to incompatibilities between transplanted germ cells and the testicular environment (Sertoli cells and paracrine factors) of the recipient.

In avian species, chicken–Japanese quail interspecific chimeras have been used commonly to follow cell lineages using in-situ markers. However, no donor-derived offspring have been obtained from recipients when Japanese quail blastoderm cells or PGCs were transplanted to chickens. The key ability of xenotransplanted PGCs to migrate to the gonads is well conserved among avian species. To date, there are a number of anecdotal reports on the production of viable progeny derived from xenotransplanted PGCs through testcross analysis of recipients. In these reports, xenogeneic germline transmission of transplanted PGCs occurred across genera (common pheasant to chicken), families (chicken to guinea fowl) or orders (duck to chicken, and houbara bustard [*Chlamydotis undulate*] to chicken), but only in males, not in females (Kang et al. 2008; Wernery et al. 2010; Liu et al. 2012; van de Lavoir et al. 2012). Although additional histological evidence is required to confirm xenogeneic spermatogenesis, these reports open up the possibility of producing full spermatogenesis in the gonads of other avian species. There are also reports on histological observations of the existence of oogonia and oocytes in primary follicles derived from female chicken PGCs in the ovaries of Japanese quail or duck prepubertal recipients (Liu et al. 2007, 2012). It is possible that xenotransplanted female PGCs are capable of entering recipient gonads and subsequently starting follicle formation, but most interspecific follicles either disappear before maturation, or their development is suppressed in the primitive follicle reservoir pool or at a later stage. In the future, the fate of xenotransplanted PGCs in recipient testes or ovaries should be investigated in more detail. Transplantation of donor PGCs carrying a reporter gene to sterile interspecific recipient embryos would provide important information on the ability of xenogeneic gametogenesis through sequential histological analysis.

12.4.4 Intratubular Transplantation of SSCs

SSCs maintain spermatogenesis in the testis throughout life, and are characterized by the ability to self-renew and to produce differentiating cells that ultimately form spermatozoa. In 1994, Brinster and colleagues developed a technique in which transplanted mouse SSCs were able to reconstitute spermatogenesis in the seminiferous tubules of the testes of infertile recipients, and this produced viable progeny (Brinster and Zimmermann 1994; Brinster and Avarbock 1994). Application of intratubular SSC transplantation to mammalian species other than rodents has so far been reported in pig, goat (*Capra hircus*), cattle, and the rhesus monkey (*Macaca mulatta*) (Honaramooz et al. 2002, 2003, Izadyar et al. 2003; Hermann et al. 2012). In 2006, the first successful transplantation of avian SSCs was reported in chicken (Trefil et al. 2006). For this, the preparation of sterile recipient roosters is required for the successful colonization of donor SSCs through intratubular transplantation. Endogenous germ cells in the testes of roosters could be depleted following repeated irradiation with moderate doses of gamma rays (five irradiations of 8 Gy each). Transplantation of dispersed testicular cells obtained from roosters could result in the reconstitution of donor-derived complete spermatogenesis in about half of the recipients. These were able to produce viable offspring through artificial insemination, but the success rate was less than 10%. This technological development of chicken SSC transplantation provides insight into the potential preservation of male fertility and avian transgenesis using SSCs. This technique further allows a straightforward quantification of SSCs. Donor-derived spermatogenesis is reconstituted after transplantation in the form of colonies in the recipient testes. Because each colony arises from a single SSC (Zhang et al. 2003; Kanatsu-Shinohara et al. 2006), the number of spermatogenic colonies is equivalent to the number of SSCs that have successfully colonized in the recipient testis. In future, the investigation of chicken SSC populations based on spermatogenic colony formation after transplantation is anticipated. Such investigations should enable interested groups to increase colony numbers through enrichment of SSCs, thereby improving the recovery of fertility.

12.4.5 Gonadal Tissue Grafting

As described above, donor germ cells are maintained in the gonadal environment involving supporting cells and paracrine factors of recipients after the transplantation of PGCs or SSCs. It is likely that the avian testes can provide a suitable environment for xenotransplanted male germ cells to differentiate into functional spermatozoa. In contrast, xenogeneic female PGC transplantation did not result in complete ova production in the ovaries of recipients, probably because of an incompatibility between donor germ cells and the recipient ovarian environment. Co-transplantation of the donor germ cells with their surrounding ovarian tissue into recipient hens restored the ovarian environment. The possibility of ovarian transplantation was first reported in 1940 by Robertson in an inbred strain of mice maintained by forced heterozygosis. This ovarian transplantation technique has been improved and applied to other mammalian species including livestock and humans. In avian species, a surgical technique for orthotropic transplantation of allogeneic ovarian tissue obtained at the post-hatching stage was developed for the chicken by Song and Silversides (2006). Briefly, donor ovarian tissues obtained from post-hatching chicks are transferred into a recipient post-hatching chick in the original position of the ovary, and are covered with the greater abdominal air sac after surgical removal of the recipient's ovary. More than 70% of manipulated recipients were able to recover fertility and produce eggs in the long term, but the production rate was approximately one-fifth of that in control birds. Of these, donor-derived offspring were obtained in about half of the laying recipients at a relatively high frequency.

This orthotropic transplantation of ovarian tissue was extended to Japanese quail (Song and Silversides 2008). Avian orthotropic ovarian transplantation provides a means for preserving female fertility preservation in birds. Both slow freezing and vitrification (see Sect. 12.6.1) are available for cryopreserving immature ovarian tissues and maintaining reproductive potential. Vitrification resulted in greater efficiency as measured by cell viability, tissue morphology, and overall reproductive performance of the cryopreserved transplants, compared with their slow-frozen counterparts (Liu et al. 2010).

This technique was further applied to evaluate the feasibility of xenogeneic transplantation of avian ovaries across phylogenetic genera. Xenografting of ovarian tissue from newborn Muscovy ducks to Pekin duck recipients, with subsequent treatment with an immunosuppressant, resulted in the production of offspring derived from a Muscovy duck ovarian transplant through artificial insemination of Muscovy duck spermatozoa (Song et al. 2012). It appears that Muscovy duck spermatozoa and the reproductive tract of the Pekin duck are compatible. Thus, hybrid offspring can be obtained by artificial insemination of female ducks with Muscovy duck spermatozoa (see Sect. 12.2.1). More importantly, this successful example of avian xenografting suggests that the endocrine system can support the xenogeneic development of oocytes, at least between these two species. In a similar fashion, it was possible to produce live progeny from xenografted mouse ovarian transplants into rat recipients (Snow et al. 2002). It is also known that testicular maturation and sperm production in ectopically grafted testicular tissues of large animals is accelerated by exposure to the endocrine environment of the castrated adult mouse recipients because of accelerated maturation of the testicular somatic cells, whereas the length of the seminiferous cycle remains unchanged. These results support the possibility that the endocrine systems controlling reproduction are widely conserved across avian and mammalian species. If this is correct, it follows that the differences in ovarian somatic cells between donor and recipient breeds affects the development of xenogeneic female germ cells in birds more seriously.

Apart from immature ovarian tissues, allografting of immature testicular tissues under the back skin of roosters could result in full spermatogenesis (Song and Silversides 2007). However, it is a laborious method of producing offspring from the spermatozoa of grafted testicular transplants because it involves surgical insemination.

12.5 Genetically Modified Birds

Introducing exogenous (foreign) genetic materials to poultry breeds has never been achieved by conventional breeding methods such as genetic crossing to date. If the genome of poultry could be modified freely, such a technique would have enormous potential for industrial applications in agriculture and pharmacy, as well as in promoting studies in basic biology. The insertion of new genetic material into avian genomes would enable the production of transgenic birds. This could provide a means of improvement of poultry egg, meat, and feather production, improvements in the nutritive value of poultry eggs and meat, impartment of disease resistance to poultry, and mass production of specific proteins or antibodies that have commercial value in egg albumin. The introduction of site-specific modifications including disrupting, inserting, and mutating the poultry genome by homologous recombination, or programmable genome-editing technologies such as the use of zinc-finger nuclease (ZFN), transcription activator-like effector nuclease (TALEN), and clustered regularly interspaced short palindromic repeats (CRISPR)-associated protein 9 (CRISPR/Cas9) could generate gene-targeted poultry. Such gene disruption could provide insights into the production of hypoallergenic poultry products and the establishment of models for studying avian biology. However, the virus-free transgenesis technologies that are well established in mammalian models have hardly been applied to poultry because of the difficulty in accessing and manipulating their zygotes. Over the past 3 decades, several avian transgenesis technologies have been established, particularly in the chicken. Recent advances of these techniques will allow the efficient generation of genetically modified chickens and open the way to the applications listed above.

12.5.1 Avian Transgenesis

All methods of generating genetically modified poultry lines rely on techniques designed to manipulate the genome of the germ cells. To date, three methods have been developed that have successfully generated genetically modified poultry lines: virus-mediated transduction of embryos, DNA microinjection into zygotes, and germline chimeric intermediate through the transplantation of genetically modified PGCs.

12.5.1.1 Virus-Mediated Transduction of Embryos

The use of retroviruses including lentiviruses as gene delivery vehicles is an efficient transgenesis technology because they can integrate stably into cell genomes. The first successful development of transgenic animals was reported for the mouse using oncoretroviruses such as the Moloney murine leukemia virus (MoMLV) (Jaenisch 1976). Ten years later, the first transgenic chickens were generated using replication-competent reticuloendotheliosis virus (REV) and avian leukosis virus (ALV) (Salter et al. 1986). Injection of retroviral vectors close to the blastoderm of chicken embryos resulted in somatic mosaicism in 25%, and transmission of the provirus to the next generation at rates of 1–11%. Subsequently, the possibility of germline insertion of a specific transgene such as that encoding bovine growth hormone using a modified form of the Rous sarcoma virus (RSV) was reported (Chen et al. 1990). However, these replication-competent vectors induced viremia in infected birds and so were impossible to use practically for transgenesis. Since the effectiveness of replication-deficient REV or ALV vectors on generations of transgenic poultry was demonstrated (Bosselman et al. 1989; Thoraval et al. 1995), virus-mediated transgenesis has been the most widely used approach for establishing various transgenic poultry lines. A conventional virus-mediated transgenesis protocol is that a high titer of the vector is injected into the subgerminal cavity of stage X embryos, so blastoderm cells including PGCs are transduced genetically. Other methods currently available involve transduction of PGCs in vivo through the intravascular injection of a vector (Kamihira et al. 2005) or ex vivo after the enrichment of PGCs (Motono et al. 2010). Several types of retroviral vectors such as REV, ALV, MoMLV, spleen necrosis virus (SNV), murine stem cell virus (MSCV), and lentiviral vectors such as equine infectious anemia virus (EIAV) or human immunodeficiency virus (HIV), have been used. The efficiencies of germline transmission of a transgene varied according to the selection of vector type. Conversely, there are several inherent disadvantages of virus-mediated transduction including random integration, the limitations of packaging size of the transgene that can be introduced, and difficulties in gene disruption, as well as safety issues associated with a use of viral vectors.

12.5.1.2 DNA Microinjection into Zygotes

Microinjection of DNA into the pronucleus of a zygote is the commonly used method to produce transgenic mammals. The first successful production of transgenic animals using this method was reported for the mouse by Gordon et al. in 1980. Although the applicability of DNA microinjection was demonstrated in chicken in 1994 by Love et al. (see Sect. 12.4.1), this technique has not been used extensively to produce transgenic poultry because of technical difficulties and complications, the low efficiency of germline transmission, and the need to euthanize a large number of hens to collect zygotes.

12.5.1.3 Transgenesis of PGCs In Vitro

By combining germline chimera production technologies, poultry PGCs can be used as an attractive source for transgenesis. A major obstacle has been that conventional transfection methods such as lipofection, electroporation and sonoporation are only available for transient gene expression, but have a limited effectiveness integrating foreign genes into the genome of intact PGCs after purification. In the first report concerning the generation of transgenic chickens via germline chimeras, non-purified dissociated germinal crescent cells were infected ex vivo with replication-defective ALV or SNV vectors (Vick et al. 1993). Recently established avian PGC enrichment techniques (see Sect. 12.4.2) enabled more efficient virus-mediated transduction of PGCs in chickens and Japanese quail.

Thus far, the long-term culture of PGCs that remain restricted to the germline has not been established for any species. In 2006, an innovative method was established for the long-term culture of male chicken PGCs with maintenance of their germline competency by van de Lavoir et al. (see Chap. 1). This PGC culture system made a major breakthrough in manipulation of the chicken genome, with subsequent cloning in a manner similar to mouse ES cells (Fig. 12.4). It was demonstrated that in-vitro culture of PGCs could be used for transgenesis by conventional electroporation of plasmid DNA and subsequent generation of transgenic chickens via germline chimeras. However, the apparently low frequency of stable integration of foreign genes into the genome of cultured PGCs remain inefficient for the practical establishment of transgenic chicken lines using conventional virus-independent methods. This might in part result from the transcriptional repression of integrated transgenes by the genome-wide epigenetic modifications present in the PGCs of many organisms. These problems in the use of non-viral transgenesis systems could be overcome by using transposons. DNA transposons are naturally occurring mobile genetic elements that transpose by a "cut and paste" process involving the mobilization of a transposon from one location on the genome and subsequent reintegration to a novel site. To date, among several transposon elements that are found in vertebrates, the transposable element of *Oryzias latipes*, number 2 (Tol2), *piggyBac*, and *Sleeping beauty* have been used predominantly for transgenesis in several species. The *piggyBac* transposon was found to efficiently integrate a foreign gene to intronic and intergenic regions of the genome in chicken PGCs cultured in vitro, and these were transmitted efficiently to the next generation without silencing (Macdonald et al. 2012; Park and Han 2012). This transgenesis protocol would be the best means currently available for establishment of transgenic chicken lines, although a precise integration site cannot be targeted.

In contrast to the studies performed to establish transgenic poultry lines using the methods listed above, only a few reports are available on

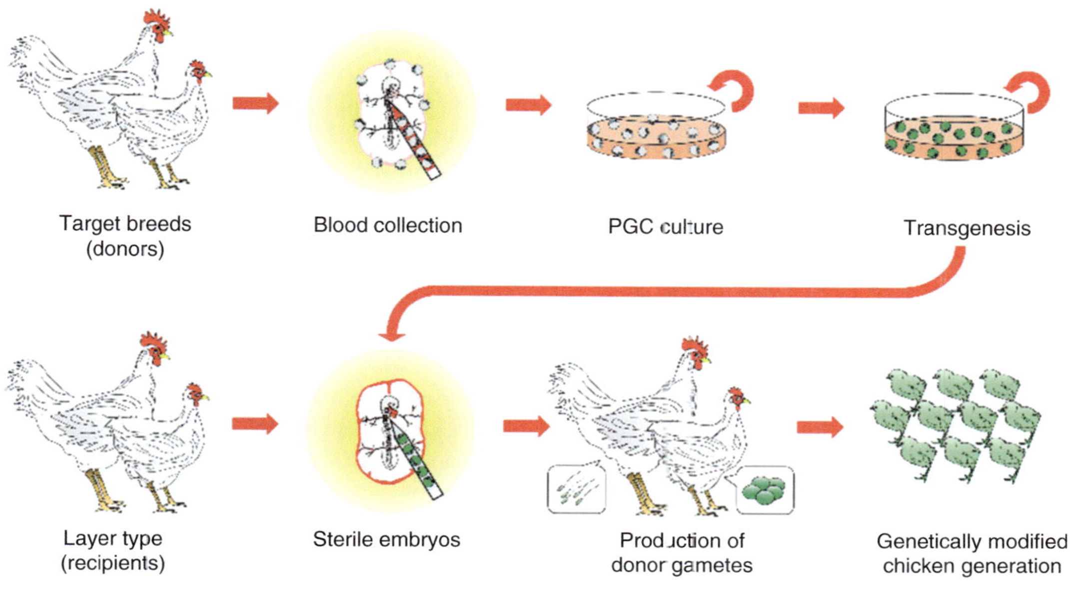

Fig. 12.4 Principle of generation of genetically modified chickens via germline chimeras. Chicken PGCs are isolated from the blood of embryos at stages 14–16, and cultured in vitro. PGC transfection is carried out; then, genetically modified PGCs are selected and propagated in vitro. Genetically modified PGCs are transplanted intravascularly to sterile recipient chicken embryos at stage 14–16 To obtain genetically modified progeny, male and female germline chimeric chickens are fertilized using artificial insemination

the generation of gene knockout poultry. In 2013, the first knockout chicken model disrupting the gene encoding heavy chain immunoglobulin was established by Schusser et al. They demonstrated that homologous recombination could be executed through in-vitro PGC culture to generate these chickens. One year later, generation of ovalbumin heterozygous gene knockout chickens was reported to induce mutations in cultured PGCs using the TALEN system (Park et al. 2014). They were the first to show the effectiveness of programmable genome-editing technology in generating gene knockout chickens. More recently, the CRISPR/Cas9 system was used to produce ovomucoid gene knockout chickens (Oishi et al. 2016). These studies have shown the availability of programmable genome-editing technologies for the efficient induction of targeted mutagenesis in cultured chicken PGCs. It seems that the CRISPR/Cas9 system is advantageous for generating gene knockout chickens because of the much higher mutational frequency in cultured PGCs. It has become possible to apply programmable genome-editing technology to produce gene knock-in in several organisms. A potential disadvantage of such technologies, in particular for CRISPR/Cas9, is the possibility of off-target effects. In future, precise evaluation of such effects in several knockout chicken lines will be required by genome-wide analysis.

The rapid generation of homozygous mutant chickens is important because interbreeding is a time-consuming process and it takes about 6 months for the birds to reach sexual maturity. Null mutant chickens have been generated by interbreeding of the G1 generation of heterozygous mutant male and female chickens, because long-term culture is only available for male PGCs. Recently defined serum-free, feeder-free, and chemically permissive culture conditions have enabled efficient derivation and propagation of chicken PGCs for both sexes (Whyte et al. 2015). This defined culture system allows gene targeting both in male and female chicken PGCs. By interbreeding both G0 male and female germline chimeras, homozygous mutants can be generated one generation earlier than is achieved with currently available approaches.

12.5.2 Genetically Modified Poultry in Agriculture

Poultry breeds are among the most popular form of livestock in the world, and their products, such as eggs and meat, are considered as among the main sources of protein. In 2012, the world's total poultry meat and egg production stood at 105.4 and 65.0 million tons, respectively. In particular, poultry meats constituted about 34.6% of the world's total meat production, representing the second largest proportion. In accordance with an increase in the world's human population, the demand for food will increase. In particular, the global demand for poultry eggs and meat is expected to increase exponentially in the next several decades. The updated poultry germline modification technologies might be able to contribute to such demands through poultry productivity improvements, including egg modification, and the impartment of disease resistance. For example, overexpression of growth hormone, a positive regulator of body growth, in transgenic mice and Atlantic salmon (*Salmo salar*) could result in increased growth rates and total body mass (Palmiter et al. 1982; Du et al. 1992). In 2015, the United States Food and Drug Administration (FDA) approved this transgenic salmon for commercial production, sale, and consumption, making it the first transgenic animal to be approved for human consumption. Congenital mutations or knockout of the gene encoding myostatin, a negative regulator of skeletal muscle growth, in various animal species, including mice, cattle, dogs, pigs, and humans, produced greater muscle mass (McPherron and Lee 1997; McPherron et al. 1997; Mosher et al. 2007; Rao et al. 2016; Schuelke et al. 2004). Overproduction of growth hormone or induction of mutagenesis in the gene encoding myostatin might apply to increase growth rates and body composition, thereby reducing feeding costs in poultry meat production.

A modern genetically selected layer hen produces up to 330 eggs per year, each containing over 6.5 g of protein. Approximately 3.6 g of the total protein is egg albumin, of which about 80% comprises three major proteins: ovalbumin, ovotransferrin, and ovomucoid. Among various egg albumins, ovalbumin and ovomucoid are major potentially allergenic proteins. Ovalbumin alone accounts for 2 g of total egg albumin protein, while ovomucoid accounts for 0.4 g. Poultry germline modification technologies would have potential applications to regulate the composition of egg albumin for such outcomes as reducing allergenicity, increasing nutrient levels and producing antimicrobial substances. Mutant chicken lines for both ovalbumin and ovomucoid have now been established using programmable genome-editing technologies (Park et al. 2014; Oishi et al. 2016). Further studies will show the component of albumin of these null mutant chicken lines.

Infection of poultry with avian influenza virus poses a global threat to both poultry production and human health; this cannot be adequately controlled by vaccination or by biosecurity measures. Recently, an experimental challenge to this problem was performed by using transgenic chickens expressing a short-hairpin interfering RNA specifically targeting viral RNA polymerase activity. This resulted in suppressed transmission of the avian influenza virus (Lyall et al. 2011).

12.5.3 Genetic Modified Poultry in Pharmacy

Mammalian germline modification technologies have been applied to the production of recombinant biologically active proteins in the mammary glands of several farm animals such as rabbits, goats, sheep (*Ovis aries*), pigs, and cows. The use of transgenic farm animals for such gene "pharming" has advanced to the stage of commercial application. Poultry, and in particular chickens, have many advantages over the mammalian bioreactors thanks to their short generation and continuous production of eggs year round. Thus, chicken eggs are attractive candidates for the mass production of biologically active human proteins or antibodies that have commercial value. A promising advantage of the use of poultry eggs for gene pharming is that chicken proteins, unlike those in other genetically modified animals, have a similar glycosylation profile of immunoglobulin G (IgG) to that of the human protein (Raju et al. 2000). Using poultry eggs as an alternative to cell-culture systems would also provide advantages in reducing production costs, progress in the quality of protein produced, and the ability to scale up. It was first shown in 2002 that secretion of a biological active bacterial protein (β-lactamase) into the egg albumin of transgenic chickens was possible using an ubiquitous promoter (Harvey et al.). Subsequently, this strategy was applied to the production of a biologically active glycosylated human protein, interferon α-2b in egg albumin (Rapp et al. 2003). However, the ubiquitous expression of biologically active proteins will sometimes lead to failures, which, in the worst-case scenario, might result in embryonic death. Tissue-specific production of a functional recombinant pharmaceutical protein would contribute in part to this issue. For example, the ovalbumin promoter enables the oviduct-specific synthesis of biologically active proteins in genetically modified hens (Lillico et al. 2007). This strategy has been widely used for gene pharming in chickens. In 2015, the FDA approved the use of genetically modified chickens laying drug-carrying eggs for pharmaceutical purposes. Increasing the expression levels of transgenes and decreasing total amount of endogenous albumin proteins would contribute to a more efficient pharmaceutical production in chicken eggs. In the former situation, several potential approaches are available, such as modification of the ovalbumin promoter, insertion of multiple copies of transgenes using piggyBac-mediated transgenesis, or the use of oviduct cell-specific inducible expression of transgenes under the control of an enhanced ubiquitous promoter using the Cre/loxP system. For the latter case, CRISPR/Cas9-mediated knockout of genes encoding egg albumins would be useful, as described above.

12.5.4 Genetic Modified Poultry in Basic Biology

The chicken has been used extensively as a model organism throughout the history of biology. For example, the chicken embryo was one of the first ever studied. Avian embryonic development is highly reproducible and exquisitely regulated. In particular, the morphology of chicken embryos is very similar to those of humans in the early stages: both are amniotes and their development is well conserved. Chicken embryos have many experimental advantages over mammalian experimental models, including their easy accessibility, manipulation, and observation throughout development because they develop in ovo but not in utero. Hence, the chicken embryo is an attractive model for studying the developmental processes of birds as well as humans and mammals. The recent development of germline modification methods described above ensure that the chicken will remains an attractive model system that will continue to make major contributions to our understanding of molecular and cellular mechanisms controlling developmental processes.

12.6 Conservation of Avian Genetic Resources

Technological developments in avian PGC transplantation, immature ovarian grafting, and artificial insemination provide insights into the ex-situ conservation of avian genetic resources. With these techniques as a background, cryopreservation of avian PGCs, ovarian tissues, and semen has been established. Cryopreservation of these cells or tissues allows the conservation of currently available avian species for future use. Regeneration of live birds and recovery of extinct species from frozen germplasm has enormous significance to pass on the present avian genetic resources to the future.

Breeds of agriculturally prized animals have been modified throughout history. Because poultry are economically valuable, the preferences of consumers and directivities of producers have driven the generation of suitable poultry breeds or lines. For the chicken, the recent need for high production efficiency, combined with the complexity and the cost of running effective breeding programs, has resulted in commercially selected lines of layers and broilers replacing several of the breeds previously reared for productive purposes. However, the increased global use of highly productive layer and broiler lines has been accompanied by losses in genetic diversity. The less popular breeds including indigenous breeds are often maintained only locally and in small populations. Consequently, these chicken breeds are at risk of becoming extinct or may suffer from inbreeding. In addition, recent concern about the highly pathogenic avian influenza virus poses a constant threat to the global poultry industry.

Experimental birds are also important genetic resources. Recent advances in avian genetics have enabled researchers to generate several experimental models, particularly in the chicken and Japanese quail. Cryopreservation of germplasm allows the establishment of research environments such as economical storage and easy distribution of experimental models and this will facilitate global development in avian research.

To date, the cryopreservation of poultry germplasm concerns mainly semen. However, the fertility of post-thawing poultry semen lags behind other species and requires significant improvements for practical use. Additionally, a major disadvantage in avian semen cryopreservation is that genes on the W chromosome and mitochondrial DNA cannot be maintained because the rooster is a ZZ homogametic sex. In contrast, cryopreservation of avian PGCs enables capture of the entire genetics of a poultry stock. Therefore, cryopreservation of PGCs together with the maintenance of frozen stocks of semen and immature ovaries will facilitate more reliable and efficient recovery of live birds in the future.

Transportation of live birds involves many risks. There are several possibilities regarding this issue: for example, transported birds can be carriers of infectious disease and subject to casualties from transport accidents. As an alternative, frozen PGCs can be used as a means of safe, economic, and hygienic transportation. Indeed, international shipments of live birds or fertilized eggs are generally restricted by quarantine regulations, whereas appropriately processed frozen materials can be transported legally. Thus, a practical trial has been reported involving successful transport of a chicken breed by shipment of cryopreserved PGCs using a dry shipper (Nakamura et al. 2016).

12.6.1 Cryopreservation of Avian PGCs

In 1949, the cryoprotective effect of glycerol was discovered accidentally by Polge and colleagues. Since then, increased understanding of the fundamental principles of cryobiology has continually helped to develop efficient methods for the conservation of a very wide range of cells and tissues. Today, these methods have found widespread application in biology, agriculture, and medicine. The two most commonly used ultra-low temperature conservation methods for animal germplasm are slow freezing and vitrification.

For cryopreservation of chicken PGCs, the slow-freezing method is commonly used. In this, cells or tissues are suspended a medium containing 1–1.5 M cryoprotectants and are cooled at a rate of −0.3 to −1 °C/min until reaching −80 °C. In this condition, the spontaneous freezing point of the cryomedium is −5 to −7 °C. Lethal intercellular ice crystal formation can be minimized if cooling is slow enough to allow sufficient efflux of water from the cells during the progressive freezing of extracellular fluid. The most widely used cryopreservation protocol for freezing chicken PGCs is a serum containing medium supplemented with 10% DMSO as a cryoprotectant at a cooling rate of −1 °C/min until reaching −80 °C. This cryopreservation protocol yields a recovery rate of approximately 50%, and over 85% viability of frozen–thawed chicken PGCs (Naito et al. 1994a; Setioko et al. 2007; Kohara et al. 2008). Several comparative studies have examined types of cryoprotectants including their concentrations and the cooling rates to develop more efficient protocols for cryopreserving chicken PGCs (Moore et al. 2006; Setioko et al. 2007; Tonus et al. 2014; Sawicka et al. 2015). Taken together, placing chicken PGCs in a medium containing more than 10% serum and either 5–10% DMSO or 10% EG as cryoprotectants at a cooling rate of −2 °C/min allows better recovery rates and viability of post-thawed chicken PGCs than classic protocols. In addition, a commercially available serum-containing DMSO-based cryomedium is also available for the cryopreservation of avian PGCs, with higher recovery rates and viability than classical freezing protocols (Nakamura et al. 2010b, 2011, 2013b).

Vitrification is a cryopreservation method in which not only cells but also the whole solution is solidified to a glassy supercooled state without the crystallization of ice. In this, cells or tissues are suspended in a medium with about 40% (v/v) cryoprotectants and are cooled at a rate of about −100,000 °C/min until reaching −196 °C. The very high concentrations of cryoprotectants needed for vitrification can cause damage to cells through osmotic stress and chemical toxicity. Therefore, the most recent vitrification methods make use of high cooling rates to reduce the concentration of cryoprotectants needed. To recover viable cells, it is also essential to warm vitrified samples at very high rates. The vitrification method has advantages over slow freezing. For example, it has enabled the cryopreservation of porcine oocytes and embryos, which are very sensitive to chilling because of their high lipid contents. Although one study reported lower recovery and viable rates of vitrified PGCs than in slow frozen–thawed PGCs (Kohara et al. 2008), vitrification should improve the recovery rates and viability of chicken PGCs after exploration of various protocols.

Storage of freeze-dried germplasm is extremely cost efficient, as no expensive and bulky liquid nitrogen containers are necessary. Since the development of a method to preserve mouse spermatozoa by freeze-drying (Wakayama and Yanagimachi 1998), this approach has been applied widely to other animal species. However, standard insemination procedures are generally not available for freeze-dried sperm samples. Sperm freeze-drying combined with ICSI and subsequent ex-vivo culture might offer possible conservation of avian genetic resources, but this requires further improvements and optimization in each procedure.

12.6.2 Gene Banking of Poultry Genetic Resources

Recent advances in cryobiological and reproductive technologies in avian species are now offering the establishment of poultry gene banking programs. At present, PGCs, immature ovaries, and semen are useful for gene banking of poultry genetic resources, with the objectives of collecting, cryopreserving, depositing, and distributing currently available breeds or lines. Here, the concepts of poultry gene banking are introduced. As shown in Fig. 12.5, the flow of ex-situ conservation of poultry genetic resources consists of five steps: (1) collection of PGCs, immature ovarian tissues and semen from target lines or breeds, (2) cryopreservation of them in liquid nitrogen, (3) transplantation of PGCs or immature ovarian tissues to recipients, (4) artificial insemination of female recipients with frozen semen or mating with male germline chimeras, and (5) recovery of live bird populations. This foreshadowed gene-banking program potentially enables the recovery of a desired breed or line within two generations. The practicality of such a gene-banking program is affected by the success of each of the cryobiological and reproductive techniques involved. In particular, the overall success rate in producing donor-derived hatchlings is critical to ensure an adequate effective population size for genetic recovery. Allogeneic transplantation of post-thawed PGCs to sterile recipients of a commercial layer line would allow them to be multiplied rapidly and significantly, recovering a viable population size from only a small flock of recipients. Therefore, this potentially enables the efficient propagation of the populations of endangered breeds, as well as efficient production of broiler chickens from layer hens. Although further studies are necessary, xenogeneic transplantation of PGCs or immature ovaries to phylogenetically distant domestic poultry might be applicable for the ex-itu conservation of endangered species of wild birds.

The most important aspect of gene banking is the collection of germplasm without losing the present wealth of genetic diversity. To recover a desired population while maintaining sustainable genetic diversity from the germplasm repository, it is recommended to preserve the germplasm from at least 13—or ideally 25—individuals of each sex (FAO 1998). From an industrial point of view, cryopreservation of PGCs, immature ovaries, and semen can be used as back-up for commercially or industrially important poultry lines or breeds that have been selected for a long time, in case they are lost as a consequence of pathogen outbreaks, genetic problems, breeding cessation, or natural disasters. Furthermore, the cost of maintaining these germplasms in liquid nitrogen is much lower than maintaining populations of live birds.

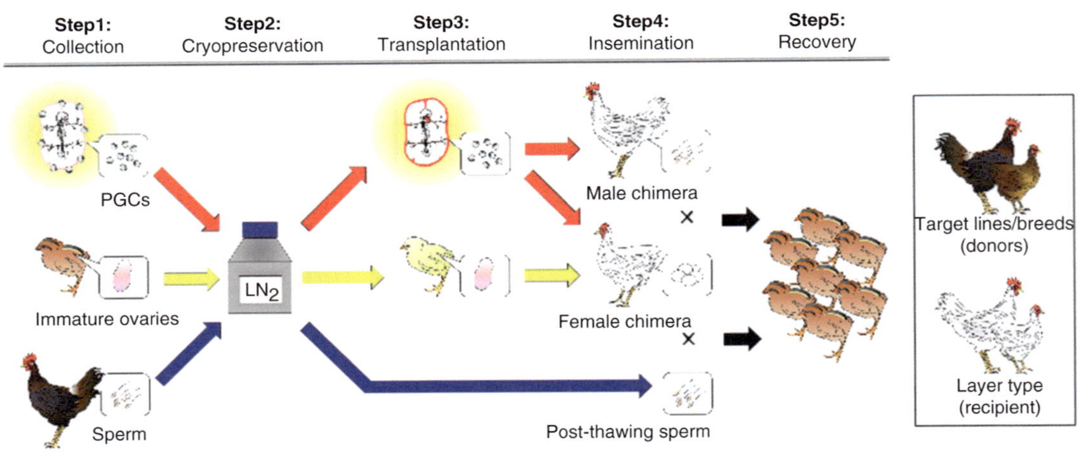

Fig. 12.5 Principle of gene banking of poultry genetic resources. PGCs, immature ovaries, and spermatozoa are collected from target lines or breeds of poultry species (*Step 1*). These genetic resources are cryopreserved in liquid nitrogen until future use (*Step 2*). Frozen-thawed PGCs or immature ovaries are allogeneically transplanted to recipients with high prolific capacity (*Step 3*). Female germline chimeras are artificially inseminated with post-thaw semen or mated with male germline chimeras (*Step 4*). Live bird populations can be recovered efficiently (*Step 5*). LN_2 liquid nitrogen (−196 °C)

In most noncommercial breeds, such as indigenous, rare, and/or endangered ones, the population size might be small, so the availability of fertilized eggs might also be restricted. In such cases, collection of germplasm stocks without influencing reproductive performance is critically important for gene banking. Semen can be collected repeatedly to cryopreserve it through successive reproductive seasons. Conversely, surgery is needed to collect immature ovaries without sacrifice. Although hens sometimes regain fertility after surgery for ovarian removal, this is not suitable for the conservation of small populations. However, a unique biological property of avian PGCs that temporarily circulate thorough the vascular system provides the opportunity to combine the reproduction of live birds with cryopreservation of PGCs (Nakamura et al. 2010b). Thus, PGCs can be collected from embryonic blood, then frozen while the residual donor embryos capable of hatching out continue to mature with normal reproductive performance. In this method, drawing blood from donor embryos involves simple windowing procedures (see Sect. 12.3.1). To date, several research groups have succeeded in the long-term culture of male PGCs from various chicken breeds including indigenous breeds. Defined culture conditions enabled the efficient derivation and propagation of both male and female chicken PGCs (Whyte et al. 2015; Nandi et al. 2016: and see Chap. 1). Therefore, at least in the chicken, the major advantage of PGCs for use as a source of gene banking is that they can be obtained from a single donor embryo and then propagated significantly in vitro prior to freezing. The principles of efficient conservation and propagation of endangered chicken breeds are shown in Fig. 12.6.

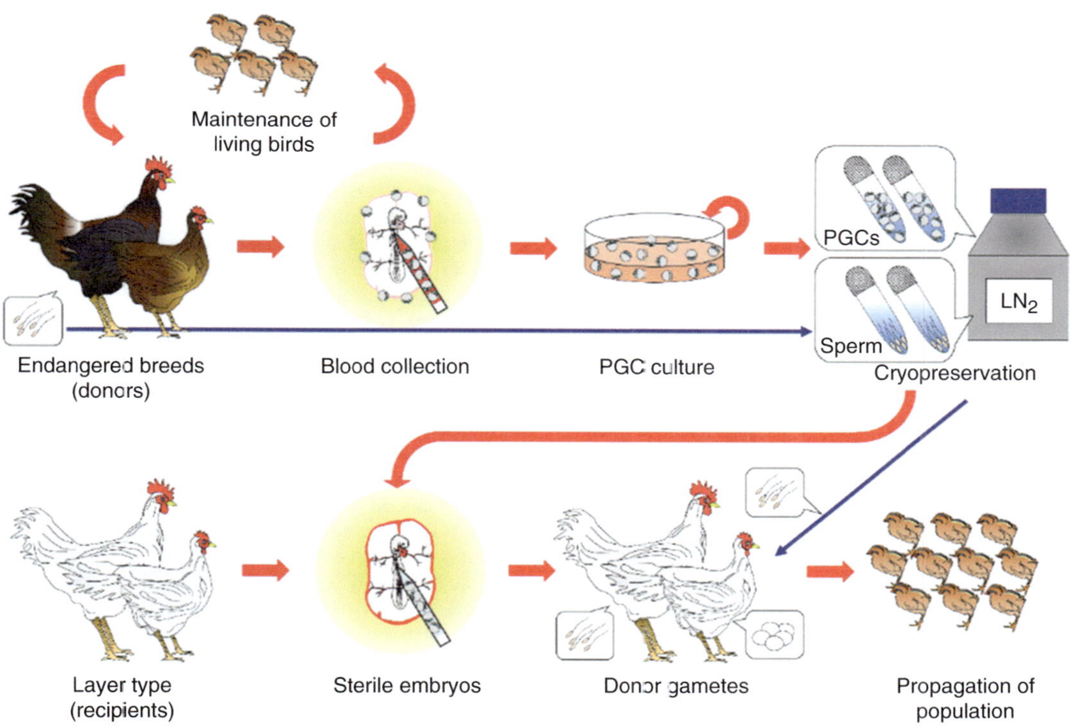

Fig. 12.6 Principle of conservation and propagation of endangered chicken breeds. Blood is collected from embryos of endangered chicken breeds at stages 14–16; these donor embryos are continued to incubate until hatching for live collection. PGCs isolated from a single embryo are propagated in vitro. Cultured chicken PGCs are cryopreserved in liquid nitrogen (LN_2) for gene banking. Frozen–thawed cultured PGCs are converted to functional spermatozoa or ova via transplantation to sterile recipient embryos of layer-type chickens. Endangered chicken populations can be propagated efficiently by artificially inseminating female recipients with frozen–thawed semen from male birds, even after breed extinction

Unlike in fish, avian PGCs hardly possess any sexual bipotency (see Chap. 1). For gene banking purposes, male and female PGCs need to be collected separately and cryopreserved, because the efficient germline transmission of PGCs requires that the sex of the donor must be matched to that of the recipient.

References

Aige-Gil V, Simkiss K. Sterilisation of avian embryos with busulphan. Res Vet Sci. 1991;50:139–44.

Alexander A, Graham J, Hammerstedt RH, et al. Effect of genotype and cryopreservation of avian semen on fertility and number of perivitelline spermatozoa. Br Poultry Sci. 1993;34:757–64.

Aoyama H, Asamoto K, Nojyo Y, et al. Monoclonal antibodies specific to quail embryo tissues: their epitopes in the developing quail embryo and their application to identification of quail cells in quail-chicken chimeras. J Histochem Cytochem. 1992;40:1769–77.

Blanco JM, Gee G, Wildt DE, et al. Species variation in osmotic, cryoprotectant, and cooling rate tolerance in poultry, eagle, and peregrine falcon spermatozoa. Biol Reprod. 2000;63:1164–71.

Blesbois E, Grasseau I, Seigneurin F. Membrane fluidity and the ability of domestic bird spermatozoa to survive cryopreservation. Reproduction. 2005;129:371–8.

Bosselman RA, Hsu RY, Boggs T, et al. Germline transmission of exogenous genes in the chicken. Science. 1989;243:533–5.

Brinster RL, Avarbock MR. Germline transmission of donor haplotype following spermatogonial transplantation. Proc Natl Acad Sci USA. 1994;91:11303–7.

Brinster RL, Zimmermann JW. Spermatogenesis following male germ-cell transplantation. Proc Natl Acad Sci USA. 1994;91:11298–302.

Carsience RS, Clark ME, Verrinder Gibbins AM, et al. Germline chimeric chickens from dispersed donor blastodermal cells and compromised recipient embryos. Development. 1993;117:669–75.

Catzeflis FM, Dickerman AW, Michaux J, et al. DNA hybridization and rodent phylogeny. In: Szalay FS, Novacek MJ, McKenna MC, editors. Mammal phylogeny (placentals), vol. 2. New York: Springer Verlag; 1993. p. 159–72.

Chalah T, Seigneurin F, Blesbois E, et al. In vitro comparison of fowl sperm viability in ejaculates frozen by three different techniques and relationship with subsequent fertility in vivo. Cryobiology. 1999;39:185–91.

Chen HY, Garber EA, Mills E, et al. Vectors, promoters, and expression of genes in chick embryos. J Reprod Fertil. 1990;41:173–82.

Clouthier DE, Avarbock MR, Maika SD, et al. Rat spermatogenesis in mouse testis. Nature. 1996;381: 418–21.

Dobrinski I, Avarbock MR, Brinster RL. Transplantation of germ cells from rabbits and dogs into mouse testes. Biol Reprod. 1999;61:1331–9.

Dobrinski I, Avarbock MR, Brinster RL. Germ cell transplantation from large domestic animals into mouse testes. Mol Reprod Dev. 2000;57:270–9.

Du SJ, Gong ZY, Fletcher GL, et al. Growth enhancement in transgenic Atlantic salmon by the use of an "all fish" chimeric growth hormone gene construct. Biotechnology. 1992;10:176–81.

Eyal-Giladi H, Kochav S. From cleavage to primitive streak formation: a complementary normal table and a new look at the first stages of the development of the chick. I. General morphology. Dev Biol. 1976;49:321–37.

FAO. Secondary guidelines for development of national farm animal genetic resources management plans: management of small populations at risk. UNEP: Nairobi, Kenya; 1998, pp. 1–210.

Gordon JW, Scangos GA, Plotkin DJ, et al. Genetic transformation of mouse embryos by microinjection of purified DNA. Proc Natl Acad Sci USA. 1980;77:7380–4.

Hamburger V, Hamilton HL. A series of normal stages in the development of the chick embryo. J Morphol. 1951;88:49–92.

Hanzawa S, Niinomi T, Miyata T, et al. Cryopreservation of chicken semen using N-methylacetamide as cryoprotective agent. Nippon Kakin Gakkaishi (Jpn J Poult Sci). 2010;47:J27–32. (in Japanese)

Harvey AJ, Speksnijder G, Baugh LR, et al. Expression of exogenous protein in the egg white of transgenic chickens. Nat Biotechnol. 2002;20:396–9.

Hermann BP, Sukhwani M, Winkler F, et al. Spermatogonial stem cell transplantation into rhesus testes regenerates spermatogenesis producing functional sperm. Cell Stem Cell. 2012;11:715–26.

Honaramooz A, Megee SO, Dobrinski I. Germ cell transplantation in pig. Biol Reprod. 2002;66:21–8.

Honaramooz A, Behboodi E, Blash S, et al. Germ cell transplantation in goats. Mol Reprod Dev. 2003;64:422–8.

Izadyar F, Den Ouden K, Stout TA, et al. Autologous and homologous transplantation of bovine spermatogonial stem cells. Reproduction. 2003;126:765–74.

Jaenisch R. Germ line integration and Mendelian transmission of the exogenous Moloney leukemia virus. Proc Natl Acad Sci USA. 1976;73:1260–4.

Jung JG, Lee YM, Kim JN, et al. The reversible developmental unipotency of germ cells in chicken. Reproduction. 2010;139:113–9.

Kagami H, Tagami T, Matsubara Y, et al. The developmental origin of primordial germ cells and the transmission of the donor-derived gametes in mixed-sex germline chimeras to the offspring in the chicken. Mol Reprod Dev. 1997;48:501–10.

Kamihira M, Oguchi S, Tachibana A, et al. Improved hatching for in vitro quail embryo culture using surrogate eggshell and artificial vessel. Dev Growth Differ. 1998;40:449–55.

Kamihira M, Ono K, Esaka K, et al. High-level expression of single-chain Fv-Fc fusion protein in serum and egg white of genetically manipulated chickens by using a retroviral vector. J Virol. 2005;79:10864–74.

Kanatsu-Shinohara M, Inoue K, Miki H, et al. Clonal origin of germ cell colonies after spermatogonial transplantation in mice. Biol Reprod. 2006;75: 68–74.

Kang SJ, Choi JW, Kim SY, et al. Reproduction of wild birds via interspecies germ cell transplantation. Biol Reprod. 2008;79:931–7.

Kang KS, Park TS, Rengaraj D, et al. Fertilisation of cryopreserved sperm and unfertilised quail ovum by intracytoplasmic sperm injection. Reprod Fertil Dev. 2016;28:1974–1981.

Karagenç L, Cinnamon Y, Ginsburg M, et al. Origin of primordial germ cells in the prestreak chick embryo. Dev Genet. 1996;19:290–301.

Kato A, Miyahara D, Kagami H, et al. Culture system for bobwhite quail embryos from the blastoderm stage to hatc hing. J Poult Sci. 2013;50:155–8.

Kohara Y, Kanai Y, Tajima A. Cryopreservation of gonadal germ cells (GGCs) from the domestic chicken using vitrification. J Poult Sci. 2008;45:57–61.

Lake PE, Ravie O, McAdam J. Preservation of fowl semen in liquid nitrogen: application to breeding programmes. Br Poult Sci. 1981;22:71–7.

Lillico SG, Sherman A, McGrew MJ, et al. Oviduct-specific expression of two therapeutic proteins in transgenic hens. Proc Natl Acad Sci USA. 2007;104: 1771–6.

Lin M, Thorne MH, Martin IC, Scheldon BL, Jones RC. Development of the gonads in the triploid (ZZW and ZZZ) fowl, Gallus domesticus, and comparison with normal diploid males (ZZ) and females (ZW). Reprod fertil Dev. 1995;7:1185–97.

Liu CH, Chang IK, Sasse J, et al. Xenogenic oogenesis of chicken (*Gallus domesticus*) female primordial germ cells in germline chimeric quail (*Coturnix japonica*) ovary. Anim Reprod Sci. 2007;101:344–50.

Liu J, Song Y, Cheng KM, et al. Production of donor-derived offspring from cryopreserved ovarian tissue in Japanese quail (*Coturnix japonica*). Biol Reprod. 2010;83:15–9.

Liu C, Khazanehdari KA, Baskar V, et al. Production of chicken progeny (*Gallus gallus domesticus*) from interspecies germline chimeric duck (*Anas domesticus*) by primordial germ cell transfer. Biol Reprod. 2012;86:1–8.

Love J, Gribbin C, Mather C, et al. Transgenic birds by DNA microinjection. Biotechnology. 1994;12:60–3.

Lyall J, Irvine RM, Sherman A, et al. Suppression of avian influenza transmission in genetically modified chickens. Science. 2011;331:223–6.

Macdonald J, Taylor L, Sherman A, et al. Efficient genetic modification and germ-line transmission of primordial germ cells using *piggyBac* and *Tol2* transposons. Proc Natl Acad Sci USA. 2012;109:E1466–72.

Mak SS, Wrabel A, Nagai H, et al. Zebra finch as a developmental model. Genesis. 2015;53:669–77.

McPherron AC, Lee SJ. Double muscling in cattle due to mutations in the myostatin gene. Proc Natl Acad Sci USA. 1997;94:12457–61.

McPherron AC, Lawler AM, Lee SJ. Regulation of skeletal muscle mass in mice by a new TGF-beta superfamily member. Nature. 1997;387:83–90.

Minvielle F, Gourichon D, Monvoisin JL. Testing homology of loci for two plumage colors, "lavender" and "recessive white," with chicken and Japanese quail hybrids. J Hered. 2002;93:73–6.

Miyahara D, Oishi I, Makino R, et al. Chicken stem cell factor enhances primordial germ cell proliferation cooperatively with fibroblast growth factor 2. J Reprod Dev. 2016;62:143–9.

Mizushima S, Hiyama G, Shiba K, et al. The birth of quail chicks after intracytoplasmic sperm injection. Development. 2014;141:3799–806.

Mocé E, Grasseau I, Blesbois E. Cryoprotectant and freezing-process alter the ability of chicken sperm to acrosome react. Anim Reprod Sci. 2010;122:359–66.

Moore DT, Purdy PH, Blackburn HD. A method for cryopreserving chicken primordial germ cells. Poult Sci. 2006:85:1784–90.

Mosher DS, Quignon P, Bustamante CD, et al. A mutation in the myostatin gene increases muscle mass and enhances racing performance in heterozygote dogs. PLoS Genet. 2007;3:e79.

Motono M, Yamada Y, Hattori Y, et al. Production of transgenic chickens from purified primordial germ cells infected with a lentiviral vector. J Biosci Bioeng. 2010;109:315–21.

Mozdziak PE, Wysocki R, Angerman-Stewart J, et al Production of chick germline chimeras from fluorescence-activated cell-sorted gonocytes. Poult Sci. 2006;85:1764–8.

Nagano M, Patrizio P, Brinster RL. Long-term survival of human spermatogonial stem cells in mouse testes. Fertil Steril. 2002;78:1225–33.

Naito M, Nirasawa K, Oishi T. Development in culture of the chick embryo from fertilized ovum to hatching. J Exp Zool. 1990;254:322–6.

Naito M, Tajima A, Tagami T, et al. Preservation of chick primordial germ cells in liquid nitrogen and subsequent production of viable offspring. J Reprod Fertil. 1994a;102:321–5.

Naito M, Tajima A, Yasuda Y, et al. Production of germline chimeric chickens, with high transmission rate of donor-derived gametes, produced by transfer of primordial germ cells. Mol Reprod Dev. 1994b;39:153–61.

Naito M, Matsubara Y, Harumi T, et al. Differentiation of donor primordial germ cells into functional gametes in the gonads of mixed-sex germline chimaeric chickens produced by transfer of primordial germ cells isolated from embryonic blood. J Reprod Fertil. 1999;117:291–8.

Naito M, Minematsu T, Harumi T, et al. Testicular and ovarian gonocytes from 20-day incubated chicken embryos contribute to germline lineage after transfer into bloodstream of recipient embryos. Reproduction. 2007;134:577–84.

Nakajima Y, Minematsu T, Naito M, et al. A new method for isolating viable gonadal germ cells from 7-day-old chick embryos. J Poult Sci. 2011;48:106–11.

Nakajima Y, Hattori T, Asano A, et al. Migration and differentiation of gonadal germ cells under cross-sex germline chimeras condition in domestic chickens. J Reprod Dev. 2014;60:406–10.

Nakamura Y. Poultry genetic resource conservation using primordial germ cells. J Reprod Dev. 2016;62(5):431–7. (Published online)

Nakamura Y, Yamamoto Y, Usui F, et al. Migration and proliferation of primordial germ cells in the early chicken embryo. Poult Sci. 2007;86:2182–93.

Nakamura Y, Yamamoto Y, Usui F, et al. Increased proportion of donor primordial germ cells in chimeric gonads by sterilisation of recipient embryos using busulfan sustained-release emulsion in chicken. Reprod Fertil Dev. 2008;20:900–7.

Nakamura Y, Usui F, Ono T, et al. Germline replacement by transfer of primordial germ cells into partially sterilized embryos in the chicken. Biol Reprod. 2010a;83:130–7.

Nakamura Y, Usui F, Miyahara D, et al. Efficient system for preservation and regeneration of genetic resources in chicken: concurrent storage of primordial germ cells and live animals from early embryos of a rare indigenous fowl (Gifujidori). Reprod Fertil Dev. 2010b;22:1237–46.

Nakamura Y, Usui F, Miyahara D, et al. Viability and functionality of primordial germ cells after freeze-thaw in chickens. J Poult Sci. 2011;48:57–63.

Nakamura Y, Usui F, Miyahara D, et al. X-irradiation removes endogenous primordial germ cells (PGCs) and increases germline transmission of donor PGCs in chimeric chickens. J Reprod Dev. 2012;58:432–7.

Nakamura Y, Kagami H, Tagami T. Development, differentiation and manipulation of chicken germ cells. Dev Growth Differ. 2013a;55:20–40.

Nakamura Y, Tasai M, Takeda K, et al. Production of functional gametes from cryopreserved primordial germ cells of the Japanese quail. J Reprod Dev. 2013b;59:580–7.

Nakamura Y, Rikimaru K, Takahashi D, et al. Production of functional gametes following transfer of frozen-thawed primordial germ cells of Hinai-dori fowl after long distance transportation for diversification of the risk to outbreaks of highly pathogenic avian influenza. Nippon Kakin Gakkaishi (Jpn J Poult Sci). 2016;53:J7–J14. (In Japanese)

Nandi S, Whyte J, Taylor L, et al. Cryopreservation of specialized chicken lines using cultured primordial germ cells. Poult Sci. 2016;95:1905–11.

Ogawa T, Dobrinski I, Avarbock MR, et al. Xenogeneic spermatogenesis following transplantation of hamster germ cells to mouse testes. Biol Reprod. 1999;60:515–21.

Oishi I. Improvement of transfection efficiency in cultured chicken primordial germ cells by Percoll density gradient centrifugation. Biosci Biotechnol Biochem. 2010;74:2426–30.

Oishi I, Yoshii K, Miyahara D, et al. Targeted mutagenesis in chicken using CRISPR/Cas9 system. Sci Rep. 2016;6:23980.

Okutsu T, Suzuki K, Takeuchi Y, et al. Testicular germ cells can colonize sexually undifferentiated embryonic gonad and produce functional eggs in fish. Proc Natl Acad Sci USA. 2006;103:2725–9.

Okutsu T, Shikina S, Kanno M, et al. Production of trout offspring from triploid salmon parents. Science. 2007;317:1517.

Ono T, Machida Y. Immunomagnetic purification of viable primordial germ cells of Japanese quail (*Coturnix japonica*). Comp Biochem Physiol A Mol Integr Physiol. 1999;122:255–9.

Ono T, Murakami T, Mochii M, et al. A complete culture system for avian transgenesis, supporting quail embryos from the single-cell stage to hatching. Dev Biol. 1994;161:126–30.

Ono T, Nakane Y, Wadayama T, Tsudzuki M, Arisawa K, Ninomiya S, Suzuki T, Mizutani M, Kagami H. Culture system for embryos of blue-breasted quail from the blastoderm stage to hatching. Exp Anim. 2005;54:7–11.

Pain B, Clark ME, Shen M, et al. Long-term in vitro culture and characterisation of avian embryonic stem cells with multiple morphogenetic potentialities. Development. 1996;122:2339–48.

Palmiter RD, Brinster RL, Hammer RE, et al. Dramatic growth of mice that develop from eggs microinjected with metallothionein-growth hormone fusion genes. Nature. 1982;300:611–5.

Park TS, Han JY. *piggyBac* transposition into primordial germ cells is an efficient tool for transgenesis in chickens. Proc Natl Acad Sci USA. 2012;109:9337–41.

Park TS, Lee HJ, Kim KH, et al. Targeted gene knockout in chickens mediated by TALENs. Proc Natl Acad Sci USA. 2014;111:12716–21.

Peláez J, Bongalhardo DC, Long JA. Characterizing the glycocalyx of poultry spermatozoa: III. Semen cryopreservation methods alter the carbohydrate component of rooster sperm membrane glycoconjugates. Poult Sci. 2011;90:435–43.

Perry MM. A complete culture system for the chick embryo. Nature. 1988;331:70–2.

Petitte JN, Clark ME, Liu G, et al. Production of somatic and germline chimeras in the chicken by transfer of blastodermal cells. Development. 1990;108:185–9.

Polge C, Smith AU, Parkes AS. Revival of spermatozoa after vitrification and dehydration at low temperatures. Nature. 1949;164:666.

Raju TS, Briggs JB, Borge SM, et al. Species-specific variation in glycosylation of IgG: evidence for the species-specific sialylation and branch-specific galactosylation and importance for engineering recombinant glycoprotein therapeutics. Glycobiology. 2000;10:477–86.

Rao S, Fujimura T, Matsunari H, et al. Efficient modification of the myostatin gene in porcine somatic cells and generation of knockout piglets. Mol Reprod Dev. 2016;83:61–70.

Rapp JC, Harvey AJ, Speksnijder GL, et al. Biologically active human interferon alpha-2b produced in the egg white of transgenic hens. Transgenic Res. 2003;12:569–75.

Reynaud G. Transfert de cellules germinales primordiales de Dindon à l'embryon de Poulet par injection intravasculaire. [The transfer of Turkey primordial germ cells to chick embryos by intravascular injection]. J Embryol Exp Morphol. 1969;21:485–507.

Robertson GAG. Ovarian transplantation in the house mouse. Proc Soc Exp Biol Med. 1940;44:302–4.

Saito T, Goto-Kazeto R, Arai K, et al. Xenogenesis in teleost fish through generation of germ-line chimeras by single primordial germ cell transplantation. Biol Reprod. 2008;78:159–66.

Salter DW, Smith EJ, Hughes SH, et al. Gene insertion into the chicken germ line by retroviruses. Poult Sci. 1986;65:1445–58.

Sawicka D, Chojnacka-Puchta L, Zielinski M, et al. Flow cytometric analysis of apoptosis in cryoconserved chicken primordial germ cells. Cell Mol Biol Lett. 2015;20:143–59.

Schuelke M, Wagner KR, Stolz LE, et al. Myostatin mutation associated with gross muscle hypertrophy in a child. N Engl J Med. 2004;350:2682–8.

Schusser B, Collarini EJ, Yi H, et al. Immunoglobulin knockout chickens via efficient homologous recombination in primordial germ cells. Proc Natl Acad Sci USA. 2013;110:20170–5.

Setioko AR, Tagami T, Tase H, et al. Cryopreservation of primordial germ cells (PGCs) from White Leghorn embryos using commercial cryoprotectants. J Poult Sci. 2007;44:73–7.

Silversides FG, Mérat P. Homology of the *s+* locus in the chicken with *Al+* in the Japanese quail. J Hered. 1991;82:245–7.

Smith CA, Roeszler KN, Bowles J, et al. Onset of meiosis in the chicken embryo; evidence of a role for retinoic acid. BMC Dev Biol. 2008;8:85.

Snow M, Cox SL, Jenkin G, et al. Generation of live young from xenografted mouse ovaries. Science. 2002;297:2227.

Song Y, Silversides FG. The technique of orthotropic ovarian transplantation in the chicken. Poult Sci. 2006;85:1104–6.

Song Y, Silversides FG. Heterotopic transplantation of testes in newly hatched chickens and subsequent production of offspring via intramagnal insemination. Biol Reprod. 2007;76:598–603.

Song Y, Silversides FG. Transplantation of ovaries in Japanese quail (*Coturnix japonica*). Anim Reprod Sci. 2008;105:430–7.

Song Y, Cheng KM, Robertson MC, et al. Production of donor-derived offspring after ovarian transplantation between Muscovy (*Cairina moschata*) and Pekin (*Anas platyrhynchos*) ducks. Poult Sci. 2012;91:197–200.

Speksnijder G, Ivarie R. A modified method of shell windowing for producing somatic or germline chimeras in fertilized chicken eggs. Poult Sci. 2000;79:1430–3.

Tagami T, Kagami H, Matsubara Y, Harumi T, Naito M, Takeda K, Hanada H, Nirasawa K. Differentiation of female primordial germ cells in the male testes of chicken (Gallus gallus domesticus). Molecular Reproduction and Development. 2007;74(1):68–75.

Tahara Y, Obara K. A novel shell-less culture system for chick embryos using a plastic film as culture vessels. J Poult Sci. 2014;51:307–12.

Tajima A, Naito M, Yasuda Y, et al. Production of germline chimeras by transfer of cryopreserved gonadal primordial germ cells (gPGCs) in chicken. J Exp Zool. 1998;280:265–7.

Tajima A, Graham EF, Shoffner RN, et al. Cryopreservation of semen from unique lines of chicken germ plasm. Poult Sci. 1990;69:999–1002.

Tajima A, Naito M, Yasuda Y, et al. Production of germ line chimera by transfer of primordial germ cells in the domestic chicken (*Gallus domesticus*). Theriogenology. 1993;40:509–19.

Takagi S, Ono T, Tsukada A, et al. Fertilization and blastoderm development of quail oocytes after intracytoplasmic injection of chicken sperm bearing the W chromosome. Poult Sci. 2007;86:937–43.

Thoraval P, Afanassieff M, Cosset FL, et al. Germline transmission of exogenous genes in chickens using helper-free ecotropic avian leukosis virus-based vectors. Transgenic Res. 1995;4:369–77.

Tonus C, Cloquette K, Ectors F, et al. Long term-cultured and cryopreserved primordial germ cells from various chicken breeds retain high proliferative potential and gonadal colonisation competency. Reprod Fertil Dev. 2014;28:628–39.

Trefil P, Micáková A, Mucksová J, et al. Restoration of spermatogenesis and male fertility by transplantation of dispersed testicular cells in the chicken. Biol Reprod. 2006;75:575–81.

Tselutin K, Seigneurin F, Blesbois E. Comparison of cryoprotectants and methods of cryopreservation of fowl spermatozoa. Poult Sci. 1999;78:586–90.

van de Lavoir MC, Diamond JH, Leighton PA, et al. Germline transmission of genetically modified primordial germ cells. Nature. 2006;441:766–9.

van de Lavoir MC, Collarini EJ, Leighton PA, et al. Interspecific germline transmission of cultured primordial germ cells. PLoS One. 2012;7:e35664.

Váradi É, Végi B, Liptói K, et al. Methods for cryopreservation of guinea fowl sperm. PLoS One. 2013;8:e62759.

Vick L, Li Y, Simkiss K. Transgenic birds from transformed primordial germ cells. Proc R Soc B Biol Sci. 1993;251:179–82.

Wakayama T, Yanagimachi R. Development of normal mice from oocytes injected with freeze-dried spermatozoa. Nat Biotechnol. 1998;16:639–41.

Wernery U, Liu C, Baskar V, et al. Primordial germ cell-mediated chimera technology produces viable pure-line Houbara bustard offspring: potential for repopulating an endangered species. PLoS One. 2010;5: e15824.

Whishart GJ. Quantitation of the fertilising ability of fresh compared with frozen and thawed fowl spermatozoa. Br Poult Sci. 1985;26:375–80.

Whyte J, Glover JD, Woodcock M, et al. FGF, insulin, and SMAD signaling cooperate for avian primordial germ cell self-renewal. Stem Cell Reports. 2015;5: 1171–82.

Yamamoto Y, Usui F, Nakamura Y, et al. A novel method to isolate primordial germ cells and its use for the generation of germline chimeras in chicken. Biol Reprod. 2007;77:115–9.

Yoshizaki G, Ichikawa M, Hayashi M, et al. Sexual plasticity of ovarian germ cells in rainbow trout. Development. 2010;137:1227–30.

Zhang X, Ebata KT, Nagano MC. Genetic analysis of the clonal origin of regenerating mouse spermatogenesis following transplantation. Biol Reprod. 2003;69:1872–8.

Zhao DF, Kuwana T. Purification of avian circulating primordial germ cells by Nycodenz density gradient centrifugation. Br Poult Sci. 2003;44:30–5.

MIX
Papier aus verantwortungsvollen Quellen
Paper from responsible sources
FSC® C105338

If you have any concerns about our products,
you can contact us on
ProductSafety@springernature.com

In case Publisher is established outside the EU,
the EU authorized representative is:
Springer Nature Customer Service Center GmbH
Europaplatz 3, 69115 Heidelberg, Germany

Printed by Libri Plureos GmbH
in Hamburg, Germany